Biosynthesis and Manipulation of Plant Products

PLANT BIOTECHNOLOGY SERIES

Edited by Don Grierson, B.Sc., Ph.D., C.Biol., F.I. Biol., Professor of Plant Physiology and Head of AFRC Research Group in Plant Gene Regulation, Department of Physiology and Environmental Science, University of Nottingham.

Most important phases of plant growth, development, and reproduction that affect food production and quality involve changes in plant gene expression. These include processes such as germination, flowering, ripening, seed development, formation of storage organs, senescence, and responses to alterations in the environment and to pathogens. Recent developments in plant physiology, biochemistry and molecular biology mean that we are beginning to understand these processes in molecular terms. Most importantly, the availability of plant genetic engineering techniques means that we can do experiments *in planta*. This makes it possible to provide new scientific information about macro-molecular interactions and control mechanisms, and to alter the properties of plants in a directed and controlled way. We are now on the threshold of a new era, poised to exploit these techniques in manipulating growth, development, and productivity of plants, making them more suitable for our needs.

This series reviews recent developments in plant biotechnology, shows how scientific understanding leads to commercial applications, and discusses opportunities and problems in this rapidly developing field of molecular breeding.

PLANT BIOTECHNOLOGY

Volume Three

Biosynthesis and Manipulation of Plant Products

Edited by

DON GRIERSON B.Sc., Ph.D., C.Biol., F.I.Biol.
Professor of Plant Physiology
and Head of ARFC Research Group in Plant Gene Regulation
Department of Physiology and Environmental Science
University of Nottingham

BLACKIE ACADEMIC & PROFESSIONAL
An Imprint of Chapman & Hall
London · Glasgow · New York · Tokyo · Melbourne · Madras

Published by
Blackie Academic & Professional, an imprint of Chapman & Hall,
Wester Cleddens Road, Bishopbriggs, Glasgow G64 2NZ

Chapman & Hall, 2–6 Boundary Row, London SE1 8HN, UK

Blackie Academic & Professional, Wester Cleddens Road, Bishopbriggs, Glasgow G64 2NZ, UK

Chapman & Hall, 29 West 35th Street, New York NY10001, USA

Chapman & Hall Japan, Thomson Publishing Japan, Hirakawacho Nemoto Building, 6F, 1-7-11 Hirakawa-cho, Chiyoda-ku, Tokyo 102, Japan

DA Book (Aust.) Pty Ltd., 648 Whitehorse Road, Mitcham 3132, Victoria, Australia

Chapman & Hall India, R. Seshadri, 32 Second Main Road, CIT East, Madras 600 035, India

First edition 1993

Typeset in 10/12 pt Times New Roman by ROM Data Corporation Ltd., Falmouth, Cornwall.

Printed in Great Britain at the University Press, Cambridge

ISBN 0 7514 0060 2

A catalogue record for this book is available from the British Library

Library of Congress Cataloging-in-Publication data available

Printed on permanent acid-free text paper, manufactured in accordance with the proposed ANSI/NISO Z 39.48-199X and ANSI Z 39.48-1984

Preface

Volumes 1 and 2 of this Plant Biotechnology series reviewed fundamental aspects of plant molecular biology and discussed production and analysis of the first generation of transgenic plants of potential use in agriculture and horticulture. These included plants resistant to insects, viruses and herbicides, which were produced by adding genes from other organisms. Realisation of the potential of plant breeding has led to a resurgence of interest in methods of altering the structure, composition and function of plant constituents, which represents an even greater challenge and offers scope for improving the quality of a wide range of agricultural products. This, in turn, has resulted in a re-evaluation of priorities and targets by industry.

Volume 3 of this series considers the biochemical and gentic basis of the biosynthesis of plant products such as starch, lipids, carotenoids and cell walls, and evaluates the ways in which biosynthesis of these products can be modified for use in the food industries. Authors also cover the biosynthesis of rare secondary products and the function and application of proteins for plant protection and therapeutic use. The emphasis throughout is on the relationship between fundamental aspects of biosynthesis and structure–function relationships, and application of this knowledge to the redesigning and altering of plant products by molecular genetics.

Don Grierson

Contributors

Colin R. Bird ICI Seeds, Jealott's Hill Research Station, Bracknell, Berks, RG12 6EY, UK

Peter M. Bramley Department of Biochemistry, Royal Holloway and Bedford New College (University of London), Egham, Surrey, TW20 OEX, UK

Tony Fawcett Plant Molecular Biology Group, Department of Biological Sciences, University of Durham, South Road, Durham DH1 3LE, UK

Gareth Griffiths Department of Botany, School of Biological Sciences, University of Bristol, Woodland Road, Bristol BS8 1UG, UK

John D. Hamill Department of Genetics and Development Biology, Monash University, Clayton, Melbourne, Victoria 3168, Australia

Martin Hartley Department of Biological Sciences, University of Warwick, Coventry CV4 7AL, UK

J. Michael Lord Department of Biological Sciences, University of Warwick, Coventry CV4 7AL, UK

Cathie Martin John Innes Institute, Colney Lane, Norwich NR4 7UH, UK

John Mitchell Department of Applied Biochemistry and Food Science, University of Nottingham School of Agriculture, Sutton Bonington, Loughborough LE12 5RD, UK

Michael J.C. Rhodes AFRC Institute of Food Research, Norwich Laboratory, Cloney Lane, Norwich NR4 7UA, UK

Wolfgang Schuch ICI Seeds, Jealott's Hill Research Station, Bracknell, Berks RG12 6EY, UK

Antoni R. Slabas Plant Molecular Biology Group, Department of Biological Sciences, University of Durham, South Road, Durham DH1 3LE, UK

Alison M. Smith John Innes Institute, Colney Lane, Norwich NR4 7UH, UK

Keith Stobard Department of Botany, School of Biological Sciences, University of Bristol, Woodland Road, Bristol BS8 1UG, UK

Gregory A. Tucker Department of Applied Biochemistry and Food Science, University of Nottingham School of Agriculture, Sutton Bonington, Loughborough LE12 5RD, UK

Contents

1 Starch biosynthesis and the potential for its manipulation

A. M. SMITH and C. MARTIN

1.1 Introduction

Starch is the major form of carbon reserve in plants. Almost all plant organs accumulate it at some stage in their development, and it constitutes half or more of the dry weight of many storage organs, for example tubers, storage roots, and the seeds of cereals and some legumes. Perhaps because it is such an abundant natural product, there has until recently been little interest in how its production in plants might be manipulated. In the last five years, however, starch synthesis has received increasing attention for two main reasons. First, there is a desire to manipulate the overall composition of the harvested parts of plants in nutritionally and commercially useful ways. Starch is a major component of many of these harvested parts, and an understanding of the regulation of its accumulation will aid attempts to make directed changes in composition. Second, there is a desire to produce a range of cultivars with starches of differing properties within single crop species. Starch has many food and industrial uses, and these require different sorts of starch with distinct physical and chemical properties. The required properties are usually produced by various chemical modifications of extracted starch. Manipulation to create cultivars which produce starches with the required properties would reduce dependence on this chemical processing. Such manipulation requires a knowledge of the way in which the properties of starch are determined during its synthesis.

The demand for information about starch synthesis has revealed large gaps in our understanding of this fundamentally important process. Characterisation of the gene products that catalyse the pathway of starch synthesis is far from complete, and for non-photosynthetic organs the precise nature of the pathway is still the subject of controversy. The models proposed to explain the way in which the structure of starch is determined during its synthesis do not account for the chemical and physical complexity of the starch granule. The origins of the genetic, developmental and environmental variation in starch content and starch structure in plant organs, which must be understood if content and structure are to be manipulated, have hardly started to be explored.

In this chapter we shall describe briefly the occurrence and structure of starch in plants, discuss the current state of knowledge of the nature and regulation of the pathway by which it is synthesised, and suggest how it may be manipulated in the future.

1.2 Occurrence of starch in plants

1.2.1 Distribution in plant organs

The role of starch as a store of carbon in plants is enormously flexible. It serves as a reserve over a matter of hours in leaves, over periods of days at particular stages of organ development, through adverse seasonal conditions in perennating organs, and over periods which may extend to years in mature reproductive structures. This flexibility is reflected in the almost universal occurrence of starch in plants and its accumulation at some stage or stages of development of almost all plant organs. The following are examples of the diverse circumstances in which starch may accumulate.

1.2.1.1 Leaves Starch in leaves is regarded as a transient reserve which serves to balance the demand for sucrose by the non-photosynthetic parts of the plant (sinks) with the rate of carbon assimilation of the leaf (Stitt, 1985). The starch content of leaves generally follows a strong diurnal rhythm. Starch accumulates as a product of photosynthesis during the day, and is mobilised to sucrose to maintain a supply of carbon to the sinks during the night. There is enormous quantitative variation in this basic pattern. First, the extent to which assimilated carbon is stored in the leaf as starch shows great interspecific variation (Huber *et al.*, 1985; Pollock and Chatterton, 1988). Second, environmental conditions—for example temperature, photon flux density, carbon dioxide concentration, day length, mineral nutrition and disease—influence the extent of starch accumulation, either through direct effects on the leaf or indirect effects on sink organs (Stitt and Quick, 1989; Chatterton and Silvius, 1980; Geiger *et al.*, 1985; Marschner, 1986; Goodman *et al.*, 1986). Third, the extent of starch accumulation in an individual leaf shows a pronounced developmental pattern. Generally speaking, a young leaf at its sink-source transition synthesises little starch, a mature leaf has a strong diurnal rhythm of starch content, a leaf approaching senescence may contain large amounts of starch and display a less pronounced diurnal rhythm, and a senescing leaf contains little or no starch (Matheson and Wheatley, 1962; Lewis, 1984). The pallisade mesophyll of the leaf generally contains more starch than the spongy mesophyll, and mature leaves of some C4 species accumulate starch in the bundle sheath but not the mesophyll cells (Lewis, 1984; Laetsch, 1968). Stomatal guard cells have a different pattern of starch accumulation from other photosynthetic cells, and may actually synthesise starch at night and mobilise it during the day (Outlaw and Manchester, 1979).

1.2.1.2 Meristems. Starch accumulates transiently in cells of most meristems, usually in a zone immediately outside the zone of cell division. This pattern is seen, for example, in the meristematic regions of both monocot and dicot leaves (Leech and Baker, 1983) and in the root cap (Barlow, 1975). Starch in the root cap is believed to play a central role in the gravitropic response of roots. It is suggested that the positioning of starch granules—called statoliths—in the cells of the root cap in response to gravity generates signals which control the differential extension of cells in a zone higher up the root (Moore and Evans, 1986). This 'statolith' model has been called into question by the generation of mutations in *Arabidopsis* which apparently eliminate starch synthesis in the plant without abolishing the gravitropic response (Caspar and Pickard, 1989). However, these plants do display reduced sensitivity to gravity, and they may still contain a small amount of starch in the root cap (Kiss *et al.*, 1989; Saether and Iversen, 1991). The precise role of starch in the gravitropic response remains to be elucidated.

1.2.1.3 Fruits, seeds and perennating organs. Fruits which have high sugar contents when mature characteristically undergo a phase of starch accumulation and mobilisation during development. In apples, pears, strawberries and tomatoes this phase occurs early in development, prior to ripening (Davies and Cocking, 1965; Knee *et al.*, 1977; Tucker and Grierson, 1987). In many tropical fruits, for example banana, starch content remains high into the ripening period, when a rapid mobilisation to sucrose, glucose and fructose coincides with the onset of the climacteric (Tucker and Grierson, 1987).

Starch is the major carbon reserve of the seeds of a very wide range of species. The part of the seed in which the deposition of starch occurs may be either the embryo (e.g. legume seeds) or the endosperm or perisperm (e.g. cereal seeds), and starch often comprises up to 50% of the final dry weight of the seed. Other seeds which contain little or no starch at maturity—for example *Sinapis alba* and soybean— have a transient phase of starch accumulation during their development (Norton and Harris, 1975; Adams *et al.*, 1980). Many of the seeds which have brief phases of starch accumulation during their development also have some potential for photosynthesis, and it is not clear for most whether starch is synthesised from sucrose imported into the seed or as a direct product of photosynthesis.

Starch is the major carbon reserve of many specialised perennating organs, including tubers (e.g. potato, yam and sweet potato), storage roots (e.g. cassava), rhizomes (e.g. arrowroots: *Maranta arundinacea, Canna edulis* and species of *Iris*), corms (e.g. taro: *Colocasia antiquorum* and cocoyam: *Xanthosoma sagittifolium*) and the turions of aquatic plants (Lewis, 1984; Harrison *et al.*, 1969; Sculthorpe, 1967). Starch also accumulates prior to and persists through adverse seasonal conditions in less specialised vegetative organs, for example the buds, medullary rays and roots of deciduous trees (Priestley, 1970).

1.2.1.4 Thermogenic organs. A few specialised reproductive organs of higher plants—for example the spadix of *Arum* species (Meeuse, 1975)—are capable of attaining temperatures several degrees above ambient for a few hours at the end of their development. The rise in temperature is brought about by extremely high rates of uncoupled respiration which occur at the expense of stored starch. The starch content of the spadix of *Arum maculatum* rises to 20–30% of the fresh weight during its development, then declines to half or less of this value in 5–10 hours during thermogenesis (ap Rees *et al.*, 1976, 1977).

1.2.2 Intracellular location of starch

Starch is synthesised exclusively inside plastids in higher plants. The great diversity of tissues and circumstances in which starch accumulates is reflected in the wide range of types of plastid in which it is found. The widespread idea that starch synthesis and storage is a function of two particular, distinct types of plastid—chloroplasts and amyloplasts—is probably misleading. Although starch synthesis is the major metabolic function of some plastids, starch is frequently found in plastids with other important metabolic functions. For example, starch-containing plastids from pea roots are capable of high rates of nitrite reduction and amino acid synthesis (Bowsher *et al.*, 1989). The plastids of oil seeds contain starch during the early stages of lipid accumulation, when their rates of fatty acid synthesis must be high (Rest and Vaughan, 1972; Bils and Howell, 1963; Yazdi-Samadi *et al.*, 1977; Norton and Harris, 1975). Chromoplasts of carrot roots contain starch throughout the period of carotene synthesis (Kirk and Tilney-Basset, 1967).

The plastids of many organs pass through phases during their differentiation in which they synthesise and store starch. During the early stage of leaf development the proplastid contains starch for a short period after cell division ceases, loses it as it passes through an amoeboid phase and thylakoids start to differentiate, then regains it subject to a diurnal rhythm as full photosynthetic capacity is attained (Leech and Baker, 1983). The plastids of the developing pea embryo differentiate from chloroplasts with little starch into non-photosynthetic amyloplasts (Smith *et al.*, 1990a), whereas the chloroplast-like plastids of *Sinapis alba* embryos accumulate starch early in their development, then lose it as the lipid content of the embryo increases (Rest and Vaughan, 1972).

1.3 Structure and composition of starch

Starch occurs as dense, water-insoluble granules, ranging in size from less than 1 μm to over 100 μm. The composition and structure of starch granules has been studied almost exclusively on starches from storage organs—seeds and perennating organs—in which the main storage product is starch. This is because these are the commercially important sources of starch and because large amounts of starch free of other cellular components can easily be obtained from these organs. Relatively little is known about leaf starch and the transient reserve starches of non-photosynthetic vegetative organs.

1.3.1 Storage organs

1.3.1.1 The composition of granules The starch of mature storage organs is composed of two classes of glucose polymer. About 70% of the weight of the starch consists of a highly branched molecule, amylopectin (Figure 1.1). The branches are composed of α-1, 4-linked chains of glucose residues, with an average degree of polymerisation of about twenty. These have free non-reducing ends and are joined via the 1-position of the residue at their reducing ends to the 6-position of a residue within another chain. The outer branches of the molecule, which are not substituted at their 6-positions, are called A-chains, and inner branches, substituted at one or more 6-positions, are called B-chains. There is one primary or C-chain per molecule, which is the only chain with a free reducing end. A typical amylopectin molecule is about 200–400 nm long and 10–15 nm wide, and has a molecular weight of 10^7 to 10^8 (Kainuma, 1988; French, 1984; Guilbot and Mercier, 1985) The currently accepted 'cluster' model of its structure proposes that the branches are not randomly distributed along the axis of the molecule but occur in discrete clusters at intervals of about 7–10 nm, separated by relatively unbranched regions (Manners and Matheson, 1981; French, 1984; Kainuma, 1988).

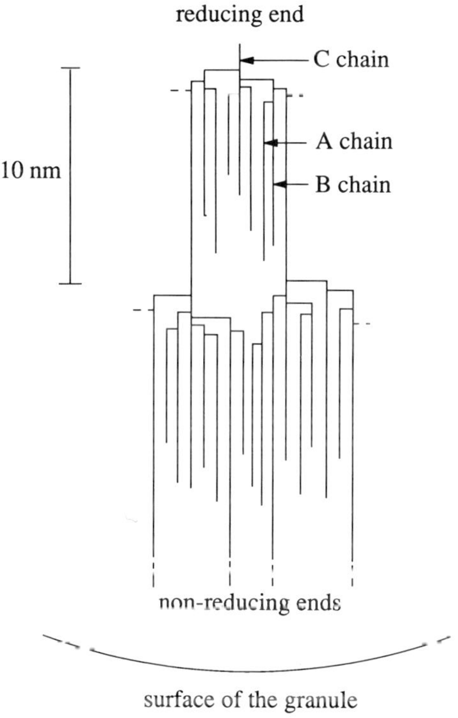

Figure 1.1 Structure, dimensions and location in the granule of part of a generalised amylopectin molecule. Redrawn from Gidley and Bociek (1985).

Within the granule, the amylopectin molecules are radially arranged with their non-reducing ends pointing outwards (Kainuma, 1988; Kassenbeck, 1978; Figure 1.1). Adjacent branches within the clusters of the molecule form double helices which are regularly packed, giving rise to long-range order (crystallinity) within the granule (French, 1984; Kainuma, 1988). About 40–50% of the granule is made up of these ordered regions (Gidley and Bociek, 1985).

The second type of starch polymer is amylose, a considerably smaller and much less branched molecule than amylopectin. It seems likely that the component of starch defined as amylose consists of both linear molecules of α-1,4-linked glucose residues and molecules comprising a small number of long chains of α-1, 4-linked residues joined by α-1,6-linkages, with degrees of polymerisation in the range of 1000–6000 (Banks and Greenwood, 1975; Hizukuri *et al.*, 1981; Kainuma, 1988; Guilbot and Mercier, 1985).

Amylose molecules are believed to exist as single, randomly-organised helices in an amorphous phase within the granule. The location of this phase relative to the crystalline regions of the granule is not known (Gidley and Bociek, 1988; Kassenbeck, 1978).

1.3.1.2 Genetic effects on composition. The generalised picture of granule composition given above is subject to great interspecific and developmental variation. The amylose to amylopectin ratio and the structures of these components in terms of branch length, frequency of branching and overall size show a wide range of variation between species and between cultivars within a species. Some of the apparent variation may result from the use of different methods and growth conditions in different studies. The fractions of starch conventionally defined as amylose and amylopectin both consist of a range of different sizes and structures of molecule, and starches frequently also contain a significant proportion of molecules which are intermediate in size and properties between these two fractions (Shannon and Garwood, 1984; Guilbot and Mercier, 1985). Because of this, different methods of starch fractionation can yield quite different values for the ratio of amylose to amylopectin.

In spite of these problems, several important generalisations about genetic variation in starch structure can be made. First, variation in amylose to amylopectin ratios shows no obvious correlation with taxonomic grouping or type of storage organ. Variation is as great between cultivars within a species as it is between species. For example, large-scale surveys of the ratio in wheat and maize endosperm and potato tuber starches gave ranges of 17–29%, 20–30% and 18–23%, respectively (Shannon and Garwood, 1984).

Second, the structure of amylopectin, and hence the structure of the crystalline regions of the granule, differs considerably between endosperm, tuber and embryo starches. The A type of crystalline structure, characteristic of the starches of cereal endosperms, reflects a closely packed array of double helices, whereas the B type of structure, characteristic of maize endosperm and tuber starches, reflects a more open array with considerably more water included in it. The third, C type, found in the starches of legume embryos, is a mixture of the A and B types of structure (Guilbot and Mercier, 1985). These different structures are related to the branch length of the amylopectin from which they are formed. The A structure arises where the average branch length is less than 20 glucose residues, and the B structure where it is more than 22 glucose residues (Hizukuri *et al.*, 1983; Gidley, 1987; Gidley and Bulpin, 1987).

Third, there is considerable variation between one type of starch-storing organ and another in the nature and amounts of non-starch components of the granule. All starch granules contain small amounts of protein and lipid, and some also contain significant amounts of phosphate. Cereal starch contains about 0.5% protein and 1% lipid, whereas tuber starch contains only 0.05% protein and 0.1% lipid (Guilbot and Mercier, 1985). The lipid of cereal starches consists mainly of lysophospholipids, which are very uncommon elsewhere in the plant. There is a strong correlation between the lipid and amylose contents of cereal starches, and it is likely that at least some of the lipid is present as inclusions within helical amylose molecules (Gidley and Bociek, 1988; Morrison *et al.*, 1984; Morrison and Gadan, 1987). An association between the amylose fraction and specific types of lipid has not been observed in starches from other sources (Morrison *et al.*, 1984). Potato starch is unusual in that one glucose unit in 300 in its amylopectin fraction is phosphorylated, on either the 6- or the 3-position (Takeda and Hizukuri, 1982).

1.3.1.3 Developmental effects on composition. The composition of starch granules changes during granule development. Precise definition of the sequence of granule development is rendered difficult by the existence of developmental gradients within immature storage organs (Shannon and Garwood, 1984). However, some developmental changes appear to be common to the granules of many different organs (Banks and Muir, 1980). First, the amylose to amylopectin ratio increases as the granule matures. For example, the amylose content of the starch increases from 9 to 27% between 8 and 28 days post-anthesis in maize endosperm, and from 12 to 20% during growth of the potato tuber from 1 to 16 cm (Tsai *et al.*, 1970; Geddes *et al.*, 1965). Second, the molecular size of both amylose and amylopectin increases during development, and the degree of branching of the amylose fraction, which is essentially linear in early development, increases (Banks and Muir, 1980).

1.3.1.4 Granule anatomy and morphology. Many starch granules show internal 'growth rings' when observed by light microscopy, by transmission electron microscopy after chemical treatment and staining, and by scanning electron microscopy after attack by α-amylases (French, 1984). The rings are concentrically arranged layers of alternating density, crystallinity, and susceptibility to attack by acids and α amylases (Buttrose, 1963a, French, 1984). The denser layers probably contain crystallites, lying tangentially to the radius of the granule and formed by the ordered packing of the branching clusters of many parallel amylopectin molecules (Kassenbeck, 1978; Kainuma, 1988; Guilbot and Mercier, 1985). The less dense layers may be boundaries at which many amylopectin molecules terminate, and crystallites are scarce (French, 1984).

The developmental origins of growth rings are unclear. In some organs—for example, wheat endosperm—the formation of rings follows a clear diurnal pattern. The denser, more crystalline layers are formed during the day, and the less dense layers at night. Imposition of continuous light upon the plant abolishes the rings. It is inferred that they reflect diurnal variations in the availability of substrate for starch synthesis (Buttrose, 1962; French, 1984). In other organs—for example, potato tuber — growth rings are not abolished in continuous light (Buttrose, 1962).

Starch granule size and shape are enormously variable. The granules of mature storage organs range from 1 to over 100 μm in length, and may be round, ovoid, reniform, elongated, compound, irregular, polyhedral or lenticular (Banks and Muir, 1980; French, 1984). There is variation between species, between cultivars, between parts of a single organ, and within individual cells. The extent of variation is well-illustrated by the starches of cereal endosperms. The granules of rice and oats are compound. Multiple starch particles, apparently separated by a layer of stroma, form in the same amyloplast. They are initially rounded, but become angular as they increase in size and pack together (Shannon and Garwood, 1984). Wheat and barley endosperm contain two distinct sizes and shapes of granule. The larger granules are lenticular with an equatorial groove, while the smaller ones are spherical (Banks and Muir, 1980). The larger granules are believed to originate from spheres. These become elaborated as two adjacent plates which grow from a point in both directions around a circumference, resulting in a flattened shape and an equatorial groove which represents the junction between the two plates (Evers, 1971). The mechanism responsible for this highly directed granule growth is not known, although clusters of tubules, which might represent an equatorial band, have been observed in the stroma adjacent to the developing granule (Briarty *et al.*, 1979; Duffus, 1984). The spherical granules are initiated later in development than the lenticular ones (May and Buttrose, 1959; Buttrose, 1963b). In barley they constitute about 6% of the weight of starch and about 90% of the number of granules at maturity (Banks and Muir, 1980; Evers *et al.*, 1974), and in wheat endosperm they can constitute more than 30% of the weight of starch (Evers and Lindley, 1977).

The mechanism by which starch granules are initiated in storage organs is not known. It has been proposed that starch molecules accumulate in specific regions of the stroma in an unorganised 'droplet' form, then spontaneously crystallise to form a nucleus around which the granule develops (Salema and Badenhuizen, 1967; French, 1984; Badenhuizen, 1969). Various techniques have revealed the presence in the granule of a core of material different in structure from the remainder of the granule (Guilbot and Mercier, 1985; McDonald *et al.*, 1991), but there is little other evidence to support the 'droplet' hypothesis. It is not known what determines whether one or several granules are initiated in an amyloplast, and whether new granules are initiated or previously formed granules continue to enlarge during a phase of starch accumulation.

Proposals about the mechanism of granule growth range from synthesis of complete molecules in the stroma followed by their deposition onto the granule surface, to elongation of projecting chains of molecules which are incorporated into the granule as they grow (Geddes and Greenwood, 1969; French, 1984;

Shannon *et al.*, 1970; Badenhuizen, 1969). The truth may lie somewhere between these two extremes. There may be no sharp boundary between the stroma and the growing granule, but rather a zone in which molecules are being both elaborated and organised into the granule. Although the lenticular granules of the cereal endosperm represent an extreme case of directed granule growth, the non-spherical shapes of most granules indicate that deposition of starch is not uniform over the surface throughout development. The general mechanisms by which preferential growth of particular regions of the surface is directed remain to be elucidated.

1.3.2. Leaves

In contrast to the granules of storage organs, starch granules in chloroplasts show little species-specific variation. They are generally disc-shaped, and smaller than mature granules of storage starch (Shannon and Garwood, 1984). Little is known of their structure and composition, but they are likely to be substantially different in these respects from granules of storage starch. The latter are formed at developmentally determined rates over relatively long periods during which little or no degradation occurs. Chloroplast starch granules are synthesised at a rate which may vary enormously in the space of a few minutes and they are usually substantially degraded and then resynthesised in a diurnal cycle.

Starch granules of pea and spinach leaves are believed to consist of a distinct, mainly crystalline core surrounded by a pasty, more amorphous mantle. The mantle may contain molecules which are on average less branched than those in the core. Degradation and resynthesis appear to occur mainly in the mantle portion of the granule (Beck, 1985; Steup *et al.*, 1983).

Starch granules from tobacco leaves have internal rings similar in appearance to those of storage starches (Buttrose, 1963a). As with potato tuber starch, the rings are not abolished by growth in continuous light. The effect of leaf age and time during the diurnal cycle on the appearance of the rings has not been reported, and the relationship of the rings to the pasty mantle and core structure of the granules of pea and spinach leaves is unclear.

1.4 The pathway of starch synthesis

In this section we shall discuss the nature of the pathway by which starch is synthesised, and the enzymes that catalyse it. The regulation of the pathway will be discussed in subsequent sections.

1.4.1 The supply of ADPglucose

There is general agreement that starch polymers in all plastids are synthesised from ADPglucose via two enzymes, starch synthase (ADPglucose:1,4-α-D-glucan glucosyltransferase, EC 2.4.1.21) and starch branching enzyme (1,4-α-D glucan:1,4-α-D-glucan 6-glycosyltransferase, EC 2.4.1.18). Evidence that this is the case will be presented in section 1.4.3. The pathway of synthesis of ADPglucose from the primary carbohydrate supply of the starch-synthesising cell is well-established for photosynthetic cells, but not for non-photosynthetic organs generally.

1.4.1.1 ADPglucose synthesis in photosynthetic cells. ADPglucose is synthesised inside chloroplasts from fructose 6-phosphate (F6P), an intermediate of the reductive pentose phosphate pathway, via three enzymes, phosphoglucose isomerase (EC 5.3.1.9), phosphoglucomutase (EC 2.7.5.1), and ADPglucose pyrophosphorylase (EC 2.2.7.27) (Figure 1.2).

Evidence that this pathway is the sole route of ADPglucose synthesis comes from studies of plants with mutations that affect the activities of the three enzymes. A reduction of 50% in phosphoglucose isomerase activity in *Clarkia xantiana* chloroplasts reduces the rate of starch synthesis in leaves by up to 50% in high light (Jones *et al.*, 1986; Kruckeberg *et al.*, 1989: mutant phenotype *Pgi1*1**, $Pgi2^a2^a$, $Pgi3^a3^a$). The elimination of plastidial phosphoglucomutase activity from leaves of *Arabidopsis thaliana* (Caspar *et al.*, 1985: line TC7) and its dramatic reduction in *Nicotiana* (Hanson and McHale, 1988; Kiss and Sack, 1990: line NS 458) results in almost starchless plants. The elimination of ADPglucose pyrophosphorylase

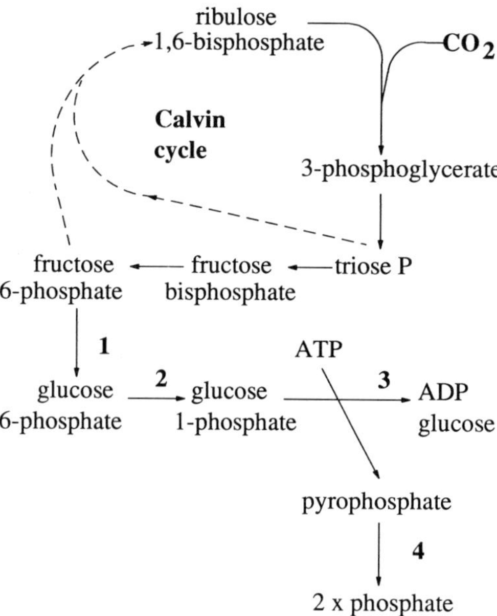

Figure 1.2 Pathway of ADPglucose synthesis in the chloroplast. Enzymes catalysing the pathway: 1, phosphoglucose isomerase; 2, phosphoglucomutase; 3, ADPglucose pyrophosphorylase; 4, alkaline inorganic pyrophosphatase.

activity from *Arabidopsis* leaves prevents starch synthesis (Lin *et al.*, 1988a: mutation at *adg1* locus, line TL25).

Phosphoglucose isomerase and phosphoglucomutase catalyse reactions central to hexose phosphate metabolism in plants. They are found almost universally in both the cytosol and the stroma, and there do not appear to be isoforms specifically associated with starch synthesis. In contrast, ADPglucose pyrophosphorylase in leaves is confined to the chloroplast (Okita *et al.*, 1979; Mares *et al.*, 1978) and is exclusively associated with starch synthesis.

Although not directly involved in ADPglucose synthesis, alkaline inorganic pyrophosphatase (EC 3.6.1.1) is important in the flux of carbon through this pathway (Figure 1.2). The reaction catalysed by ADPglucose pyrophosphorylase is far removed from equilibrium *in vivo* in favour of ADPglucose synthesis because alkaline pyrophosphatase hydrolyses the pyrophosphate that it produces (Neuhaus and Stitt, 1990; Weiner *et al.*, 1987). Alkaline pyrophosphatase in leaves is probably confined to the chloroplast (Gould and Winget, 1973; Weiner *et al.*, 1987; Bucke, 1970).

1.4.1.2 ADPglucose synthesis in non-photosynthetic organs. In heterotrophic cells, in which the primary source of carbon is sucrose, ADPglucose for starch synthesis must be derived from a product of sucrose catabolism in the cytosol. There is continuing controversy about which cytosolic metabolite crosses the amyloplast envelope as the substrate for ADPglucose synthesis. Three candidates have received serious attention: triose phosphate, hexose phosphate and ADP glucose itself (Figure 1.3).

Until relatively recently, triose phosphate was held to be the form in which carbon from sucrose crossed the amyloplast envelope (Figure 1.3, mechanism C), for the following reasons. First, triose phosphate is the major form in which carbon crosses the chloroplast envelope (Flügge and Heldt, 1984). It does so via a triose phosphate–phosphate exchange translocator in the inner membrane (Heldt and Rapley, 1970; Fliege et al., 1978). Several lines of evidence indicate that a translocator with these properties is also present in the envelope of non-photosynthetic plastids (Liedvogel and Kleinig, 1980; Emes and Traska, 1987; Alban *et al.*, 1988; Ngernprasirtsiri *et al.*, 1988). Since chloroplasts and amyloplasts are developmentally interconvertible, it seemed likely that the major flux of carbon across the amyloplast envelope was as triose phosphate (Shannon and Garwood, 1984; Jenner, 1976).

Second, isolated amyloplasts from maize endosperm and potato tuber were reported to take up triose phosphate and convert it to insoluble material, believed to be starch (Mohabir and John, 1988; Echeverria *et al.*, 1988). However, no evidence was provided that the insoluble material was indeed starch. In addition, the amyloplast preparations were significantly contaminated with cytosolic enzymes, and the possibility that triose phosphate was converted to some other metabolite prior to uptake by the amyloplast cannot be ruled out.

Third, non-photosynthetic plastids from several sources were claimed to contain

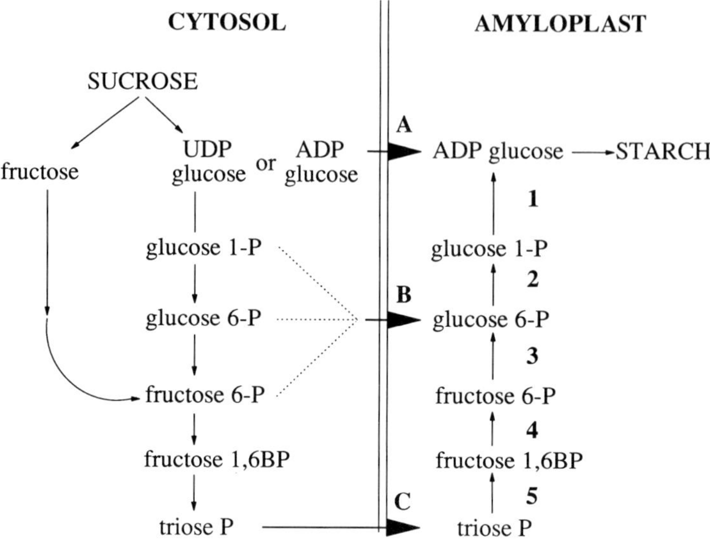

Figure 1.3 Three proposed mechanisms for entry of cytosolic metabolites into the amyloplast. **A**, adenylate translocator; **B**, putative hexose or hexose phosphate translocator; **C**, triose phosphate–phosphate exchange translocator. Enzymes catalysing the interconversion of metabolites in the amyloplast: 1, ADPglucose pyrophosphorylase; 2, phosphoglucomutase; 3, phosphoglucose isomerase; 4, fructose 1,6-bisphosphatase; 5, aldolase.

all of the enzymes required to convert triose phosphate to starch (for example cauliflower florets: Journet and Douce, 1985; soybean suspension cultures: Macdonald and ap Rees, 1983; potato tubers: Mohabir and John, 1988; rice endosperm: Nakamura *et al.*, 1989). However, it is likely that one of the required enzymes—fructose 1,6-bisphosphatase (EC 3.1.3.11)—is in fact absent from these plastids (Entwistle and ap Rees, 1988, 1990). Its apparent activity was almost certainly due to contaminating cytosolic pyrophosphate: F6P 1-phosphotransferase (EC 2.7.1.90).

Elegant experiments in the last three years with intact tissues have provided strong evidence that hexose units, rather than triose phosphate, are taken up by amyloplasts as the substrate for starch synthesis (Figure1.3, mechanism B). Intact developing wheat ears and isolated endosperms were incubated with glucose labelled in the 1-position with ^{13}C. The distribution of label in the carbon atoms of glucose produced from hydrolysis of starch from the endosperm was then determined (Keeling *et al.*, 1988). Since the two triose phosphates, dihydroxyacetone phosphate and glyceraldehyde 3-phosphate, are in equilibrium via the enzyme triose phosphate isomerase, metabolism of the supplied glucose to starch through triose phosphate would result in randomisation of the 1- and 6-positions and an equal labelling of these positions in the glucose moieties of starch (Figure 1.4). Metabolism of glucose to starch via a route in which only hexose units were involved would not lead to labelling of the 6-position in starch. In fact, most of the label in starch was retained in the 1-position. It is thus highly unlikely that triose

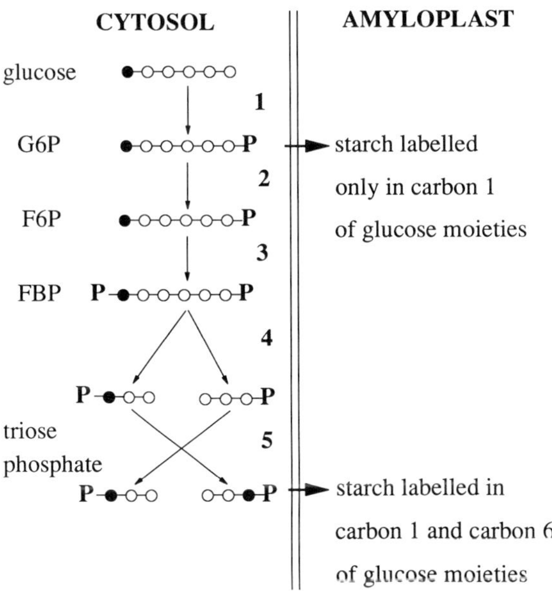

Figure 1.4 Experiment of Keeling *et al.* (1988) to discover whether hexose or triose units enter the amyloplasts of wheat endosperm. Enzymes catalysing the interconversion of cytosolic metabolites: 1, hexokinase; 2, phosphoglucose isomerase; 3, phosphofructokinase/pyrophosphate : F6P 1-phosphotransferase; 4, aldolase; 5, triose phosphate isomerase.

phosphate is imported into amyloplast as the substrate for starch synthesis in wheat endosperm. Similar experiments on developing potato tubers, maize endosperm, cotyledons of *Vicia faba* and cultured, heterotrophic *Chenopodium* cells have also shown only very limited redistribution of label between the 1- and 6-positions of hexose during starch synthesis (Hatzfeld and Stitt, 1990; Viola *et al.*, 1991).

Two further sorts of evidence support the idea that hexose units rather than triose phosphate enter the amyloplast. First, almost all non-photosynthetic plastids lack both fructose 1,6-bisphosphatase and pyrophosphate : F6P 1-phosphotransferase, the two enzymes which would allow conversion of triose to hexose phosphate (Entwistle and ap Rees, 1990; Frehner *et al.*, 1990). Second, experiments with amyloplasts isolated from pea embryos and wheat endosperm show that hexose phosphates can be taken up and made into starch. Intact amyloplasts from pea embryos can take up glucose 6-phosphate (G6P) and incorporate carbon from it into starch at a rate comparable with the rate of starch synthesis in the intact seed (Hill and Smith, 1991; Figure 1.5). Starch synthesis is dependent upon the provision of ATP. This is expected since the synthesis of ADPglucose from glucose 1-phosphate (G1P) requires ATP. Free hexose and other hexose and triose phosphates do not support physiologically significant rates of starch synthesis by these amyloplasts. Amyloplasts from wheat endosperm can take up G1P and incorporate carbon from it into starch (Tyson and ap Rees, 1988). The ability of these amyloplasts to take up and metabolise G6P in an ATP-dependent manner has not

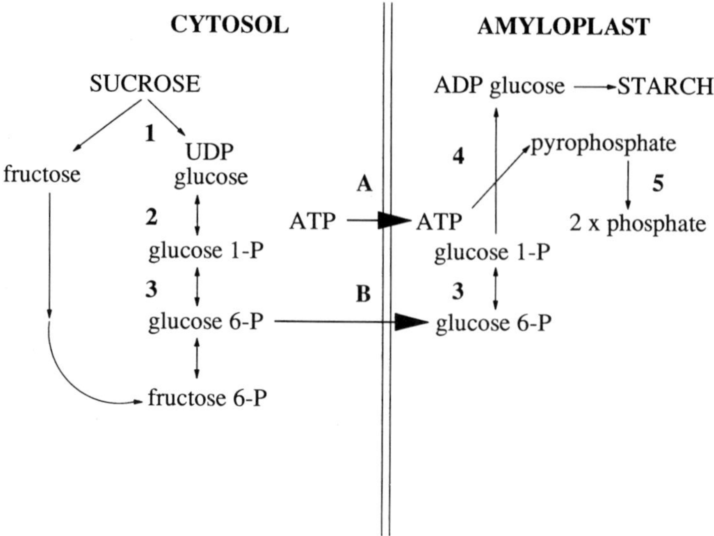

Figure 1.5 Proposed pathway of ADPglucose synthesis in developing pea embryos (Hill and Smith, 1991). **A,** adenylate translocator ;**B,** putative glucose 6-phosphate translocator. Enzymes catalysing the pathway: 1, sucrose synthase; 2, UDPglucose pyrophosphorylase; 3, phosphoglucomutase; 4, ADP glucose pyrophosphorylase; 5, alkaline inorganic pyrophosphatase.

been investigated, and it is possible that they can transport both G1P and G6P as substrates for starch synthesis. A mutant of *Nicotiana* which lacks plastidial phosphoglucomutase—and hence cannot interconvert G6P and G1P—is grossly deficient in starch in all its organs, indicating that G1P is not the transported metabolite in this species (Kiss and Sack, 1990).

The evidence presented above indicates strongly that the source of carbon for ADPglucose synthesis in non-photosynthetic plastids generally is hexose monophosphate imported from the cytosol (Figure 1.5). It is established that chloroplasts generally cannot transport hexose phosphates (Borchert *et al.*, 1989), hence uptake of G6P or G1P by non-photosynthetic plastids is likely to be via a translocator unique to these plastids. Plastids from pea roots are reported to take up G6P (Bowsher *et al.*, 1989). The mechanism may be a phosphate-exchange translocator which also recognises 3-phosphoglycerate and dihydroxyacetone phosphate, but not G1P or 2-phosphoglycerate (Borchert *et al.*, 1989). Two molecules of phosphate are generated in the amyloplast for each molecule of hexose phosphate converted to starch (Figure 1.5). The existence of a hexose phosphate–phosphate exchange translocator would allow the export to the cytosol of one of these phosphate molecules. The second molecule must presumably also be returned to the cytosol, but the mechanism by which this occurs is not known.

It is not known how amyloplasts obtain ATP for the synthesis of ADPglucose from G1P *in vivo*. However, the ability of exogenous ATP to support physiological rates of starch synthesis in amyloplasts from pea embryos indicates that direct uptake of ATP from the cytosol may be important (Hill and Smith, 1991; Figure

1.5). An adenylate translocator which might serve this function has been identified on the envelope of amyloplasts in sycamore suspension cells (Ngernprasirtsiri *et al.*, 1989).

Synthesis of ADPglucose from G1P—either imported directly or made via phosphoglucomutase from imported G6P—is catalysed, as in photosynthetic cells, by ADPglucose pyrophosphorylase. Its central importance in non-photosynthetic organs is demonstrated by the existence of mutations which specifically reduce its activity and as a consequence reduce the rate of starch synthesis. Mutations in the *shrunken-2 (sh2)* and *brittle-2 (bt2)* loci of maize, which are believed to encode the two subunits of the enzyme in the endosperm (Preiss *et al.*, 1990), result in 90% reductions of enzyme activity during endosperm development and 75% reductions in the starch content of the mature seed (Preiss *et al.*, 1989). A mutation at the *rb* locus of pea results in the loss of about 90% of the activity of the enzyme from the developing embryo, and a 40–50% reduction in the starch content of the seed at maturity (Smith *et al.*, 1989; Wang and Hedley, 1991). A mutation which essentially abolishes activity of the enzyme in *Arabidopsis* results in a dramatic reduction in the starch content of both leaves and roots (Caspar and Pickard, 1989: line TL25). Abolition of activity of the enzyme in potato tubers via genetic manipulation using antisense genes almost completely eliminates tuber starch (L. Willmitzer, personal communication).

ADPglucose pyrophosphorylase is confined to the plastid in non-photosynthetic organs (Journet and Douce, 1985; Macdonald and ap Rees, 1983; Mohabir and John, 1988; Entwistle and ap Rees, 1988), and its activity is generally correlated with phases of starch synthesis during development. For example, activity increases as starch accumulation increases in maize endosperm (Ozbun *et al.*, 1973), wheat grain (Turner, 1969), and potato tuber (Sowokinos, 1976).

As in leaves, alkaline inorganic pyrophosphatase is an important component of starch synthesis in non-photosynthetic organs. The enzyme is essentially confined to the plastid in these organs and its activity is considerably in excess of the rate of starch synthesis (Gross and ap Rees, 1986). A specific association of the enzyme with starch synthesis is suggested by its absence from onion bulb, a storage organ which does not accumulate starch (Gross and ap Rees, 1986).

The role of plastidial enzymes in the synthesis of ADPglucose has recently been questioned in a report that intact amyloplasts from cultured sycamore cells can take up ADPglucose itself as a substrate for starch synthesis (Pozueta-Romero *et al.*, 1991; Figure 1.3, mechanism A). The ADPglucose probably enters these amyloplasts via the adenylate translocator (Ngernprasirtsiri *et al.*, 1989). It has been suggested that uptake of ADPglucose from the cytosol may be the major route of its supply to starch synthesis *in vivo* (Pozueta-Romero *et al.*, 1991). ADPglucose could be synthesised directly from sucrose via sucrose synthase. However, there are strong arguments against this proposal. As discussed above, there is excellent evidence from mutant and transgenic plants that most or all of the flux of carbon into starch is via a pathway that includes ADPglucose pyrophosphorylase. In addition, the affinity of sucrose synthase for ADP is much lower than its affinity for UDP, which is believed to be its principal substrate. It is unlikely that ADP glucose synthesis can proceed at physiologically significant rates in the cytosol (ap Rees *et al.*, 1985; ap Rees, 1988).

1.4.2 ADPglucose pyrophosphorylase

ADPglucose pyrophosphorylase is the best-studied of the committed enzymes of starch synthesis, mainly because it has regulatory properties believed to be critical—in leaves at least—to the regulation of the pathway of starch synthesis as a whole. The enzyme has been purified from several photosynthetic and non-photosynthetic sources, its kinetic properties have been studied in detail, and the genes that encode it have been cloned from several sources. Detailed analysis of its expression and the structure–function relationships of its subunits are being undertaken (Preiss, 1988; Preiss *et al.*, 1989; Anderson *et al.*, 1990; Müller-Röber *et al.*, 1990). The following is a brief summary of current knowledge of the enzyme.

1.4.2.1 Kinetic properties. ADPglucose pyrophosphorylase catalyses the readily-reversible reaction ATP + G1P = ADPglucose + pyrophosphate. *In vivo*, the synthesis of ADPglucose is strongly favoured by the hydrolysis of pyrophosphate by alkaline inorganic pyrophosphatase, which renders pyrophosphate levels in the plastid virtually undetectable (Weiner *et al.*, 1987).

The most striking property of the enzyme is that it is strongly regulated by 3-phosphoglycerate and phosphate. Phosphate is a strong inhibitor of the enzyme from all the sources from which it has been examined. For example, the enzymes from potato tuber and maize endosperm are inhibited 75–95% by 1–3 mM phosphate (Sowokinos and Preiss, 1982; Plaxton and Preiss, 1987). This inhibition is relieved—to varying degrees, depending on the source of the enzyme—by 3-phosphoglycerate. The enzyme from most sources is activated by 3-phosphoglycerate. Reported activations with saturating concentrations of 3-phosphoglycerate (2–10 mM) range from 25–35-fold (potato tuber, spinach leaf and maize endosperm: Sowokinos and Preiss, 1982; Morell *et al.*, 1988; Plaxton and Preiss, 1987) to 4–6-fold (maize leaf mesophyll and tomato leaf: Spiltrano and Preiss, 1987; Sanwal *et al.*, 1968). However, the enzyme from wheat endosperm is apparently not activated by 3-phosphoglycerate (Olive *et al.*, 1989), and the enzyme from developing pea embryos is activated less than two-fold (C. Hylton and A.M. Smith, unpublished results).

Several metabolites in addition to 3-phosphoglycerate activate the enzyme. The most commonly reported of these are hexose 6-phosphates, fructose 1,6-bisphosphate and phosphoenolpyruvate. The degree of activation by these metabolites is usually much less than by 3-phosphoglycerate, but in some cases it is quite substantial. For example, the enzyme from maize endosperm is activated about 25-fold by 3-phosphoglycerate, and 14–17-fold by hexose 6-phosphates (Plaxton and Preiss, 1987).

1.4.2.2 Physical properties. ADPglucose pyrophosphorylase is a heterotetramer composed of two types of subunit with molecular weights in the range 50–60 kDa (Morell *et al.*, 1987; Lin *et al.*, 1988a,b; Okita *et al.*, 1990; Preiss *et al.*, 1990). There is good biochemical evidence from protein sequencing, tryptic digestion patterns, and antigenic relationships that the two subunits are proteins of different sequence (Morell *et al.*, 1987; Okita *et al.*, 1990; Preiss *et al.*, 1990).

Direct evidence that different genes encode the two subunits has recently been provided. Genes encoding subunits of the enzyme have been cloned and sequenced from several sources (rice endosperm: Krishnan *et al.*, 1986; Anderson *et al.*, 1989; spinach leaf: Preiss *et al.*, 1989; wheat endosperm: Olive *et al.*, 1989). These are either known to encode the smaller of the two subunits, or have not been definitely attributed to either subunit. However, genes which encode two different sorts of subunit have now been cloned from both maize and potato. Mutations at the *bt-2* and *sh-2* loci of maize almost abolish activity of ADPglucose pyrophosphorylase in the endosperm. The *bt-2* mutation causes the loss of the small and the *sh-2* mutation the loss of the large subunit of the enzyme (Preiss *et al.*, 1990). The *bt-2* and *sh-2* genes have been cloned, and there is good evidence that they encode the small and large subunits respectively (Preiss *et al.*, 1990; Bhave *et al.*, 1990; Bae *et al.*, 1990). The two genes have been used as probes to clone homologous genes from potato, and these have been partially characterised (Müller-Röber *et al.*, 1990).

There is increasing evidence that there may be more than one gene encoding each subunit of the enzyme, and that these may be differentially expressed. The *sh-2* and *bt-2* homologues in potato (called S and B respectively) show quite different patterns of expression in the plant (Müller-Röber *et al.*, 1990). The S-gene is highly expressed in the tuber, but expressed at only a very low level in leaves. However, high levels of expression are induced in leaves by conditions which cause them to accumulate sucrose. The B-gene is expressed in both tubers and leaves, and its expression is not affected by accumulation of sucrose (Müller-Röber *et al.*, 1990). These results indicate that the subunit composition of the enzyme in tubers and leaves may be different, although this has not yet been directly examined. The leaf enzyme in the absence of accumulation of sucrose could be primarily a homotetramer of the subunit encoded by the S-gene, or could contain a second type of subunit not encoded by the S- or B-gene. Its composition might be altered by conditions which cause accumulation of sucrose in the leaf.

It has been suggested that members of gene families encoding the subunits of the enzyme may be differentially expressed in the leaf and endosperm of wheat (Olive *et al.*, 1989). However, it is not clear which subunit(s) the clones represent. The enzyme in wheat leaves is claimed to differ in size and kinetic properties from that in endosperm, again indicating that it may have a different subunit composition (Olive *et al.*, 1989), but data to substantiate this have not been published.

Several other lines of evidence are consistent with the idea that the enzyme in different organs and tissues of the plant may be encoded by different genes. First, the enzymes from the mesophyll and the bundle sheath of maize leaves have substantially different properties (Spiltrano and Preiss, 1987). Second, the enzyme activity in the embryos of developing maize seeds is affected by the *bt-2* but not the *sh-2* mutation, whereas both affect the enzyme activity in the endosperm (Dickinson and Preiss, 1969). Third, Western-blotting experiments indicate that the subunit sizes of the enzyme in the leaves of cereals may be different from those in the endosperm (Krishnan *et al.*, 1986).

1.4.2.3 The roles of the subunits. Both sorts of subunit are required for full activity of ADPglucose pyrophosphorylase. Mutations that result in the loss of one of the two subunits of the enzyme from *Arabidopsis* leaves (Lin *et al.*, 1988b) and either of the two subunits of the enzyme from maize endosperm (Preiss *et al.*, 1990), result in a dramatic loss of activity. The residual activity in these cases could be due to the formation of an active enzyme by the remaining subunit, the existence of a third type of subunit or a minor, distinct form of the enzyme, or a very low level of expression of the missing subunit.

The requirement for two sorts of subunit for full activity of the plant enzyme is interesting in that the ADPglucose pyrophosphorylase from bacteria is a tetramer of only one sort of subunit, encoded by a single gene (Preiss, 1984). The predicted amino acid sequences for the large and small subunits of the plant enzymes show considerable similarity both to each other and to the enzyme from bacteria. At present many of these comparisons can be made only for incomplete sequences. However, it is clear that the small subunits form a group of proteins which are more similar to each other than to the large subunits (Müller-Röber *et al.*, 1990; Bae *et al.*, 1990; Okita *et al.*, 1990; Preiss *et al.*, 1990). It has been suggested that the two subunits of the plant enzyme arose from a duplication of the gene in an organism with a single type of subunit. The two products were initially interchangeable, but sequence divergence between the two genes has subsequently occurred to the extent that their products have become complementary (Bae *et al.*, 1990; Bhave *et al.*, 1990)

The functional differences between the two sorts of subunit are not understood. Functional analysis of the enzyme from spinach leaf indicates that both subunits can bind the activator, 3-phosphoglycerate. However, only the large subunit appears to have the capacity to bind the substrate, ATP, although the substrate site may actually be shared between the two sorts of subunit (Morell *et al.*, 1988; Preiss *et al.*, 1989).

1.4.3 The synthesis of α-1,4-glucan

Although it is now generally accepted that the α-1,4-linkages of starch are synthesised via an ADPglucose starch synthase, two older ideas persist in the literature. The first is that at least part of the flux of carbon into starch is catalysed by starch phosphorylase, and the second is that UDPglucose rather than ADPglucose is the main nucleotide sugar used by starch synthases.

Phosphorylase catalyses the readily-reversible reaction

$$\text{G1P} + (\text{glucosyl})_n = (\text{glucosyl})_{n+1} + \text{phosphate}.$$

It is present at sufficient activity to account for the observed net flux of carbon into starch in many starch-synthesising organs (Stitt and Steup, 1985; Hawker *et al.*, 1979; Tsay *et al.*, 1983), although a major portion of the activity may not be inside the plastid (Steup, 1988). However, the high K_m for G1P (in excess of 1 mM), and the very unfavourable ratio of phosphate and G1P concentrations that is believed to exist in plastids, make it highly unlikely that net synthesis of starch occurs via this enzyme (Preiss and Levi, 1979, 1982). It is likely that the major, if not the only, function of the plastidial isoforms of starch phosphorylase is in starch degradation (Steup, 1988).

The view that UDPglucose might be the primary sugar nucleotide involved in starch synthesis arose because granule-bound forms of starch synthase can utilise both UDPglucose and ADPglucose, and UDPglucose and enzymes that might generate it—sucrose synthase and UDPglucose pyrophosphorylase—are present in starch-synthesing organs. UDPglucose is almost invariably present at higher levels than ADPglucose (Jenner, 1982; ap Rees *et al.*, 1984) and activities of UDPglucose pyrophosphorylase are usually considerably greater than those of ADPglucose pyrophosphorylase (Hawker *et al.*, 1979; Sowokinos, 1976; Smith *et al.*, 1989). However, evidence that UDPglucose makes little or no contribution to starch synthesis is now overwhelming (ap Rees *et al.*, 1984; ap Rees, 1988). First, the extent to which starch synthesis can utilise UDPglucose is very limited. Soluble starch synthases do not do so, and the affinity of granule-bound starch synthases for UDPglucose is generally much lower than for ADPglucose (Ghosh and Preiss, 1966; Preiss and Levi, 1979; Smith, 1990a; ap Rees *et al.*, 1984; Macdonald and ap Rees, 1983). Some of the reported UDPglucose-dependent starch synthesis may have been artifactual (ap Rees *et al.*, 1984). Second, the fact that mutant and transgenic plants which have very low activities of ADPglucose pyrophosphorylase also have greatly reduced rates of starch synthesis (sections 1.4.1.1 and 1.4.1.2) indicates strongly that most or all of the flux of carbon into starch is through ADPglucose. Third, there is no good evidence that UDPglucose is synthesised inside amyloplasts. The enzymes which catalyse its synthesis from sucrose and its entry into glycolysis, sucrose synthase and UDPglucose pyrophosphorylase (ap Rees, 1988) are located outside the amyloplast (Entwistle and ap Rees, 1988; Macdonald and ap Rees, 1983; Mohabir and John, 1988). In spinach leaves most or all of the UDPglucose is in the cytosol (Gerhardt and Heldt, 1984), and this may well be true for plant tissues generally.

1.4.3.1 Starch synthases. The starch synthase activity of plants is located both in the soluble fraction of the amyloplast and bound onto the starch granule. The maximum catalytic activities of the granule-bound and soluble starch synthases are difficult to assess in starch-synthesising organs generally, for two main reasons. First, many organs also contain substantial activities of starch-degrading enzymes, which interfere with the assay for starch synthase (Pollock and Preiss, 1980). Second, granule-bound starch synthase activity is often increased by mechanical damage to the granules (Frydman and Cardini, 1967). It is not known whether the activity revealed on mechanical damage is part of the active fraction of the enzyme *in vivo*.

The organs in which starch synthases are best understood are the developing maize endosperm, the potato tuber, and the developing pea embryo. This section is a survey of the current state of knowledge of the proteins responsible for starch synthase activity in plants. The roles of these proteins will be discussed in detail in section 1.5.

Maize endosperm. Attempts to assess the maximum catalytic activity of the starch synthases in maize endosperm are subject to the problems described above. About half of the total starch synthase activity in crude extracts is associated with starch granules (Macdonald and Preiss, 1985). The activities of both the soluble and granule-bound activities are probably in excess of the rate of starch synthesis in the endosperm, although not greatly so. For example, in kernels 16–22 days after pollination, in which the starch content increased by about 20 nmol glucose units per min per kernel, the soluble starch synthase activity was reported to be 30 nmol per min per kernel (Ozbun *et al.*, 1973).

The soluble starch synthase activity of the endosperm can be separated by ion-exchange chromatography into two fractions with distinct properties (Boyer and Preiss, 1979; Ozbun *et al.*, 1971a; Pollock and Preiss, 1980). The main difference between these two forms is in their ability to synthesise starch in the absence of a 'primer'—an α-1,4-glucan added to the assay as a substrate for starch synthesis. At low salt concentrations both forms require a primer. Addition of high concentrations of certain salts—of which citrate in excess of 0.1 M is the most effective—to the assay lowers the K_m of both of the forms for primers and allows one of the two forms (SSS I), but not the other (SSS II), to display high activity in the absence of any exogenous primer (Ozbun *et al.*, 1971a; Pollock and Preiss, 1980; Macdonald and Preiss, 1983, 1985).

It is thought that the 'unprimed' activity displayed by SSS I in the presence of citrate is a function of small amounts of α-1,4-glucan bound to the enzyme protein. The high salt concentration is believed to bring about a change in the conformation of the enzyme which dramatically lowers its K_m for α-1,4-glucan, and allows it to use its bound glucan as a primer. These ideas are supported by the fact that preparations of SSS I contain significant amounts of polymerised glucose even after extensive purification (Pollock and Preiss, 1980), and treatment of partially purified SSS I with amylolytic enzymes can destroy its 'unprimed' activity without affecting its activity with exogenous primers (Schiefer *et al.*, 1978).

The physiological significance of this high-salt effect on soluble starch synthase activity is not clear, and it does not of itself establish that SSS I and SSS II are different proteins rather than a single protein with and without bound α-1,4-glucan. Investigations of the physical and antigenic properties of the two forms lend more weight to the argument that they are different proteins. First, antibodies raised to SSS I effectively immunoprecipitated the activity of this form from purified preparations, but were ineffective against the activity of SSS II (Macdonald and

Preiss, 1985). Second, SSS I and SSS II were shown by sucrose density gradient centrifugation to have molecular weights of 70 and about 95 kDa, respectively (Macdonald and Preiss, 1985). However, weights determined by this method could be affected by modifications of the protein such as the presence of bound glucan. Third, the ratio of SSS I to SSS II activity in partially purified preparations is greater from endosperm with a mutation at the *Dull* locus than from wildtype endosperm (Boyer and Preiss, 1981). This has been taken as evidence that the two forms are under independent genetic control, but does not prove that they are different proteins. Thus although the balance of evidence suggests that the soluble starch synthase of the endosperm consists of two different proteins, definitive proof of this is lacking.

Proof that at least part of the granule-bound starch synthase activity of the maize endosperm is a different protein from the soluble starch synthase comes from studies of mutations at the *waxy* locus. These mutations result in the loss of the amylose component of the starch, most of the granule-bound starch synthase (GBSS) activity (Nelson and Rines, 1962; Tsai, 1974; Nelson *et al.*, 1978) and a major, 58 kDa protein tightly bound to the starch granule (Echt and Schwartz, 1981). However, they have no effect on the amount of starch in the endosperm or on the activity of soluble starch synthase (Ozbun *et al.*, 1971a). In experiments which utilised an unstable mutation at the *waxy* locus caused by a transposable element, the 58 kDa protein has been shown to be the product of a structural gene at this locus. A cDNA clone for the 58 kDa protein hybridised to a single DNA fragment in digests of genomic DNA. The hybridising fragments in wildtype and phenotypically wildtype revertant plants were the same size, but the fragment in line with the transposon inserted at the *waxy* locus was larger (Shure *et al.*, 1983). This indicates that the transposon was within the structural gene encoding the 58 kDa protein.

There is a clear relationship between the size and amount of the 58 kDa protein and the alleles at the *waxy* locus. For example, the amount of the protein is correlated with the dosage of the wildtype allele (Echt and Schwartz, 1981), and revertant lines with partial excision of the transposon and intermediate phenotypes may have granule-bound proteins immunologically related to the 58 kDa protein but of a larger size (Shure *et al.*, 1983; Schwartz and Echt, 1982; Wessler *et al.*, 1986).

There is excellent molecular and biochemical evidence that the 58 kDa 'waxy' protein is the major GBSS during at least part of the development of the maize endosperm. First, there is a good correlation between the amount of the protein and the GBSS activity in gene-dosage experiments (Echt and Schwartz, 1981). Second, small mutations in the *waxy* gene, which do not affect its expression, can give rise to reduced GBSS activity (Wessler *et al.*, 1986). Third, purification of GBSS solubilised from starch granules with amylolytic enzymes yields a major form of the enzyme of about 60 kDa. This form is absent from starch in *waxy* mutant endosperms (Macdonald and Preiss, 1983, 1985). Fourth, the amino acid sequence predicted from the sequence of the *waxy* gene shows strong similarities to that predicted from the *glgA* gene of *Escherichia coli* (Klösgen *et al*., 1986; Kumar *et al*., 1986; Preiss, 1990). This gene encodes an ADPglucose-dependent α-1,4-glucan synthase involved in glycogen synthesis (Preiss, 1984). The amino acid sequence predicted for the waxy protein contains a small motif (–KTGG–) which is conserved in all glycogen synthases and is believed to lie at the ADP/ADPglucose binding site (Furukawa *et al*., 1990)

In addition to the waxy protein, there may be at least one other GBSS in the maize endosperm. A second, minor form of the enzyme, with an apparent molecular weight of 93 kDa, is present in protein solubilised from starch granules (Macdonald and Preiss, 1983). The starch of *waxy* mutant endosperms retains some starch synthase activity, and this has different kinetic properties from the forms identified in wildtype endosperm (Nelson *et al*., 1978; Macdonald and Preiss, 1985). At least a part of this non-waxy activity may be attributable to the binding of a soluble form of the enzyme to the granule (Macdonald and Preiss, 1985). However, the fact that this activity is not affected by alleles at the *Waxy* locus and appears earlier in development than the wildtype GBSS activity indicates that it is distinct from the waxy protein, and may at least in part be due to a GBSS normally masked by the waxy form during most of the development of wildtype endosperm (Nelson *et al*., 1978).

Potato tuber. Starch synthase activity is greater in the granule-bound than the soluble fraction of potato tubers throughout most of their development (Hawker *et al*., 1979; Tsay and Kuo, 1980). Reported activities of both enzymes (20–100 nmol min^{-1} g^{-1} fresh weight, Hawker *et al*., 1979) are comparable with the rate of starch synthesis of intact tubers (about 40 nmol glucose units min^{-1} g^{-1} fresh weight, Morrell and ap Rees, 1986).

Tuber starch contains only one major protein, of 60 kDa, and there is excellent evidence that it is a GBSS involved in the synthesis of amylose. First, it is antigenically closely related to the waxy protein of maize (Vos-Scheperkeuter *et al*., 1986). Second, an amylose-free mutant (*amf*) of potato lacks this protein, and its starch has only 4% of the GBSS activity of wildtype starch (Jacobsen *et al*., 1989; Hovenkamp-Hermelink *et al*., 1987). The mutation responsible for this phenotype is in the coding sequence of a structural gene encoding the 60 kDa granule-bound protein (van der Leij *et al*., 1991a). Third, the predicted amino acid sequence for the 60 kDa protein shows strong similarities to that for the waxy protein of maize and the *glgA* gene product of *E. coli* (van der Leij *et al*., 1991a). Fourth, the introduction into wildtype potatoes of antisense constructs derived from a cDNA for the 60 kDa protein results in a loss or reduction in the amylose content of the starch, the GBSS activity, and the amount of the 60 kDa protein (Visser *et al*., 1991).

The solubilisation of significant amounts of the GBSS activity of potato tubers has so far proved impossible (Vos-Scheperkeuter *et al*., 1986; Frydman and Cardini, 1967). However, the existence on the granule of a minor protein of about 95 kDa with strong antigenic similarity to a GBSS of pea embryos (Smith, 1990a, see below) indicates that there may be more than one form of GBSS in potato tubers.

There are conflicting reports about the soluble starch synthase of potato tubers. Either one (Ponstein, 1990) or three (Hawker *et al.*, 1972) forms have been separated by ion exchange chromatography. Three different purifications have yielded two polypeptides, of 78 and 85 kDa (Ponstein, 1990), a single polypeptide of 90 kDa (Hawker *et al.*, 1972), and a single polypeptide of 70 kDa (Baba *et al.*, 1990). There are many possible reasons for the discrepancies between these three studies: for example, they used different cultivars of potato at different stages of development and extracted them in different media. It seems likely, however, that the 60 kDa protein identified as a GBSS is not responsible for the soluble activity.

Pea embryo. Pea embryos contain considerably less soluble starch synthase than GBSS activity during their development. The GBSS activity in crude extracts is always several-fold in excess of that required to account for the rate of starch synthesis, whereas the soluble activity is only comparable with the rate of starch synthesis (Smith *et al.*, 1989).

The starch granules of the embryo contain a major, 59 kDa waxy protein (GBSS I) antigenically closely related to the waxy protein of maize, potato and other species (Smith, 1990a and unpublished data). However, starch synthase activity solubilised from the granule is associated not with this protein but with a protein of 77 kDa (GBSS II), which is antigenically only weakly related to the waxy protein (Smith, 1990a). Recent work in our laboratory has shown that GBSS I and II are the products of different genes, but that both are related to α-1,4-glucan synthases (Dry *et al.*, 1992). The amino-acid sequences of both proteins—predicted from full-length cDNA clones—are homologous to the waxy protein of maize, potato and other species, and to the *glgA* gene product of *E. coli*. Both proteins contain the –KTGG– motif thought to lie at the binding site for ADPglucose (Furukawa *et al.*, 1990). In GBSS I, as in the waxy proteins of other species, it is close to the N-terminus of the mature protein. However, in GBSS II this motif is about 17 kDa in from the N-terminus of the mature protein. Whereas the C-terminal 60 kDa of GBSS II shows strong similarity to waxy proteins along its entire length, the 17 kDa N-terminal portion is not found in waxy proteins or glycogen synthases.

At present it is not known why the solubilised GBSS I is inactive. There are no losses of starch synthase activity during extraction and solubilisation which would indicate the loss of a major form of the enzyme (Smith, 1990a). On the other hand, there are no obvious non-conservative differences in sequence between this protein and other waxy proteins which might indicate the presence of a mutation that has led to loss of activity (Dry *et al.*, 1992). The possibility that the protein is inactive due to damage during extraction from the granule cannot be ruled out.

Two forms of soluble starch synthase have been purified from developing pea embryos (Denyer and Smith, 1992) which together probably account for all of the soluble starch synthase activity of the granule. They are proteins of 60 (SSS I) and 77 kDa (SSS II). They have similar kinetic properties and neither displays significant citrate-stimulated, unprimed activity. Although the relationship between these forms and the GBSS of the embryo is not yet established, our data indicate that SSS I and II are different proteins and that SSS II is very similar, and possibly identical, to GBSS II (Denyer and Smith, 1992). Neither of the soluble starch synthases is significantly antigenically related to GBSS I (Denyer and Smith, 1992).

General features of starch synthases. Many other storage organs possess a granule-bound protein closely related to the waxy proteins of maize, potato and pea. The predicted amino acid sequences of the examples for which the genes have been cloned show strong similarities to each other (barley: Rohde *et al.*, 1988; rice: Wang *et al.*, 1990; and references above). Many organs, particularly the endosperms of grass and cereal seeds, are affected by mutations which, like the *waxy* mutation of maize and the *amf* mutation of potato, eliminate the amylose component of starch, the waxy protein and the GBSS activity (Hseih, 1988; Sano, 1984; Okuno and Sakaguchi, 1982).

The extent of occurrence of a waxy-like protein in leaves is less clear. In potato the gene encoding the GBSS is expressed throughout the plant, including the leaves, and the *amf* mutation eliminates amylose from starch in all organs (Visser *et al.*, 1989). However, the *waxy* mutation of maize is reported to affect only the endosperm, pollen and embryo sac (Echt and Schwartz, 1981), and no mRNA for the waxy protein could be detected in maize leaves (Klösgen *et al.*, 1986). The gene encoding GBSS I of pea is highly expressed in embryos, but expression can barely be detected in leaves (Dry *et al.*, 1992). It is not possible at present to judge whether there are qualitative differences in the nature of GBSS proteins between leaves and storage organs in some species and not others, or whether there are simply quantitative differences between species in the relative amounts of mRNA for the waxy protein in different organs. The extensive degradation of leaf starch at night presumably results in the release of much of the leaf GBSS from the granule. The fate of this released protein has not yet been addressed.

We believe that most if not all storage organs have one or more forms of GBSS in addition to the waxy protein. Evidence is presented above that this is the case for the pea embryo, and may be the case for maize and potato. Starches from a range of other storage organs, including embryos, tubers and endosperms, contain a protein of about 77 kDa which is antigenically closely related to the 77 kDa GBSS of the pea embryo (A. Smith, unpublished results).

Evidence concerning the nature of soluble starch synthases in plants generally is scarce. The enzyme from several sources has been partially purified and shown to contain one to three forms after ion exchange chromatography. Where multiple forms are present, these usually differ in the ratio of their primed to unprimed activities and their affinities for various primers (Ozbun *et al.*, 1971b; Hawker *et al.*, 1974; Schulman and Ahokas, 1990; Dang and Boyer, 1988; Pisigan and del Rosario, 1976). The reported presence of one form of the enzyme in maize leaves and two in endosperm has been taken to indicate that different soluble starch synthase proteins are expressed in different organs of the plant (Dang and Boyer, 1988). However, another study of maize leaves reported resolution of the soluble starch synthase activity into two forms (Hawker and Downton, 1974). Evidence that multiple forms of the enzyme are different gene products is lacking in all of these cases.

The relationship between granule-bound and soluble forms of starch synthase is not fully understood for any starch-synthesising organ. There is general agreement that the waxy family of starch synthases is confined to the granule. In contrast, our work on pea embryos indicates that the novel form of GBSS which we have characterised, GBSS II, may be present in both a soluble and a granule-bound form. This issue cannot be resolved until more proteins with starch synthase activity have been fully characterised.

1.4.4 The branching of starch

1.4.4.1 Branching enzymes. Starch-branching enzyme catalyses the formation of branches in starch molecules through the formation of α-1,6-linkages. The precise mechanism of action of the enzyme is not fully understood. The enzyme catalyses the cleavage of a linear, α-1,4-linked chain of glucoses, probably within 15–20 glucose residues of the non-reducing end. The cleaved, non-reducing terminal portion is then rejoined to the molecule from which it was cleaved, or perhaps an adjacent molecule, at an internal glucose residue via an α-1,6-linkage. Enzymes from different sources differ in their specificities for glucan substrates. The best-studied enzyme—from potato tubers—can branch both linear and branched substrates. It will not cleave linear substrates of less than 30–40 glucose residues, but can cleave A-chains of only 10–14 residues in branched molecules. This has been interpreted as a requirement for a substrate with a stable, double-helical conformation. It is likely that the enzyme cannot cleave A-chains within about six glucose residues of a branch point (Manners, 1985; Borovsky *et al.*, 1976, 1979; Guilbot and Mercier, 1985).

Branching-enzyme activity is confined to plastids (Smith, 1988; Echeverria *et al.*, 1985, 1988) but its location within the plastid has not been widely studied. It has been assumed that the enzyme is exclusively soluble, but this may be because it cannot be assayed in the presence of large amounts of starch (Edwards *et al.*, 1988; Smith, 1990b), rather than because it is absent from the starch granule. We have been able to purify branching enzyme activity from protein solubilised from washed starch granules of pea embryos. This activity is associated with a protein of the same size as one of the two soluble forms in the embryo (see below), and it seems likely that it is a granule-bound fraction of that form (A. Smith, unpublished data). The occurrence of granule-bound branching enzyme needs further investigation.

Relatively little is known about changes in the activity of the enzyme in relation to starch synthesis in plant organs. Both of the commonly used assays are prone to interference by several components of crude extracts of plants (Edwards *et al.*, 1988; Smith, 1988; Smith, 1990b; Vos-Scheperkeuter *et al.*, 1989) and estimates of its maximum catalytic activity are frequently unreliable. However, there are indications that its activity is related to the rate of starch synthesis. For example, its activity increases markedly in the early stages of starch deposition in developing pea embryos (Edwards *et al.*, 1988; Smith, 1988; Smith *et al.*, 1989).

The enzyme from most of the sources from which it has been purified can be resolved into two or three forms by ion exchange chromatography (maize kernels: Boyer and Preiss, 1978a; sorghum seed: Boyer, 1985; spinach leaf: Hawker *et al.*, 1974; rice endosperm: Smyth, 1988; pea embryos: Matters and Boyer, 1981; Smith, 1988). In most of these examples the forms differ in their substrate affinities, but it is not known whether they are different proteins, modifications of a single protein, or fractions of a single protein with and without associated glucan (Hawker *et al.*, 1974). The enzymes from maize endosperm, potato tubers and pea embryos have been studied in most detail.

Maize endosperm. The enzyme from developing endosperm can be resolved into three forms (I, IIa and IIb) by ion-exchange and aminobutyl Sepharose chromatography (Boyer and Preiss, 1978a). All three forms have molecular weights of about 80 kDa (Boyer and Preiss, 1978a; Singh and Preiss, 1985). Forms IIa and b differ only in that IIb has a higher activity with amylopectin than IIa, but they differ considerably from form I in respect of substrate affinity and the branching patterns that they produce *in vitro* (Boyer and Preiss, 1978a). Forms I and II are only weakly antigenically related (Singh and Preiss, 1985). Antibodies to the branching enzyme from potato tuber and the glycogen-branching enzyme from *E. coli* inhibited form I much more strongly than forms IIa and b (Singh and Preiss, 1985; Vos-Scheperkeuter *et al.*, 1989). The amino acid composition and tryptic digest pattern of form I are significantly different from those of forms IIa and b (Singh and Preiss, 1985). It seems likely that form I is a different protein from forms IIa and IIb.

Genetic evidence indicates that forms IIa and IIb are different proteins. Mutations at the *amylose extender* (*ae*) locus of maize, which dramatically reduce the level of branching of the starch in the endosperm (Deatherage *et al.*, 1954), cause a large reduction in the activity of branching enzyme in crude extracts of endosperm, and loss specifically of form IIb (Boyer and Preiss, 1978b). Forms I and IIa remain unaltered in activity, kinetic properties and size (Hedman and Boyer, 1982, 1983). There is a good correlation between the dosage of the mutant allele at *ae* and the activity of form IIb (Fergason *et al.*, 1966; Hedman and Boyer, 1983). However, detailed studies of forms IIa and IIb have revealed that the proteins responsible for their activities are essentially identical. Reciprocal analysis with monoclonal antibodies, tryptic digests and amino acid analysis all failed to distinguish between them (Singh and Preiss, 1985). It is possible that the effect of the *ae* mutation on form II is indirect. For example, forms IIa and IIb could be a single protein, resolving into two fractions with somewhat different properties because of variable amounts of glucan bound to the protein molecules. A mutation which altered the nature of the glucan in the amyloplast might then affect the chromatographic behaviour and activity of the protein through a change in the amount or nature of the glucan bound to it.

Potato tuber. Only one form of starch-branching enzyme has been found in potato tubers, but there is a lack of agreement about its nature. Independent reports give the size of the purified protein as 79 kDa (Vos-Scheperkeuter *et al.*, 1989), 85 kDa (Borovsky *et al.*, 1975) and 103 kDa (Blennow and Johansson, 1991). Extraction of enzymes from potato tubers is particularly difficult, and smaller forms of the enzyme may be products of the hydrolysis of the 103 kDa form. This large form is reported to be labile during extraction unless strenuous precautions are taken to inhibit proteases (Blennow and Johansson, 1991). Its primary breakdown product is a protein of 86 kDa, which retains branching-enzyme activity. The 103 kDa form is the only branching enzyme in preparations from freshly harvested, developing tubers, but the 86 kDa form and other forms of intermediate size are always present in preparations from stored tubers (Blennow and Johansson, 1991). This suggests that the 103 kDa form may be labile *in vivo* during dormancy. Reports that tubers have a single branching enzyme of 79 or 86 kDa were based on purifications from dormant tubers (Vos-Scheperkeuter *et al.*, 1989) or made without protease inhibitors (Borovsky *et al.*, 1975).

Pea embryo. Two forms of branching enzyme, which differ in both kinetic and physical properties, have been purified to homogeneity from pea embryos (Smith, 1988). These probably account for all of the activity of branching enzyme in the embryo. Form I consists of two proteins, of 108 and 114 kDa, whereas form II is a protein of 100 kDa. The proteins of form I are antigenically indistinguishable, and are probably products of the same gene (Smith, 1988; Bhattacharyya *et al.*, 1990). They are only weakly antigenically related to form II. Form I has much lower activity with amylose than form II, and the polymer that it produces in the phosphorylase-stimulation assay has a different branching pattern from that produced by form II (Smith, 1988).

Unequivocal proof that forms I and II are encoded by different genes is provided by a mutation (at the *rugosus* (*r*) locus) which dramatically reduces the branching of starch in the embryo. The mutation causes the loss of both the activity and the proteins of form I, while form II is unaffected (Smith, 1988). We have shown that that the *r* locus is actually a structural gene that encodes form I. In mutant plants, the gene is disrupted by a transposon-like insertion of DNA, reduced levels of an enlarged transcript are produced, and no protein is synthesised (Bhattacharyya *et al.*, 1990). The predicted amino-acid sequence of form I shows regions of marked similarity to that of the glycogen-branching enzyme of *E. coli* (Baecker *et al.*, 1986). We have recently cloned the gene that encodes form II of the enzyme, and this too shows similarities to the glycogen-branching enzyme (D. Bewley and C. Martin, unpublished results).

It is likely that there is variation in the nature of branching enzymes between organs within a plant, as well as between species. Form I of branching enzyme in peas is present in both leaves and embryos, but form II appears to be absent from leaves. In spite of this, leaves of mutant plants which lack form I have branching enzyme activity. This activity may be due to a third isoform of the enzyme and, consistent with this, antibodies to form I specifically recognise a protein of about 80 kDa in extracts of leaves but not of embryos (Smith *et al.*, 1990b).

1.4.4.2 Debranching enzymes. Debranching enzymes, which catalyse the hydrolysis of α-1,6-linkages of glucans, are involved in starch degradation in plants (Steup, 1988). They are present in starch-synthesising as well as starch-degrading organs, for example in both developing and germinating cereal grains, but they are generally thought to be in an inactive form during periods of starch synthesis (Steup, 1988). However, study of the *sugary-1* (*su-1*) mutation of maize endosperm indicates that they may play a role in determining starch structure during its synthesis. The mutation causes the accumulation of a highly-branched, water-soluble glucan in the endosperm (Sumner and Somers, 1944). This phytoglycogen comprises up to 25% of the final dry weight of the kernel, and is formed after the appearance of starch during development (Preiss, 1988). The mechanism of its formation is controversial, and two hypotheses have been put forward. First, it has been reported that *su-1* endosperms have a modified form of starch-branching enzyme, capable of creating a more highly branched polymer than is present in normal starch (Boyer *et al.*, 1982). However, this finding has not been subsequently confirmed. Second, *su-1* endosperms are reported to have a much lower activity of debranching enzyme than wildtype endosperms (Pan and Nelson, 1984). Three forms of the enzyme can be resolved by hydroxylapatite chromatography from preparations from wildtype endosperms. One of these is missing and the other two severely reduced in activity in *su-1* endosperms. This observation has led to the suggestion that the structure of amylopectin in normal starches is the result of an equilibrium between branching and debranching activity during the formation of the granule (Preiss, 1988). A deficiency in debranching activity would then be expected to result in the formation of highly branched glucans.

Insufficient information is available about either the branching or the debranching enzymes of wildtype and *su-1* endosperms to resolve this problem. The latter hypothesis has profound implications for the mechanism of starch synthesis, and the synthesis of phytoglycogen clearly requires further study.

1.5 Regulation of starch synthesis

The preceeding sections illustrate the vast potential for variation in the structure and amount of starch in plant organs, and in the nature of the pathway of starch synthesis. It is clear that regulation of the pathway is likely to be complex, to occur at several different levels, and to operate in different ways in different organs. In this section we shall consider first the ways in which the rate of starch synthesis may be regulated in different organs, and second how the structure of starch is determined.

1.5.1 The rate of starch synthesis

1.5.1.1 Turnover of starch There are good *a priori* reasons for assuming that the rate of starch accumulation in plant organs represents an equilibrium between its synthesis and degradation. Enzymes of starch degradation are active in all starch-synthesising organs. Much of this activity may be extraplastidial, or even in a different part of the organ from the site of starch accumulation (Stitt *et al.*, 1978; Okita and Preiss, 1980; Kakefuda *et al.*, 1986; Ziegler, 1988) but in many organs at least part of the activity is known to be in the plastids in which starch is accumulating. Starch phosphorylase is present in starch-synthesising chloroplasts, and amylase may also be present (Stitt *et al.*, 1978; Okita and Preiss, 1980; Echeverria and Boyer, 1986; ap Rees, 1988; Ziegler, 1988). Amylase and phosphorylase are present in the starch-containing plastids of cauliflower buds and phosphorylase is present in the amyloplasts of soybean suspension cultures (Journet and Douce, 1985; Macdonald and ap Rees, 1983). The enzymes of starch degradation are not known to have regulatory properties which would prevent their activity during periods of starch synthesis (Stitt and Steup, 1985; Steup, 1988).

In spite of the apparent lack of a mechanism to prevent starch degradation during its synthesis, there is almost no good evidence that significant starch degradation occurs at the same time as starch synthesis in either leaves or storage organs. Although starch turnover can occur in leaves held in the light for very protracted periods, and for a short period at the onset of the light period, the bulk of evidence indicates that there is almost no degradation of leaf starch through most of the normal light period (Kruger *et al.*, 1983; Fondy and Geiger, 1985). Less is known about the degradation of starch during periods of net synthesis and storage in non-photosynthetic organs. However, starch is metabolically stable in rapidly growing pea roots, and metabolic stability may be a feature of reserves in non-photosynthetic organs generally (Hargreaves and ap Rees, 1988).

These limited observations indicate that, during their synthesis, starch reserves in intact plants do not generally undergo significant degradation to the level of hexose units. We shall therefore consider the regulation of the rate of synthesis simply in terms of a unidirectional flux. However, the occurrence of limited internal hydrolysis or rearrangement of newly synthesised starch polymers, for example via endoamylase, debranching enzyme and disproportionating enzyme (D enzyme: Jones and Whelan, 1969; Lin and Preiss, 1988), cannot be ruled out. This could have consequences for the structure of the starch, and might also affect the availability of glucan substrates for starch synthases and starch-branching enzymes.

1.5.1.2 Leaves Regulation of starch synthesis in leaves presents problems entirely different from those in non-photosynthetic organs. The whole of photosynthetic metabolism is integrated to allow close interaction between the operation of the Calvin cycle, the synthesis of ATP, the poising of electron transport, and the control of starch and sucrose synthesis. The rate of starch synthesis must respond in a rapid and modulated manner to changes in the demand for photosynthate by sink organs and environmentally induced changes in the rate of photosynthetic carbon assimilation. The mechanisms by which starch synthesis is regulated are inevitably complex. They are described in detail in recent publications by Stitt and colleagues (Stitt, 1990; Neuhaus and Stitt, 1990; Stitt and Quick, 1989). The following is a much-simplified summary of the current state of knowledge.

Under most physiological conditions, control of the partitioning of assimilate between starch and sucrose occurs primarily in the cytosol, in the pathway of sucrose synthesis (Figure 1.6). The activities of two enzymes on the pathway, fructose 1,6-bisphosphatase and sucrose phosphate synthase, are subject to complex metabolic regulation. They respond extremely sensitively and in parallel to changes in the demand for sucrose by the sink organs and the supply of assimilate from the chloroplast. Changes in the flux through the pathway can be brought about very rapidly and without large and potentially damaging changes in cytosolic metabolite levels.

Changes in the rate of sucrose synthesis alter the phosphate status of the chloroplast. This is because assimilate is exported from the chloroplast as triose phosphate in a one-for-one exchange with phosphate released during the synthesis

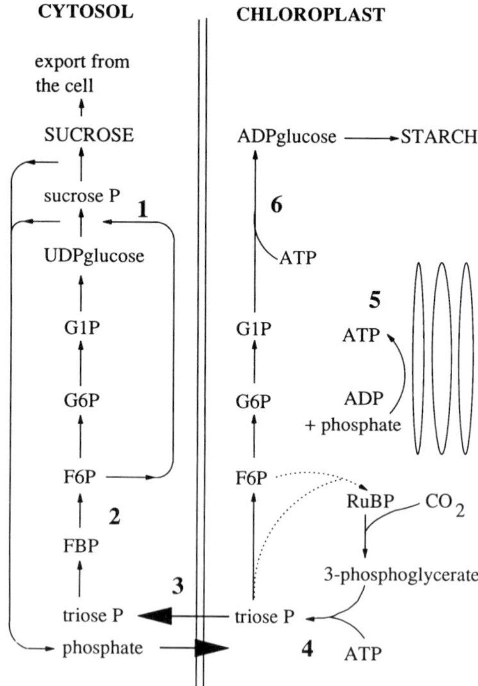

Figure 1.6 Important steps in the regulation of starch synthesis in chloroplasts. 1, sucrose phosphate synthase; 2, fructose 1,6-bisphosphatase; 3, triose phosphate–phosphate exchange translocator; 4, phosphoglycerate kinase and glyceraldehyde 3-phosphate dehydrogenase; 5, ATP synthesis via photophosphorylation; 6, ADPglucose pyrophosphorylase.

of sucrose in the cytosol. High phosphate levels in the chloroplast promote ATP synthesis via photophosphorylation. This in turn promotes the ATP-requiring conversion of 3-phosphoglycerate to triose phosphate in the Calvin cycle, thus lowering the concentration of 3-phosphoglycerate in the chloroplast (Figure 1.6). Conversely, low concentrations of phosphate in the chloroplast reduce the rate of ATP synthesis and hence allow the level of 3-phosphoglycerate to increase. These changes in the phosphate : 3-phosphoglycerate ratio in the chloroplast brought about by changes in the rate of sucrose synthesis allow the generation of the following simple model for the regulation of starch synthesis.

The response of the first committed enzyme of starch synthesis, ADPglucose pyrophosphorylase, to changes in the phosphate : 3-phosphoglycerate ratio (section 1.4.2.1) make it an ideal candidate for the regulation of starch synthesis. When the ratio rises the enzyme will be activated, allowing more starch synthesis to occur. Conversely, when the ratio falls the activation state of the enzyme and hence the rate of starch synthesis will fall. Thus, for example, if the rate of sucrose synthesis declines because of a fall in sink demand, the phosphate : 3-phosphoglycerate ratio in the chloroplast, and hence the rate of starch synthesis, will increase.

The evidence that this is the primary mechanism of regulation of starch synthesis is not complete. Although changes in the 3-phosphoglycerate levels in chloroplasts can be measured, it is extremely difficult to make reliable estimates of changes in the levels of phosphate in chloroplasts under normal physiological conditions *in vivo*. It is thus not possible to assess the relationship between changes in the activation state of ADPglucose pyrophosphorylase and the rate of starch synthesis *in vivo* by direct measurements. However, recent elegant experiments by Stitt and colleagues have allowed the importance of ADPglucose pyrophosphorylase in the regulation of starch synthesis to be estimated through the application of metabolic control theory (Torres *et al.*, 1986).

The rates of starch and sucrose synthesis in wildtype plants and mutant plants with reduced activities of four enzymes of the pathway of starch synthesis (plastidial phosphoglucose isomerase and phosphoglucomutase (Figure 1.2), ADPglucose pyrophosphorylase and starch-branching enzyme) were used to assess the flux control coefficients of each of these enzymes (Neuhaus and Stitt, 1990). In low light ADPglucose pyrophosphorylase was the only one of the enzymes which exerted any significant control over starch synthesis. In high light it was still the single most important enzyme, but the control was more evenly partitioned. All of the other enzymes exerted some control. Even when ADPglucose pyrophosphorylase was exerting a very high level of control, its activity *in vivo* was no more than 20% of its maximum catalytic activity. It seems that the enzyme is usually strongly inhibited *in vivo*. In low light, factors involved in light harvesting exerted the greatest influence over control. This is reflected in the fact that a doubling of light intensity in low light conditions almost doubled the rate of starch synthesis. Sucrose synthesis exerted a strong negative control on starch synthesis, consistent with the view that control of partitioning occurs primarily in the cytosol.

The extent to which these results are applicable to starch synthesis under ambient conditions is not clear. Measurements were made in saturating levels of carbon dioxide, and many additional factors concerned with the supply and assimilation of carbon dioxide are likely to exert control in ambient conditions. Of the four mutations used, one was in pea, one in *Clarkia* and two in *Arabidopsis*, hence the picture that emerges is not for a single pathway of starch synthesis. The control coefficients of starch synthase and alkaline pyrophosphatase were not assessed. Nonethless, three important conclusions from this study are likely to be generally applicable. First, within the pathway of starch synthesis itself the activity of ADPglucose pyrophosphorylase is probably the single most important factor in determining flux. Second, the contribution of other enzymes of the pathway can, under some conditions, be substantial. Third, as expected from the general complexity of the control of photosynthetic metabolism, factors outside the pathway exert strong negative and positive control. The nature and extent of this control change dramatically with changing external conditions.

An interesting fact to emerge from these experiments is that starch-branching enzyme, which does not catalyse any net synthesis of starch, can contribute significantly to the control of flux through the pathway under high light conditions (Neuhaus and Stitt, 1990). The mechanism by which low branching-enzyme activity reduces the flux of carbon into starch is probably as follows. A reduction in activity will lead to a reduced availability of non-reducing ends of starch polymers, one of the two substrates of starch synthase. Consistent with this, mutant pea leaves with reduced activity of branching enzyme have up to seven-fold greater amounts of the other substrate, ADPglucose, than wildtype leaves (Smith *et al.*, 1990b). This elevated level of ADPglucose could act to reduce overall flux through the pathway in two ways. First, ADPglucose pyrophosphorylase may be inhibited by high concentrations of ADPglucose (Preiss and Levi, 1979). Second, as much as half of the adenine-nucleotide pool of the chloroplast may be sequestered as ADPglucose in the mutant leaf (Smith *et al.*, 1990b). The lowered stromal ATP levels will restrict both ADPglucose synthesis and the operation of the Calvin cycle.

The regulation of partitioning of carbon into starch in leaves in the longer term, the basis of interspecific differences in partitioning, and the response of partitioning to environmental change, are less well understood. It is likely that in all of these cases modulation of the activity of sucrose phosphate synthase, leading to an altered capacity for sucrose synthesis and hence a change in partitioning, will play an important role (Huber *et al.*, 1985; Stitt *et al.*, 1987; Stitt and Quick, 1989). It is not known whether changes also occur in the maximum catalytic activities or regulatory properties of enzymes of starch synthesis. The demonstration that genes encoding subunits of ADPglucose pyrophosphorylase are differentially expressed in response to both developmental and metabolic signals (Müller-Röber *et al.*, 1990) raises the possibility that changes in this enzyme could contribute to longer-term changes in partitioning.

1.5.1.3 Storage organs. Regulation of starch synthesis in storage organs might be expected to be different from that described above for leaves. The storage organ receives its carbon as sucrose, at a rate which is unlikely to exhibit marked short-term fluctuations. Starch and many of the other major products of sucrose dissimilation are probably not subject to significant degradation during the developmental period. There is thus no obvious requirement in the developing organ for the sorts of regulatory mechanisms which operate in leaves to allow sensitive responses to externally imposed fluctuations in carbon supply and demand.

There is insufficient information about starch synthesis in storage organs to assess in any detail how it is regulated. In this section we review the limited evidence available, and speculate about where control may lie. We shall consider the general relationship between sucrose import and starch synthesis in starch-storing organs, and the way in which the partitioning of carbon between starch synthesis and other metabolic pathways may be controlled (Figure 1.7). We shall then discuss the extent to which individual steps on the committed pathway of starch synthesis may contribute to the control of flux through the pathway.

Control of partitioning. In starch-storing organs, a very high proportion of the sucrose entering the organ may be converted to starch. Starch often constitutes as much as half of the final dry weight of the organ. In developing pea embryos and potato tubers, over half of the sucrose metabolised during feeding experiments was converted to starch (Edwards and ap Rees, 1986a; Oparka and Wright, 1988a). Both the uptake of sucrose and the extent to which it is converted to starch rather than stored as sucrose are extremely sensitive to cell turgor in discs of growing potato tubers (Oparka and Wright, 1988a, b). The percentage of sucrose that is converted to starch shows a sharp optimum at a cell turgor of 80 kPa (Oparka and Wright, 1988b). The cause of this sensitivity is not known, but it could be of great importance in regulating the partitioning of sucrose into starch in intact tubers. Cell turgor in tubers in the field varies considerably on a diurnal basis. The lowest values, achieved during the day, are close to the optimum for maximum partitioning of sucrose into starch. There is thus likely to be a diurnal rhythm in partitioning in the field. Consistent with this prediction, tubers grown in continuously water-stressed conditions which resulted in turgor values close to optimum for starch synthesis had 87% of their dry weight as starch, whereas tubers grown in irrigated conditions had 65% of their dry weight as starch (Oparka and Wright, 1988b).

Partitioning of carbon from sucrose between starch synthesis, other biosynthetic pathways and respiration occurs primarily at two points: the hexose phosphate pools in the cytosol and the amyloplast (Figure 1.7). In the cytosol, hexose phosphate is partitioned between the hexose phosphate translocator of the amyloplast envelope and cytosolic glycolysis, which provides the substrate for respiration and a host of biosynthetic pathways. In the amyloplast, hexose phosphate is probably partitioned primarily between starch synthesis and the oxidative pentose phosphate pathway. Preparations of amyloplasts from developing pea embryos are capable of metabolising supplied glucose 6-phosphate via both of these pathways (Hill and Smith, 1991; J. Foster and A. Smith, unpublished results). The extent to

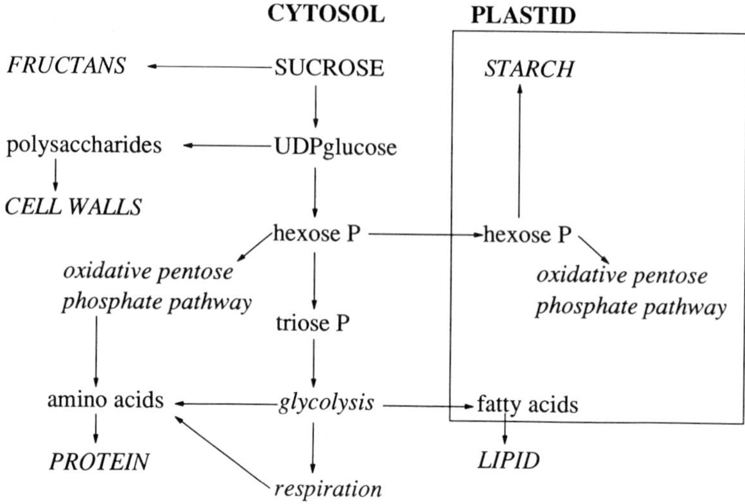

Figure 1.7 Possible fates of sucrose in the cells of developing, non-photosynthetic storage organs.

which hexose phosphate in the amyloplast is metabolised via glycolysis is not known.

Two major areas of ignorance about non-photosynthetic starch synthesis limit our understanding of partitioning in the cytosol. First, the mechanism of hexose phosphate uptake is unknown (section 1.4.1.2). Starch synthesis releases—via alkaline pyrophosphatase—the phosphate imported as hexose phosphate, and it seems likely that hexose phosphate might enter in exchange for this phosphate. This mechanism has been suggested for the hexose phosphate translocator of pea root plastids (Borchert *et al.*, 1989).

Second, the means of supply of ATP for starch synthesis is not known. It could be supplied by direct import of ATP (Figure 1.5) or a metabolite shuttle, for example the exchange of triose phosphate and 3-phosphoglycerate. The flux of ATP or a metabolite from which it can be generated must equal the flux of hexose phosphate into starch synthesis.

The fluxes of glycolytic metabolites across the amyloplast envelope during starch synthesis could allow the interaction of starch synthesis with processes which regulate the rate of entry of sucrose into glycolysis. Sucrose probably enters glycolysis in storage organs via sucrose synthase and UDPglucose pyrophosphorylase (Morrell and ap Rees, 1986; ap Rees, 1988; Edwards and ap Rees, 1986b). Synthesis of GlP from UDPglucose requires pyrophosphate. Control of pyrophosphate levels in the cytosol may rest with pyrophosphate-dependent phosphofructokinase, which generates pyrophosphate in the gluconeogenic direction (ap Rees *et al.*, 1985). This enzyme allows rapid cycling between pools of hexose phosphates and triose phosphates in the cytosol, such that the two pools are close to equilibrium (Hatzfeld and Stitt, 1990). Processes such as starch synthesis which affect levels of metabolites involved in this equilibrium could potentially affect the rate of entry of sucrose into glycolysis.

Information about the control of partitioning can be gained from plants with mutations that reduce starch synthesis through effects on the committed pathway. The extent to which the flux of carbon through other pathways is affected by decreases in the flux into starch varies from one mutation, and one storage organ, to another. However, the total increase in fluxes of carbon through other pathways of hexose phosphate metabolism is usually much less than is required to account for the decrease in the flux into starch. The remainder of the decrease is accounted for by one or a combination of sucrose accumulation, synthesis of fructans or polysaccharides derived from UDPglucose (Figure 1.7), and a decline in the rate of dry-matter accumulation of the organ.

Accumulation of sucrose is almost universal in organs in which a mutation has resulted in a decrease in starch synthesis. It is caused, for example, by the mutation at the *r* and *rb* loci of pea (Wang and Hedley, 1991), the *amy-1* and *shx* loci of barley (Shannon and Garwood, 1984; Schulman and Ahokas, 1990), and the *ae, du, su, sh-2*, and *bt-2* loci of maize (Shannon and Garwood 1984). Genetically manipulated potatoes which almost completely lack ADPglucose pyrophosphorylase activity and hence starch, have very high sucrose contents (L. Willmitzer, personal communication). The mutation at the *r* locus of peas causes a large increase in hexose phosphate as well as sucrose in the developing embryo (Edwards and ap Rees, 1986b), and this may be true of the other mutations as well. These results indicate that there is little coordinate regulation of fluxes through pathways of hexose-phosphate metabolism. The most important factors in the control of pathways such as lipid and amino acid biosynthesis and respiration must be distant from the hexose and triose phosphate pools of the cytosol and amyloplast (Figure 1.7).

Many observations indicate that there may be complex indirect interactions between sucrose levels, starch synthesis, and the synthesis of specific storage proteins. For example, the mutation at the *r* locus of peas, which reduces the rate of starch synthesis via a lesion in starch-branching enzyme, specifically reduces the amount of one of the three sorts of storage globulin in the embryo (Davies, 1980). Several mutations which affect the endosperm of barley ('high-lysine' mutants) reduce both its starch content and the levels of prolamins, the major storage proteins (Doll, 1984). In maize, mutations in genes known to encode enzymes of the pathway of starch synthesis (*sh-2, bt-2*) and mutations where the primary effect is believed to be on storage proteins (*opaque, floury*), all decrease the synthesis of both starch and zeins (DiFonzo *et al.*, 1978; Tsai *et al.*, 1970).

The precise explanations for the above examples are not known. However, mechanisms which might underlie the indirect linkage between sucrose, starch and other storage products can be suggested. First, the expression of genes that encode storage proteins and enzymes of starch synthesis may be regulated by sucrose. High levels of sucrose induce expression of genes that encode one of the subunits of ADPglucose pyrophosphorylase and the storage protein patatin in potatoes (Müller-Röber *et al.*, 1990; Rocha-Sosa *et al.*, 1989). It is tempting to speculate that expression of other proteins involved in starch synthesis might be regulated in this way. Coarse control of the pathway of starch synthesis and of the synthesis of storage proteins by sucrose might provide a means of linking the availability of photosynthate to the initiation of storage organs in plants.

Second, high osmotic potential may selectively affect the stability of some mRNAs. The mutation at the *r* locus of peas decreases the levels of legumin and its transcript in developing embryos, but appears to have no effect on the rate of transcription of legumin genes (Turner *et al.*, 1990). The legumin mRNA may thus be unstable at the higher levels of sucrose, and hence of osmotic pressure, in the mutant embryo.

Control within the pathway. There is no quantitative information about the partitioning of control of starch synthesis in non-photosynthetic organs. The effects of the mutations that alter the rate of starch synthesis do, however, allow some suggestions to be made about the importance of some of the individual steps on the pathway.

ADPglucose pyrophosphorylase. Although regulation of this enzyme is the key factor in the control of starch synthesis in leaves, there is no good evidence that this is so in non-photosynthetic organs. There is no requirement in these organs for a rapid response to the sorts of changes in the phosphate : 3-phosphoglycerate ratio which signal fluctuations in carbon supply and demand in leaves. However, the activity of all reported ADPglucose pyrophosphorylases is modulated by phosphate : 3-phosphoglycerate ratios. Thus although the activity of ADPglucose pyrophosphorylase in non-photosynthetic organs usually greatly exceeds the rate of starch synthesis, it may well be substantially inhibited *in vivo*, just as it is in leaves.

Evidence that the subunit composition and regulatory properties of ADPglucose pyrophosphorylase may be different in leaves and storage organs was presented in section 1.4.2.2. This could be taken to indicate that the regulatory role of the enzyme differs in photosynthetic and non-photosynthetic organs. However, the evidence currently available is not conclusive.

The existence of mutations which specifically reduce the activity of ADP glucose pyrophosphorylase in maize endosperm and pea embryos, and of genetically manipulated potato tubers with reduced activity of the enzyme, should allow its control coefficient to be assessed for non-photosynthetic organs in the same way as for leaves (section 1.5.1.2). However, in all of these cases the reduction in activity of the enzyme is apparently due to the loss of a putative subunit. This may result in the alteration of the kinetic properties as well as the total activity of the enzyme. This is certainly the case in pea embryo: the residual activity in mutant pea embryos is much more sensitive to phosphate : 3-phosphoglycerate ratios than is the wildtype enzyme (C. Hylton and A. Smith, unpublished data). The *sh-2* and *bt-2* mutations that reduce ADPglucose pyrophosphorylase activity in maize endosperm also increase the K_m of the enzyme for GIP (Hannah and Nelson, 1976). This renders interpretation of control coefficients measured with these mutants very difficult. This criticism may also apply to the *Arabidopsis* mutant used to measure the control coefficient of the enzyme in leaves (Neuhaus and Stitt, 1990).

Starch synthase. The fact that *waxy* mutants have only low activities of granule-bound starch synthase, yet have wildtype levels of starch in their storage organs (section 1.4.3.1) might be taken to indicate that granule-bound starch synthase exerts no control over the rate of starch synthesis, and hence that any control at this step is solely due to the soluble enzyme. However, interpretation of the effects of *waxy* mutations is complicated by the existence of granule-bound isoforms of starch synthase which are not encoded by the *waxy* genes, and which may also be present in the soluble fraction (section 1.4.3.1). The roles of these isoforms in starch synthesis and their fates in *waxy* mutants are not known (see section 1.5.2.1).

There are no known mutations which specifically affect the activity of soluble starch synthase. Mutations at the *Dull* locus of maize and the *Shx* locus of barley both cause substantial reductions in the activity of soluble starch synthase, but also affect activities of other enzymes (Boyer and Preiss, 1981; Schulman and Ahokas, 1990). The primary lesions caused by these mutations are not known.

The activity of soluble starch synthase in non-photosynthetic organs is low. The difficulties of assaying this enzyme in crude extracts may lead to underestimates of its activity, but it is likely that its maximum catalytic activity does not greatly exceed the rate of starch synthesis (section 1.4.3.1). Limited evidence that significant control might be exerted at this step under some conditions comes from studies of the effects of high temperatures on starch synthesis in storage organs. The rate of starch synthesis in many organs does not increase, or actually declines, with increasing temperature above about 25–30°C (Krauss and Marschner, 1984; Bhullar and Jenner, 1986). This unusual response has been attributed, at least in part, to thermal lability of soluble starch synthase. Soluble starch synthase activity in extracts of developing wheat grains declines as the temperature at which the ears are treated prior to extraction is increased above about 30°C (Rivjen, 1986; Caley *et al.*, 1990). The Q_{10} of both isoforms of soluble starch synthase from pea embryos is low and falls slightly in the temperature range 20–35°C. The activity is rapidly destroyed at 40–45°C (Denyer and Smith, 1992). However, the importance of this enzyme in the general effect of temperature on starch synthesis is not yet clear. The effects of temperature on the activities of other components of the pathway have not yet been fully examined.

Starch-branching enzyme. Reductions in the activity of branching enzyme can cause substantial reductions in the rate of starch synthesis in non-photosynthetic organs. The mutation at the *r* locus of peas, which causes the loss of one of two isoforms of the enzyme from the embryo, reduces the rate of starch synthesis during development such that the final starch content of the seed is 40% lower than the wildtype. The mechanism by which reduced branching-enzyme activity reduces the overall rate of starch synthesis is probably similar to that proposed for the mutation in leaves (section 1.5.1.2). Consistent with this, the level of ADPglucose in mutant embryos is considerably greater than in wildtype embryos (Edwards *et al.*, 1988). No quantitative assessment of the role of this enzyme in controlling flux in non-photosynthetic organs has yet been made. There are problems with the use of mutations which reduce the activity of one of two or more isoforms in making such assessments, akin to the problems described for mutations affecting the subunits of ADPglucose pyrophosphorylase. Because isoforms of starch-branching enzyme differ in their kinetic properties (section 1.4.4.1), the activity in mutant plants is likely to be qualitatively different from that in wildtype plants. This will render interpretation of control coefficients difficult. This problem also applies to the use of the mutant at the *r* locus to measure the control coefficient of branching enzyme in leaves (section 1.5.1.2).

1.5.2 Determination of starch structure

Our current understanding of the pathway of starch synthesis is insufficient to allow satisfactory explanations of how the structure of starch is determined. We have no clues about the biochemical origins of several features of starch structure. Most types of starch granule have several integral proteins of no known function, which may be involved in the determination of structure. The only aspect of structure which has received sustained attention from biochemists is the amylose to amylopectin ratio. This is in part because it is superficially one of the easiest aspects of structure to quantify, and also because it is affected by well-characterised mutations in several storage organs. In this section we shall review current understanding of the amylose to amylopectin ratio, and speculate on the origins of some other aspects of starch structure.

1.5.2.1 The amylose to amylopectin ratio. The amylose to amylopectin ratio is likely to be determined largely by the two enzymes responsible for polymer synthesis, i.e. starch synthase and starch-branching enzyme. Most of the available information about amylose synthesis comes from study of *waxy* mutations, which result in the loss of the amylose component of starch.

As discussed in section 1.4.3.1, the *waxy* mutations are in genes that encode a GBSS. They result in the loss of most of the GBSS activity but have no effect on soluble starch synthase activity. This suggests strongly that GBSS is responsible for amylose synthesis. The mechanism by which this occurs is not understood. It has been suggested that the location of the *waxy* gene product on the granule allows the polymer it synthesises to escape the action of branching enzyme. It was argued that soluble starch synthase and starch-branching enzyme (also held to be soluble) act together to elongate amylopectin molecules. Their product crystallises into the surface of the granule. The waxy protein binds onto the crystalline matrix where its product, formed within the matrix, is essentially unavailable to the branching enzymes. The product could be physically separated from the branching enzymes, or, because of the environment in which it is located in the matrix, it could take on a conformation unfavourable to branching-enzyme activity. Branching enzyme is believed to act preferentially on double-helical structures, whereas amylose within the granule appears not to form double helices (sections 1.3.1.1. and 1.4.4.1; Roybt, 1984; Guilbot and Mercier, 1985).

The existence of other forms of GBSS, not encoded by the *waxy* genes, casts some doubt on this simple model (section 1.4.3.1). It is not yet known whether these forms synthesise amylose or amylopectin in wildtype plants, and how they behave in *waxy* mutant plants. However, the granule-bound fraction of their activity may not be important in starch synthesis *in vivo*. The enzymes may be involved in polymer synthesis in a soluble layer at the granule surface, then become trapped into the granule as their product crystallises into the surface. The layer of active synthesis will be lost during purification of granules, and isolated purified granules may therefore contain only the fraction of the enzyme trapped within the crystalline matrix. The fact that the starch synthase activity of isolated granules is greatly increased by mechanical damage to the granules is consistent with the idea that much of this activity is trapped inside the granule in a way which may render it incapable of starch synthesis *in vivo*. In contrast, the waxy GBSS is not found in the soluble fraction, and must therefore carry out its physiological function while attached to the granule. The locations of the physiologically important fractions of the waxy and non-waxy forms of GBSS may therefore differ *in vivo*.

Other models to explain the role of the waxy protein in amylose synthesis can be put forward. For example, the protein might protect a part of the product of starch synthase from the action of branching enzyme, rather than itself synthesising a polymer unavailable to branching enzyme. However, the number of amylose molecules with ends at the periphery of the granule is likely to be greater than the number of waxy protein molecules in the granule. It has been estimated that 3.5×10^8 chains of amylose molecules project from the surface of a granule with a diameter of 15 μm and a weight of 2.65×10^{-9}g. This assumes a degree of polymerisation for amylose of 1000 and that the molecules lie perpendicular to the granule surface (French, 1984). We know from our work on pea starch that the waxy protein is distributed throughout the interior of the granule, and that there is about 1.5 mg of this protein per gram of starch. A granule of 2.65×10^{-9}g would therefore contain 6×10^6 molecules of waxy protein. On this basis it seems unlikely that the waxy protein prevents branching by a simple blanketing mechanism.

1.5.2.2 Other aspects of starch structure. The *waxy* mutations have almost no effect on aspects of starch structure other than the amylose to amylopectin ratio. The structure of amylopectin in *waxy* mutants, for example, is the same as that of wildtype starch (Shannon and Garwood, 1984). However, other mutations which alter the amylose to amylopectin ratio have complicated effects on polymer size and structure and on granule morphology. As discussed in section 1.3.1.1, amylose and amylopectin are convenient working definitions rather than tightly defined classes of polymer. Factors that decrease the degree of branching of amylopectin polymers may thus increase the apparent amylose content of the starch without having any direct effect upon the normal mechanism of amylose synthesis mediated by the waxy protein. The factors which determine granule morphology are not understood, but it is likely that changes in amylopectin branching can affect morphology through alterations in the nature or degree of crystallinity of the starch.

The mutations at the *ae* locus of maize and the *r* locus of pea both increase the amylose to amylopectin ratio through action on the degree of branching of amylopectin (section 1.4.4.1). Unlike the wildtype starch, mutant maize starches contain loosely branched polymers intermediate in size between amylose and amylopectin, and branched polymers of low molecular weight (Yeh *et al.*, 1981). The average polymer size in both the amylose and amylopectin fractions of mutant pea starch is reduced relative to wildtype starch, and branched polymers of low molecular weight are present (Colonna and Mercier, 1984). The simple morphology of the wildtype granule becomes elongated and irregular in the mutant maize endosperm, and deeply-fissured and compound in appearance in the mutant pea embryo (Banks and Muir, 1980).

At present, the ways in which isoforms of starch synthase and starch-branching enzyme interact to synthesise amylopectin molecules *in vivo* are not known. It seems likely, however, that the structure of amylopectin is determined by the nature of the isoforms of starch synthase and starch-branching enzyme involved in its synthesis. First, differences in starch structure between organs and through time are matched by differences in the expression of isoforms of these enzymes. Second, different isoforms of branching enzyme from the same organ have different substrate affinities and produce polymers of different structures *in vitro*. Third, the mutations at the *r* and *ae* loci, which result in the loss of one out of two or three forms of branching enzyme from the organs upon which they act, have profound effects on the structure of amylopectin.

It has been suggested that physical interactions between isoforms of starch synthase and starch-branching enzyme are important in determining the structure of amylopectin during its synthesis *in vivo* (Preiss and Levi, 1982). The fact that particular isoforms of the two enzymes copurify to a striking extent during extraction from some organs has been cited as evidence for this (Schiefer *et al.*, 1978; Boyer and Preiss, 1981). However, copurification may reflect an association of two proteins with a fraction of glucan rather than with each other (Hawker *et al.*, 1974). There is at present no good evidence of physical interactions between the two enzymes *in vivo*.

1.5.2.3 Indirect effects on structure. Mutations which have direct effects on enzymes other than starch synthase and starch-branching enzyme may result in both decreased starch content and altered amylose to amylopectin ratio in storage organs. For example, the *rb* mutation of peas, which is believed to have a specific effect on ADPglucose pyrophosphorylase (Smith *et al.*, 1989), increases the amylopectin content of the mature seed from 70 to 87% of the total starch (Wang and Hedley, 1991). There is insufficient information about the detailed effects of these mutations to allow the mechanism by which they affect polymer synthesis to be deduced. Three possible explanations are as follows.

First, changes in the rate of synthesis of ADPglucose might bring about changes in the ratio of amylose to amylopectin if, for example, the K_m for ADPglucose of a starch synthase responsible for amylose synthesis was different from that of starch synthases involved in amylopectin synthesis. If this explanation is correct and generally applicable, it has profound consequences for the synthesis of starch in leaves. The rapid and frequent fluctuations which can occur in the rate of synthesis of ADPglucose in leaves would give rise to corresponding fluctuations in the amylose to amylopectin ratio through the starch granule.

Second, if the mutations affect the rate of starch synthesis only or primarily at particular times during development, rather than throughout, they may alter the amylose to amylopectin ratio of the mature granule without affecting the relative rates of synthesis of the two sorts of polymer at any point in development. This is because the ratio changes considerably during the development of a normal granule (section 1.3.1.3). A reduction in the overall rate of synthesis at a particular time in development can thus affect the final ratio.

Third, mutations which reduce the rate of starch synthesis through action on one particular enzyme could have indirect effects on the maximum catalytic activity and/or isoform composition of starch synthase and starch-branching enzyme, and hence on the amylose to amylopectin ratio. Several mechanisms could bring this about. For example, increases in sucrose concentration resulting from the decrease in starch synthesis could induce expression of metabolically regulated genes. The change in osmotic potential of the organ as a result of sucrose accumulation could have differential effects on the stabilities of mRNAs, and hence on the activities of particular isoforms of enzymes (section 1.5.1.3).

1.5.3 Changes during development

As discussed in preceeding sections, the isoform and subunit composition of enzymes of starch synthesis probably changes during the development of all starch-storing organs. These changes will alter the partitioning of control of both the overall flux through the pathway, and the flux into particular structures of polymer. The final amount and structure of starch in a storage organ must thus be regarded as intimately linked, and an integral of a continuously changing pattern of control. The widely held view of starch as the product of a single pattern of control, established through coordinate regulation of gene expression early in development, is certainly not correct.

1.6 Potential for the genetic manipulation of starch

The existence of well-characterised mutations which alter the quality and/or quantity of starch in plant organs illustrates the potential for directed genetic manipulation of starch. In this section we shall discuss the feasibility of manipulation to increase or decrease the starch content of plant organs and to alter the quality of starch. The fact that the harvestable, starch-storing organs of plants are also usually the means by which the crop is propagated may limit the sorts of modifications which can be carried out. We shall consider the difficulties which this may present.

1.6.1 Alteration of starch content

The use of genetic manipulation to reduce the capacity for starch synthesis of plant organs may be agronomically desirable for several reasons. First, a reduced flux of carbon into starch in a storage organ may allow increased synthesis of other storage products such as lipid (see chapter 3) and protein. This could enhance the nutritional properties or industrial value of the organ, or permit simpler extraction of its products. Mutations like that at the *r* locus of peas demonstrate that this approach is feasible. The reduction in branching enzyme activity caused by the mutation results in a decrease in the starch content of the seed, an increase in its sucrose and lipid content, and an alteration in its storage protein composition (Bhattacharyya *et al.*, 1990). These changes are of commercial importance: wildtype peas are grown primarily for animal feed whereas the sweeter and more palatable mutant seeds are used for direct human consumption.

The *r* mutation also illustrates the inherent difficulties in making directed changes in the overall composition of storage organs via manipulation of starch synthesis. As with most mutations that reduce the rate of starch synthesis, carbon is diverted primarily into sucrose storage and cell wall synthesis, rather than lipid and protein (section 1.5.1.3). Many of the changes caused by the mutation are thought to be complicated secondary consequences of the osmotic effect of the increased sucrose content of the seed (Bettey and Smith, 1990; Turner *et al.*, 1990). It is thus likely to be difficult to predict the effect of a manipulated reduction in starch synthesis upon the levels of other storage products, and the effect on valuable products like lipid and protein will probably be small. The general effects of an increased sucrose content in the storage organ upon the performance of the plant in the field are also difficult to predict and could be detrimental. For example, these organs may be more liable to attack by pathogens during the early stages of germination or sprouting if they contain high levels of sucrose than if they contain starch.

Reduction or elimination of the capacity for starch synthesis in a storage organ might allow diversion of carbon from sucrose into other pathways introduced into the organ via genetic manipulation. Fructans, for example, are synthesised directly from sucrose. A starch-storing organ rendered starch-free by genetic manipulation might be converted to a fructan-storing organ by further manipulation.

Dramatic reductions in starch synthesis via genetic manipulation have already been achieved. Elimination of one of the two subunits of ADPglucose pyrophosphorylase from potato tubers by antisense transformation almost eliminates enzyme activity and starch from the tuber, and results in a very large accumulation of sucrose (L. Willmitzer, personal communication).

An ability to increase starch accumulation through genetic manipulation could potentially improve the agronomic value of starch-storing organs by increasing their dry-matter content. This would be of particular value in the potato crop, where an increase in the relatively low dry-to fresh-weight ratio would make the tubers more economical to harvest and transport, and improve their properties for the manufacture of potato-based processed foods.

In the absence of definitive information about the control of flux through the pathway of starch synthesis in storage organs, ADPglucose pyrophosphorylase is perhaps the best target for attempts to increase flux by genetic manipulation. It would probably be less difficult technically to introduce the bacterial than the plant enzyme into plants by genetic manipulation, because this enzyme has only one sort of subunit. The important regulatory sites of the bacterial enzyme have been defined (Preiss, 1990), and can be eliminated by protein engineering and mutations. Preliminary reports of a successful attempt to increase the starch content of potato tubers by manipulation of starch synthesis have recently appeared (Fraley *et al.*, 1991).

1.6.2 Alteration of starch structure

The ability to alter starch structure has many potential applications. It could reduce dependence on chemical modification of starch for industrial use, and it might generate starches with novel physical properties and hence new sorts of industrial applications. The relationship between the physical properties of starch and its underlying chemistry is poorly understood, and it is not yet clear what modifications at the structural level are required to achieve any particular set of physical properties. Equally, the relationship between structure and the pathway of synthesis is not understood in sufficient detail to allow many sorts of potentially desirable modifications to be made in a directed manner. However, the very wide genetic variation for starch structure that exists in crop plants shows that there is great potential for modification of structure through genetic manipulation in the future.

Manipulation of structure can probably be achieved in several different ways. The maximum catalytic activities of selected isoforms of starch synthase and/or starch-branching enzyme can be increased or decreased, and the timing of their appearance during development can be altered. Isoforms of these enzymes from other species of plant, or of related enzymes from other types of organism, can be introduced, with or without prior elimination of endogenous isoforms. Enzymes which modify starch but are either not normally expressed during its synthesis or not naturally found in plants, for example debranching and disproportionating enzymes and cyclodextrin glycosyltransferase, can be introduced. The availability of organ-specific, metabolically regulated and inducible promoters should make it possible to express an introduced gene construct specifically in the harvested organ of the plant and at the particular point in organ development when its effect will be greatest.

Alterations in starch structure brought about by genetic manipulation of potatoes have recently been reported. Expression in the tuber of an antisense construct derived from the *waxy* gene reduces GBSS activity and the amylose content of the starch (Visser *et al.*, 1991). Conversely, the amylose content of the tuber starch of the *amf* mutant can be increased by transformation with the wildtype *waxy* gene (van der Leij *et al.*, 1991b). Tubers with amylose contents from zero to wildtype levels have been produced by these methods. Attempts are also being made to produce cyclodextrins from starch in potato tubers by introduction of a bacterial gene encoding cyclodextrin glycosyltransferase (Shewmaker *et al.*, 1991).

Genetic manipulation of starch structure may have several commercially undesirable side effects. Alterations which reduce the branching of starch may well reduce the rate of starch synthesis and hence the yield of starch (sections 1.5.1.2 and 1.5.1.3). If reliable methods of increasing the yield of starch through genetic manipulation can be established (see above), it may be possible to use them to compensate for any loss of yield incurred through alteration of structure.

Alteration of structure may also alter granule size and/or morphology, and this may have detrimental effects on the extractability of the starch. For example, the mutation at the *r* locus of pea alters granule shape from simple to deeply fissured. The fissured granules tend to fracture into small fragments upon extraction.

The accumulation of novel starch polymers might reduce the viability of the reproductive organs of a transgenic plant. If the novel structure can be attacked only slowly, or not at all, by the starch-degrading enzymes of the plant, germination or sprouting of the propagules may be adversely affected. The solution to this type of problem may come from new methods of regulating the expression of introduced genes. It may be possible in future to allow the production on transgenic plants of propagules which either contain modified starches, or are phenotypically wildtype. The former would provide the raw material for industry, and the latter the means of propagation of the cultivar.

Acknowledgment

We are grateful to all of our colleagues for valuable discussions during the preparation of this chapter and to Professor Lothar Willmitzer for making unpublished results from his laboratory available to us.

References

Adams, C.A., Rinne, R.W. and Fjerstad, M.C. (1980) Starch deposition and carbohydrase activities in developing and germinating soya bean seeds. *Ann. Bot.* **45**: 577–582.

Alban, C., Joyard, J. and Douce, R. (1988) Preparation and characterisation of envelope membranes from non-green plastids. *Plant Physiol.* **88**: 709–717.

Anderson, J.M., Hnilo, J., Larson, R., Okita, T.W., Morell, M. and Preiss, J. (1989) The encoded primary sequence of a rice seed ADP-glucose pyrophosphorylase and its homology to the bacterial enzyme. *J. Biol. Chem.* **264**: 12238–12242.

Anderson, J.M., Okita, T.W. and Preiss, J. (1990) Enhancing carbon flow into starch: the role of ADPglucose pyrophosphorylase. In *The Molecular and Cellular Biology of the Potato* (Vayda, M.E. and Park, W.D., eds), CAB International, Wallingford, Oxford, pp. 159–179

ap Rees, T. (1988) Hexose phosphate metabolism by nonphotosynthetic tissues of higher plants. In *The Biochemistry of Plants, Vol. 14: Carbohydrates* (Preiss, J., ed.), Academic Press, San Diego, pp. 1–34.

ap Rees, T., Fuller, W.A. and Wright, B.W. (1976) Pathways of carbohydrate oxidation during thermogenesis by the spadix of *Arum maculatum*. *Biochim. Biophys. Acta* **437**: 22–35.

ap Rees, T., Wright, B.W. and Fuller, W.A. (1977) Measurements of starch breakdown as estimates of glycolysis during thermogenesis by the spadix of *Arum maculatum* L. *Planta* **134**: 53–56.

ap Rees, T., Leja, M., Macdonald, F.D. and Green, J.H. (1984) Nucleotide sugars and starch synthesis in spadix of *Arum maculatum* and suspension cultures of *Glycine max*. *Phytochem.* **23**: 2463–2468.

ap Rees, T., Morell, S., Edwards, J., Wilson, P.M. and Green, J.H. (1985) Pyrophosphate and the glycolysis of sucrose in higher plants. In *Regulation of Carbon Partitioning in Photosynthetic Tissue* (Heath, R.L., and Preiss, J., eds), Waverley Press, Baltimore, pp. 27–44.

Baba, T., Noro, M., Hiroto, M. and Arai, Y. (1990) Properties of primer-dependent starch synthesis cataylsed by starch synthase from potato tubers. *Phytochem.* **29,** 719–723.

Badenhuizen, N.P. (1969) *The Biogenesis of Starch Granules in Higher Plants*. Appleton-Century-Crofts, New York.

Bae, J.M., Giroux, M. and Hannah, L. (1990) Cloning and characterisation of the *brittle-2* gene of maize. *Maydica* **35**: 317–322.

Baecker, P.A., Greenberg, E. and Preiss, J. (1986) Biosynthesis of bacterial glycogen. Primary structure of *Escherichia coli* 1,4-α-D-glucan: 1,4-α-D-glucan 6-α-D-(1,4-α-D-glucano)-transferase as deduced from the nucleotide sequence of the *glgB* gene. *J. Biol. Chem.* **261**: 8738–8743.

Banks, W. and Greenwood, C.T. (1975) *Starch and its Components*. Edinburgh University Press, Edinburgh.

Banks, W. and Muir, D.D. (1980) Structure and chemistry of the starch granule. In *The Biochemistry of Plants, Vol. 3: Carbohydrates: structure and function* (Preiss, J., ed.), Academic Press, New York, pp. 321–369.

Barlow, P.W. (1975) The root cap. In *The Development and Function of Roots* (Torrey, J.G. and Clarkson, D.T., eds), Academic Press, London, pp. 22–54.

Beck, E. (1985) The degradation of transitory starch granules in chloroplasts. In *Regulation of Carbon Partitioning in Photosynthetic Tissue* (Heath, R.L. and Preiss, J., eds), Waverley Press, Baltimore, pp. 27–44.

Bettey, M. and Smith, A.M. (1990) Nature of the effect of the *r* locus on the lipid content of embryos of peas (*Pisum sativum* L.) *Planta* **180**: 420–428.

Bhattacharyya, M.K., Smith, A.M., Ellis, T.H.N., Hedley, C. and Martin, C. (1990) The wrinkled-seed character of peas described by Mendel is caused by a transposon-like insertion in a gene encoding starch branching enzyme. *Cell* **60**: 115–122.

Bhave, M.R., Lawrence, S., Barton, C. and Hannah, L.C. (1990) Identification and molecular characterisation of *shrunken-2* cDNA clones of maize. *Plant Cell* **2**: 581–588.

Bhullar, S.S. and Jenner, C.R. (1986) Effects of temperature on the conversion of sucrose to starch in the developing wheat endosperm. *Aust. J. Plant Physiol.* **13**: 605–615.

Bils, R.F. and Howell, R.W. (1963) Biochemical and cytological changes in developing soybean cotyledons. *Crop Sci.* **3**: 304–308.

Blennow, A. and Johansson, G. (1991) Isolation of a Q-enzyme with M_r 103 000 from potato tubers. *Phytochem.* **30**: 437–441.

Borchert, S., Grosse, H. and Heldt, H.W. (1989) Specific transport of inorganic phosphate, glucose 6-phosphate, dihydroxyacetone phosphate and 3-phosphoglycerate into amyloplasts from pea roots. *FEBS Lett.* **253**: 183–186.

Borovsky, D., Smith, E.E. and Whelan, W.J. (1975) Purification and properties of potato 1,4-α-D-glucan: 1,4-α-D-glucan 6-(1,4-α-D-glucano)-transferase. *Eur. J. Biochem.* **59**: 615–625.

Borovsky, D., Smith, E.E. and Whelan, W.J. (1976) On the mechanism of amylose branching by potato Q-enzyme. *Eur. J. Biochem.* **62**: 307–312.

Borovsky, D., Smith, E.E., Whelan, W.J., French, D. and Kikumoto, S. (1979) The mechanism of Q-enzyme action and its influence on the structure of amylopectin. *Arch. Biochem. Biophys.* **198**: 627–631.

Bowsher, C.G., Hucklesby, D.P. and Emes, M.J. (1989) Nitrite reduction and carbohydrate metabolism in plastids purified from roots of *Pisum sativum* L. *Planta* **177**: 359–366.

Boyer, C.D. (1985) Soluble starch synthases and starch-branching enzymes from developing seeds of *Sorghum*. *Phytochem.* **24**: 15–18.

Boyer, C.D. and Preiss, J. (1978a) Multiple forms of (1,4) -α-D-glucan, (1,4) -α-D-glucan-6-glycosyl transferase from developing *Zea mays* kernels. *Carbohydr. Res.* **61**: 312–334.

Boyer, C.D. and Preiss, J. (1978b) Multiple forms of starch branching enzyme of maize: evidence for independent genetic control. *Biochem. Biophys. Res. Commun.* **80**: 169–175.

Boyer, C.D. and Preiss, J. (1979) Properties of citrate-stimulated starch synthesis catalysed by starch synthase I of developing maize kernels. *Plant Physiol.* **64**: 1039–1042.

Boyer, C.D. and Preiss, J. (1981) Evidence for independent genetic control of the multiple forms of maize endosperm branching enzymes and starch synthases. *Plant Physiol.* **67**: 1141–1145.

Boyer, C.D., Simpson, E.K.G. and Damewood, P.A. (1982) The possible relationship of starch and phytoglycogen in sweet corn. II. The role of branching enzyme. *Stärke* **34**: 181–185.

Briarty, J.G., Hughes, C.E. and Evers, A.D. (1979) The developing endosperm of wheat—a stereological analysis. *Ann. Bot.* **44**: 641–658.

Bucke, C. (1970) The distribution and properties of alkaline inorganic pyrophosphatase from higher plants. *Phytochem.* **9**: 1303–1309.

Buttrose, M.S. (1962) The influence of environment on the shell structure of starch granules. *J. Cell Biol.* **14**: 159–167.

Buttrose, M.S. (1963a) Electron microscopy of acid-degraded starch granules. *Stärke* **3**: 85–92.

Buttrose, M.S. (1963b) Ultrastructure of the developing wheat endosperm. *Aust. J. Biol. Sci.* **16**: 305–317.

Caley, C.Y., Duffus, C.M. and Jeffcoat, B. (1990) Effects of elevated temperature and reduced water uptake on enzymes of starch synthesis in developing wheat grains. *Aust. J. Plant Physiol.* **17**: 431–439.

Caspar, T. and Pickard, B.G. (1989) Gravitropism in a starchless mutant of *Arabidopsis*. *Planta* **177**: 185–197.

Caspar, T., Huber, S.C. and Somerville, C. (1985) Alterations in growth, photosynthesis and respiration in a starchless mutant of *Arabidopsis thaliana* (L.) Heynh. deficient in chloroplast phosphoglucomutase activity. *Plant Physiol.* **79**: 11–17.

Chatterton, N.J. and Silvius, J.E. (1980) Photosynthetic partitioning into leaf starch as affected by daily photosynthetic period duration in six species. *Physiol. Plant.* **49**: 141–144.

Colonna, P. and Mercier, C. (1984) Macromolecular structure of wrinkled- and smooth-pea starch components. *Carbohydr. Res.* **126**: 233–247.

Dang, P.L. and Boyer, C.D. (1988) Maize leaf and kernel starch synthases and starch branching enzymes. *Phytochem.* **27**: 1255–1259.

Davies, D.R. (1980) The r_a locus and legumin synthesis in *Pisum sativum*. *Biochem. Genet.* **18**: 1207–1219.

Davies, J.W. and Cocking, E.C. (1965) Changes in carbohydrates, proteins and nucleic acids during cellular development in tomato locule tissue. *Planta* **67**: 242–253.

Deatherage, W.L., MacMasters, M.M., Vineyard, M.L. and Bear, R.P. (1954) A note on starch of high amylose content from corn with high starch content. *Cereal Chem.* **31**: 50–52.

Denyer, K. and Smith, A.M. (1992) The purification and characterisation of two forms of soluble starch synthase from pea embryos. *Planta* **186:** 607–617

Dickinson, D.B. and Preiss, J. (1969) Presence of ADP-glucose pyrophosphorylase in *shrunken-2* and *brittle-2* mutants of maize endosperm. *Plant Physiol.* **44**: 1058–1062.

Di Fonzo, N., Fornasari, E., Gentinetta, E., Salamini, F. and Soave, C. (1978) Proteins and carbohydrate accumulation in normal, *opaque-2* and *floury* maizes. In *Carbohydrate and Protein Synthesis* (Miflin, B.J. and Zoschke, M., eds), Commission of the European Communities, Brussels-Luxembourg, pp. 199–212.

Doll, H. (1984) Nutritional aspects of cereal proteins and approaches to overcome their deficiencies. *Phil. Trans. R. Soc. Lond. B* **304**: 373–380.

Dry, I., Smith, A.M., Edwards, E.A., Bhattacharyya, M., Dunn, P. and Martin, C. (1992) Characterisation of cDNAs encoding two isoforms of granule-bound starch synthase which show differential expression in developing storage organs. *Plant J.* **2**: 193–202.

Duffus, C.M. (1984) Metabolism of reserve starch. In *Storage Carbohydrates in Vascular Plants* (Lewis, D.H., ed.), Cambridge University Press, Cambridge, pp. 231–252.

Echeverria, E. and Boyer, C.D. (1986) Localization of starch biosynthetic and degradative enzymes in maize leaves. *Amer. J. Bot.* **73**: 167–171.

Echeverria, E., Boyer, C.D., Liu, K.-C. and Shannon, J. (1985) Isolation of amyloplasts from developing maize endosperm. *Plant Physiol.* **77**: 513–519.

Echeverria, E., Boyer, C.D., Thomas, P.A., Liu, K.-C. and Shannon, J.C. (1988) Enzyme activities associated with maize kernel amyloplasts. *Plant Physiol.* **86**: 786–792.

Echt, C.S. and Schwartz, D. (1981) Evidence for the inclusion of controlling elements within the structural gene at the waxy locus in maize. *Genetics* **99**: 275–284.

Edwards, J. and ap Rees, T. (1986a) Sucrose partitioning in round and wrinkled varieties of *Pisum sativum*. *Phytochem.* **25**: 2027–2032.

Edwards, J. and ap Rees, T. (1986b) Metabolism of UDP-glucose by developing embryos of round and wrinkled varieties of *Pisum sativum*. *Phytochem.* **25**: 2033–2039.

Edwards, J., Green, J.H. and ap Rees, T. (1988) Activity of branching enzyme as a cardinal feature of the r_a locus in *Pisum sativum*. *Phytochem.* **27**: 1615–1620.

Emes, M.J. and Traska, A. (1987) Uptake of inorganic phosphate by plastids purified from the roots of *Pisum sativum* L. *J. Exp. Bot.* **38**: 1781–1788.

Entwistle, G. and ap Rees, T. (1988) Enzymic capacities of amyloplasts from wheat (*Triticum aestivum*) endosperm. *Biochem. J.* **255**: 391–396.

Entwistle, G. and ap Rees, T. (1990) Lack of fructose-1,6-bisphosphatase in a range of higher plants that store starch. *Biochem. J.* **271**: 467–472.

Evers, A.D. (1971) Scanning electron microscopy of wheat starch. III: Granule development in the endosperm. *Stärke* **23**: 157–162.

Evers, A.D. and Lindley, J. (1977) The particle-size distribution in wheat endosperm starch. *J. Sci. Food Agric.* **28**: 98–102.

Evers, A.D., Greenwood, C.T., Muir, D.D. and Venables, C. (1974) Studies on the biosynthesis of starch granules. 8: A comparison of the properties of the small and large granules in mature cereal starches. *Stärke* **26**: 42–46.

Fergason, V.L., Helm, J.L. and Zuber, M.S. (1966) Gene dosage effects at the *ae* locus on amylose content of corn endosperm. *J. Hered.* **57**: 90–94.

Fliege, R., Flügge, U.-I., Werdan, K. and Heldt, H.W. (1978) Specific transport of inorganic phosphate, 3-phosphoglycerate and triosephosphates across the inner membrane of the envelope in spinach chloroplasts. *Biochim. Biophys. Acta* **502**: 232–247.

Flügge, U.-I. and Heldt, H.W. (1984) The phosphate–triose phosphate–phosphoglycerate translocator of the chloroplast. *Trends Biochem. Sci.* **9**: 530–533.

Fondy, B.R. and Geiger, D.R. (1985) Diurnal changes in allocation of newly-fixed carbon in exporting sugar beet leaves. *Plant Physiol.* **78**: 753–757.

Fraley, R.T., Perlak, F.J., Fischhoff, D.A., Turner, N., Stark, D., Barry, G. and Kishore, G. (1991) Improving potato processing and pest control through gene transfer. In *Abstracts of 2nd International Potato Molecular Biology Symposium*, St Andrews, Scotland, Potato Marketing Board.

Frehner, M., Pozueta-Romero, J. and Akazawa, T. (1990) Enzyme sets of glycolysis, gluconeogenesis, and oxidative pentose phosphate pathway are not complete in nongreen highly purified amyloplasts of sycamore (*Acer pseudoplatanus* L.) cell suspension cultures. *Plant Physiol.* **94**: 538–544.

French, D. (1984) Organisation of starch granules. In *Starch: Chemistry and Technology* (Whistler, R.L., BeMiller, J.N., and Paschall, E.F., eds), Academic Press, Orlando, pp. 183–247.

Frydman, R.B. and Cardini, C.E. (1967) Studies on the biosynthesis of starch. II. Some properties of the adenosine diphosphate glucose:starch glucosyltransferase bound to the starch granule. *J. Biol. Chem.* **242**: 312–317.

Furukawa, K., Tagaya, M., Inoye, M., Preiss, J. and Fukui, T. (1990) Identification of lysine 15 at the active site in *Escherichia coli* glycogen synthase. *J. Biol. Chem.* **265**: 2086–2090.

Geddes, R. and Greenwood, C.T. (1969) Observations on the synthesis of the starch granule. *Stärke* **6**: 148–152.

Geddes, R., Greenwood, C.T. and Mackenzie, S. (1965) Studies on the biosynthesis of starch granules. III. The properties of the components of starches from the growing potato tuber. *Carbohydr. Res.* **1**: 71–82.

Geiger, D.R., Jablonski, L.M. and Ploeger, B.J. (1985) Significance of carbon allocation to starch in growth of *Beta vulgaris* L. In *Regulation of Carbon Partitioning in Photosynthetic Tissue* (Heath, R.L. and Preiss, J., eds), Waverley Press, Baltimore, pp. 289–308.

Gerhardt, R. and Heldt, H.W. (1984) Measurement of subcellular metabolite levels in leaves by fractionation of freeze-stopped material in non-aqueous media. *Plant Physiol.* **75**: 542–547.

Ghosh, H.P. and Preiss, J. (1966) Adenosine diphosphate glucose pyrophosphorylase. A regulatory enzyme in the biosynthesis of starch in spinach leaf chloroplasts. *J. Biol. Chem.* **241**: 4491–4504.

Gidley, M. (1987) Factors affecting the crystalline type (A–C) of native starches and model compounds: a rationalisation of observed effects in terms of polymorphic structures. *Carbohydr. Res.* **161**: 301–304.

Gidley, M.J. and Bociek, S.M. (1985) Molecular organisation in starches: a ^{13}C CP/MAS NMR study. *J. Amer. Chem. Soc.* **107**: 7040–7044.

Gidley, M.J. and Bociek, S.M. (1988) ^{13}C CP/MAS NMR studies of amylose inclusion complexes, cyclodextrins, and the amorphous phase of starch granules. *J. Amer. Chem. Soc.* **110**: 3820–3829.

Gidley, M. and Bulpin, P.V. (1987) Crystallisation of malto-oligosaccharides as models of the crystalline forms of starch: minimum chain-length requirement for the formation of double helices. *Carbohydr. Res.* **161**: 291–300.

Goodman, R.N., Ziraly, Z. and Wood, K.R. (1986) *The Biochemistry and Physiology of Plant Disease.* University of Missouri Press, Columbia, Missouri.

Gould, J.M. and Winget, G.D. (1973) A membrane-bound alkaline inorganic pyrophosphatase in isolated spinach chloroplasts. *Arch. Biochem. Biophys.* **154**: 606–613.

Gross, P. and ap Rees, T. (1986) Alkaline inorganic pyrophosphatase and starch synthesis in amyloplasts. *Planta* **167**: 140–145.

Guilbot, A. and Mercier, C. (1985) Starch. In *The Polysaccharides, Vol. 3* (Aspinall, G.O., ed.), Academic Press, London, pp. 209–282.

Hannah, L.C. and Nelson, O.E. (1976) Characterisation of ADP-glucose pyrophosphorylase from *shrunken-2* and *brittle-2* mutants of maize. *Biochem. Genet.* **14**: 547–560.

Hanson, K.R. and McHale, N.A. (1988) A starchless mutant of *Nicotiana sylvestris* containing a modified plastid phosphoglucomutase. *Plant Physiol.* **88**: 838–844.

Hargreaves, J.A. and ap Rees, T. (1988) Turnover of starch and sucrose in roots of *Pisum sativum. Phytochem.* **27**: 1627–1629.

Harrison, S.G., Masefield, G.B. and Wallis, M. (1969) *The Oxford Book of Food Plants.* Oxford University Press, Oxford.

Hatzfield, W.-D. and Stitt, M. (1990) A study of the rate of recycling of triose phosphates in heterotrophic *Chenopodium rubrum* cells, potato tubers, and maize endosperm. *Planta* **180**: 198–204.

Hawker, J.S. and Downton, W.J.S. (1974) Starch synthetases from *Vitis vinifera* and *Zea mays. Phytochem.* **13**: 893–900.

Hawker, J.S., Ozbun, J.L. and Preiss, J. (1972) Unprimed starch synthesis by soluble ADPglucose-starch glucosyltransferase from potato tubers. *Phytochem.* **11**: 1287–1293.

Hawker, J.S., Ozbun, J.L., Ozaki, H., Greenberg, E. and Preiss, J. (1974) Interaction of spinach leaf adenosine diphosphate glucose α-1,4-glucan α-4-glucosyl transferase and α-1-glucan, α-1,4-glucan-6-glycosyl transferase in synthesis of branched α-glucan. *Arch. Biochem. Biophys.* **160**: 530–551.

Hawker, J.S., Marschner, H. and Krauss, A. (1979) Starch synthesis in developing potato tubers. *Physiol. Plant.* **46**: 25–30.

Hedman, K.D. and Boyer, C.D. (1982) Gene dosage at the *amylose-extender* locus of maize: effects on the levels of starch-branching enzymes. *Biochem. Genet.* **20**: 483–492.

Hedman, K.D. and Boyer, C.D. (1983) Allelic studies of the *amylose-extender* locus of *Zea mays* L.: levels of the starch branching enzymes. *Biochem. Genet.* **21**: 1217–1222.

Heldt, H.W. and Rapley, L. (1970) Unspecific permeation and specific uptake of substances in spinach chloroplasts. *FEBS Lett.* **7**: 139–142.

Hill, L.M. and Smith, A.M. (1991) Evidence that glucose 6-phosphate is imported as the substrate for starch synthesis by the plastids of developing pea embryos. *Planta* **185**: 91–96.

Hizukuri, S., Takeda, Y., Yasuda, M. and Suzuki, A. (1981) Multi-branched nature of amylose and the action of debranching enzyme. *Carbohydr. Res.* **94**: 205–213.

Hizukuri, S., Kaneko, T. and Takeda, Y. (1983) Measurement of the chain length of amylopectin and its relevance to the origin of crystalline polymorphism of starch granules. *Biochim. Biophys. Acta* **760**: 188–191.

Hovenkamp-Hermelink, J.H.M., Jacobsen, E., Ponstein, A.S., Visser, R.G.F., Vos-Scheperkeuter, G.H., Bijmolt, E.W., de Vries, J.N., Witholt, B. and Feenstra, W.J. (1987) Isolation of an amylose-free starch mutant of the potato (*Solanum tuberosum* L.). *Theor. Appl. Genet.* **57**: 217–221.

Hsieh, J.-S. (1988) Genetic studies on the *Wx* gene of sorghum [*Sorghum bicolor* (L.) Moench.] I. Examination of the protein product of the waxy locus. *Bot. Bull. Academia Sinica* **29**: 293–299.

Huber, S.C., Kerr, P.S. and Kalt-Torres, W. (1985) Regulation of sucrose formation and movement. In *Regulation of Carbon Partitioning in Photosynthetic Tissue* (Heath, R.L. and Preiss, J., eds), Waverley Press, Baltimore, pp. 199–214.

Jacobsen, E., Hovenkamp-Hermelink, J.M.H., Krigsheld, H.T., Nijdam, H., Pijnacker, L.P., Witholt, B. and Feenstra, W.J. (1989) Phenotypic and genotypic characterisation of an amylose-free starch mutant of the potato. *Euphytica* **44**: 43–48.

Jenner, C. (1976) Wheat grains and spinach leaves as accumulators of starch. In *Transport and Transfer Processes in Plants* (Wardlaw, I. and Passioura, J.B., eds), Academic Press, New York, pp. 77–83.

Jenner, C. (1982) Storage of starch. In *Encyclopedia of Plant Physiology New Series Vol. 13A: Plant carbohydrates II, Intracellular carbohydrates*. (Loewus, F.A. and Tanner, W., eds), Springer-Verlag, Berlin, pp. 700–747.

Jones, G. and Whelan, W.J. (1969) The action pattern of D-enzyme, a transmaltodextrinylase from potato. *Carbohydr. Res.* **9**: 483–490.

Jones, T.W.A., Pichersky, E. and Gottlieb, L.D. (1986) Enzyme activity in EMS-induced null mutations of duplicated genes encoding phosphoglucose isomerase in *Clarkia*. *Genetics* **113**: 101–114.

Journet, E.P. and Douce, R. (1985) Enzymic capacities of purified cauliflower bud plastids for lipid symthesis and carbohydrate metabolism. *Plant Physiol.* **79**: 458–467.

Kainuma, K. (1988) Structure and chemistry of the starch granule. In *The Biochemistry of Plants, Vol. 14: Carbohydrates* (Preiss, J., ed.), Academic Press, San Diego, pp. 141–180.

Kakefuda, G., Duke, S.H. and Hostak, M.S. (1986) Chloroplast and extrachloroplastic starch-degrading enzymes in *Pisum sativum* L. *Planta* **168**: 175–182.

Kassenbeck, P. (1978) Beitrag zur Kenntnis der Verteilung von Amylose und Amylopektin in Stärkekörnern. *Stärke* **30**: 40–46.

Keeling, P.L., Wood, J.R., Tyson, R.W. and Bridges, I.G. (1988) Starch biosynthesis in developing wheat grain. Evidence against the direct involvement of triose phosphates in the metabolic pathway. *Plant Physiol.* **78**: 311 319.

Kirk, R.T.O. and Tilney-Basset, R.A.E. (1967) *The Plastids*. W.H. Freeman and Co., London.

Kiss, J.Z. and Sack, F.D. (1990) Severely reduced gravitropism in dark-grown hypocotyls of a starch-deficient mutant of *Nicotiana sylvestris*. *Plant Physiol.* **94**: 1867–1873.

Kiss, J.Z., Hertel, R. and Sack, F.D. (1989) Amyloplasts are necessary for full gravitropic sensitivity in roots of *Arabidopsis thaliana*. *Planta* **177**: 198–206.

Klösgen, R.B., Gierl, A., Schwarz-Sommer, Z. and Saedler, H. (1986) Molecular analysis of the *waxy* locus of *Zea mays*. *Mol. Gen. Genet.* **203**: 237–244.

Knee, M., Sargent, J.A.and Osborne, D.J. (1977) Cell wall metabolism in developing strawberry fruits. *J. Exp. Bot.* **28**: 377–396.

Krauss, A. and Marschner, H. (1984) Growth rate and carbohydrate metabolism of potato tubers exposed to high temperatures. *Potato Res.* **27**: 297–303.

Krishnan, H.B., Reeves, C.D. and Okita, T.W. (1986) ADPglucose pyrophosphorylase is encoded by different transcripts in leaf and endosperm of cereals. *Plant Physiol.* **81**: 642–645.

Kruckeberg, A.L., Neuhaus, H.E., Feil, R., Gottlieb, L.D. and Stitt, M. (1989) Decreased-activity mutants of phosphoglucose isomerase in the cytosol and chloroplast of *Clarkia xantiana*. I. Impact on mass action ratios and fluxes to sucrose and starch, and estimation of flux control coefficients and elasticity coefficients. *Biochem. J.* **261**: 457–467.

Kruger, N.J., Bulpin, P.V. and ap Rees, T. (1983) The extent of starch degradation in the light in pea leaves. *Planta* **157**: 271–273.

Kumar, A., Larson, C.E. and Preiss, J. (1986) Biosynthesis of bacterial glycogen. Primary structure of *Escherichia coli* ADP-glucose: α-1,4-glucan, 4-glucosyltransferase as deduced from the nucleotide sequence of the *glgA* gene. *J. Biol. Chem.* **261**. 16256–16259.

Laetsch, W.M. (1968) Chloroplast specialisation in dicotyledons possessing the C4-dicarboxylic acid pathway of photosynthetic CO_2 fixation. *Amer. J. Bot.* **55**: 875–883.

Leech, R.M. and Baker, N.R. (1983) The development of photosynthetic capacity in leaves. In *The Growth and Functioning of Leaves* (Dale, J.E. and Milthorpe, F.L., eds), Cambridge University Press, Cambridge, pp. 273–307.

van der Leij, F.R., Visser, R.G.F., Ponstein, A.S., Jacobsen, E. and Feenstra, W.J. (1991a) Sequence of the structural gene for granule-bound starch synthase of potato (*Solanum tuberosum* L.) and evidence for a single point deletion in the *amf* allele. *Mol. Gen. Genet.* **228**: 240–248.

van der Leij, F.R., Visser, R.G.F., Oosterhaven, K., van der Kop, D.A.M., Jacobsen, E. and Feenstra, W.J. (1991b) Complementation of the amylose-free starch mutant of potato (*Solanum tuberosum* L.) by the gene encoding granule-bound starch synthase. *Theor. Appl. Genet.*: in press.

Lewis, D.H. (1984) Occurrence and distribution of storage carbohydrates. In *Storage Carbohydrates in Vascular Plants* (Lewis, D.H. ed.) Cambridge University Press, Cambridge, pp. 1–54.

Liedvogel, B. and Kleinig, H. (1980) Phosphate translocator and adenylate translocation in chromoplast membranes. *Planta* **150**: 170–173.

Lin, T.-P. and Preiss, J. (1988) Characterisation of D-enzyme (4-α-glucanotransferase) in *Arabidopsis* leaf. *Plant Physiol.* **86**: 260–265.

Lin, T.-P., Caspar, T., Somerville, C.R. and Preiss, J. (1988a) Isolation and characterisation of a starchless mutant of *Arabidopsis thaliana* (L.) Heynh. lacking ADPglucose pyrophosphorylase activity. *Plant Physiol.* **86**: 1131–1135.

Lin, T.-P., Caspar, T., Somerville, C.R. and Preiss, J. (1988b) A starch-deficient mutant of *Arabidopsis* with low ADPglucose pyrophosphorylase activity lacks one of the two subunits of the enzyme. *Plant Physiol.* **88**: 1175–1181.

Macdonald, F.D. and ap Rees, T. (1983) Enzymic properties of amyloplasts from suspension cultures of soybean. *Biochim. Biophys. Acta* **755**: 81–89.

Macdonald, F.D. and Preiss, J. (1983) Solubilisation of the starch-granule-bound starch synthase of normal maize kernels. *Plant Physiol.* **73**: 175–178.

Macdonald, F.D. and Preiss, J. (1985) Partial purification and characterisation of granule-bound starch synthases from normal and *waxy* maize. *Plant Physiol.* **78**: 849–852.

McDonald, A.L.M., Stark, J.R., Morrison, W.R. and Ellis, R.P. (1991) The composition of starch granules from developing barley genotypes. *J. Cereal Sci.* **13**: 93–112.

Manners, D.J. (1985) Starch. In *Biochemistry of Storage Carbohydrates in Green Plants* (Dey, P.M. and Dixon, R.A., eds), Academic Press, London, pp. 149–203.

Manners, D.J. and Matheson, N.K. (1981) The fine structure of amylopectin. *Carbohydr. Res.* **90**: 99–110.

Mares, D.J., Hawker, J.S. and Possingham, J.V. (1978) Starch synthesising enzymes in developing leaves of spinach (*Spinacia oleracea* L.) *J. Exp. Bot.* **29**: 829–835.

Marschner, H. (1986) *Mineral Nutrition of Higher Plants*. Academic Press, London.

Matheson, N.K. and Wheatley, J.M. (1962) Starch changes in developing and senescing tobacco leaves. *Aust. J. Biol. Sci.* **15**: 445–458.

Matters, G.L. and Boyer, C.D. (1981) Starch synthases and starch-branching enzymes from *Pisum sativum*. *Phytochem.* **20**: 1805–1809.

May, L.H. and Buttrose, M.S. (1959) Physiology of cereal grain. II: Starch granule formation in the developing barley kernel. *Aust. J. Biol. Sci.* **12**: 146–159.

Meeuse, B.J.D. (1975) Thermogenic respiration in Aroids. *Annu. Rev. Plant Physiol.* **26**: 117–126.

Mohabir, G. and John, P. (1988) Effect of temperature on starch synthesis in potato tuber tissue and amyloplasts. *Plant Physiol.* **88**: 1222–1228.

Moore, R. and Evans, M.L. (1986) How roots perceive and respond to gravity. *Amer. J. Bot.* **73**: 574–587.

Morell, M.K., Bloom, M. and Preiss, J. (1987) Subunit structure of spinach leaf ADPglucose pyrophosphorylase. *Plant Physiol.* **85**: 185–187.

Morell, M.K., Bloom, M. and Preiss, J. (1988) Affinity labelling of the allosteric activator site(s) of spinach leaf ADPglucose pyrophosphorylase. *J. Biol. Chem.* **263**: 633–637.

Morrell, S. and ap Rees, T. (1986) Sugar metabolism in developing tubers of *Solanum tuberosum*. *Phytochem.* **25**: 1579–1585.

Morrison, W.R. and Gadan, H. (1987) The amylose and lipid contents of starch granules in developing wheat endosperm. *J. Cereal Sci.* **5**: 263–275.

Morrison, W.R., Milligan, T.P. and Azudin, M.N. (1984) A relationship between the amylose and lipid contents of starches from diploid cereals. *J. Cereal Sci.* **2**: 257–271.

Müller-Röber, B.T., Kossman, J., Hannah, L.C., Willmitzer, L. and Sonnenwald, U. (1990) One of two different ADP-glucose pyrophosphorylase genes from potato responds strongly to elevated levels of sucrose. *Molec. Gen. Genet.* **224**: 136–146.

Nakamura, Y., Yuki, K., Park, S.Y. and Ohya, T. (1989) Carbohydrate metabolism in the developing endosperm of rice grains. *Plant Cell Physiol.* **30**: 833–839.

Nelson, O.E. and Rines, H.W. (1962) The enzymatic deficiency in the waxy mutant of maize. *Biochem. Biophys. Res. Commun.* **9**: 297–300.

Nelson, O.E., Chourey, P.S. and Chang, M.T. (1978) Nucleotide diphosphate sugar–starch glucosyl transferase activity of *wx* starch granules. *Plant Physiol.* **72**: 383–386.

Neuhaus, H.E. and Stitt, M. (1990) Control analysis in photosynthetic partitioning. Impact of reduced activity of ADP-glucose pyrophosphorylase or plastid phosphoglucomutase on the fluxes to starch and sucrose in *Arabidopsis thaliana* (L.) Heynh. *Planta* **182**: 445–454.

Ngernprasirtsiri, J., Harinasut, P., Macherel, D., Strzalka, K., Takabe, T., Akazawa, T. and Kojima, J.

(1988) Isolation and characterisation of the amyloplast envelope-membrane from cultured white-wild cells of sycamore (*Acer pseudoplatanua* L.). *Plant Physiol.* **87**: 371–378.
Ngernprasirtsiri, J., Takabe, T. and Akazawa, T. (1989) Immunochemical analysis shows that an ATP/ADP translocator is associated with the inner-envelope membranes of amyloplasts from *Acer pseudoplatanus* L. *Plant Physiol.* **89**: 1024–1027.
Norton, G. and Harris, J.F. (1975) Compositional changes in developing rape seed (*Brassica napus* L.). *Planta* **123**: 163–174.
Okita, T.W. and Preiss, J. (1980) Starch degradation in spinach leaves. *Plant Physiol.* **66**: 870–876.
Okita, T.W., Greenberg, E., Kuhn, D.N. and Preiss, J. (1979) Subcellular localisation of the starch degradative and biosynthetic enzymes of spinach leaves. *Plant Physiol.* **64**: 187–192.
Okita, T.W., Nakata, P.A., Anderson, J.M., Sowokinos, J., Morell, M. and Preiss, J. (1990) The subunit structure of potato tuber ADPglucose pyrophosphorylase. *Plant Physiol.* **93**: 785–790.
Okuno, K. and Sakaguchi, S. (1982) Inheritance of starch characteristics in perisperm of *Amaranthus hypochondriacus*. *J. Hered.* **73**: 467.
Olive, M.R., Ellis, J.R. and Schuch, W.W. (1989) Isolation and nucleotide sequences of cDNA clones encoding ADP-glucose pyrophosphorylase polypeptides from wheat leaf and endosperm. *Plant Molec. Biol.* **12**: 525–538.
Oparka, K. and Wright, K.M. (1988a) Osmotic regulation of starch synthesis in potato tubers? *Planta* **174**: 123–126.
Oparka, K. and Wright, K.M. (1988b) Influence of cell turgor on sucrose partitioning in potato tuber storage tissues. *Planta* **175**: 520–526.
Outlaw, J.L. and Manchester, J. (1979) Guard cell starch concentration quantitatively related to stomatal aperture. *Plant Physiol.* **64**: 79–82.
Ozbun, J.L., Hawker, J.S. and Preiss, J. (1971a) Adenosine diphosphoglucose-starch glucosyltransferases from developing kernels of waxy maize. *Plant Physiol.* **48**: 765–769.
Ozbun, J.L., Hawker, J.S. and Preiss, J. (1971b) Multiple forms of α-1,4 glucan synthetase from spinach leaves. *Biochem. Biophys. Res. Commun.* **43**: 631–636.
Ozbun, J.L., Hawker, J.S., Greenberg, E., Lammel, C., Preiss, J. and Lee, E.Y.C. (1973) Starch synthetase, phosphorylase, ADPglucose pyrophosphorylase, and UDPglucose pyrophosphorylase in developing maize kernels. *Plant Physiol.* **51**: 1–5.
Pan, D. and Nelson, O.E. (1984) A debranching enzyme deficiency in endosperms of the *sugary-1* mutants of maize. *Plant Physiol.* **74**: 324–328.
Pisigan, R.A. and del Rosario, E.J. (1976) Isoenzymes of soluble starch synthetase from *Oryza sativa* grains. *Phytochem.* **15**: 71–73.
Plaxton, W.C. and Preiss, J. (1987) Purification and properties of nonproteolytic degraded ADPglucose pyrophosphorylase from maize endosperm. *Plant Physiol.* **83**: 105–112.
Pollock, C. and Chatterton, N.J. (1988) Fructans. In *The Biochemistry of Plants, Vol. 14: Carbohydrates* (Preiss, J., ed.), Academic Press, San Diego, pp. 109–140.
Pollock, C.J. and Preiss, J. (1980) The citrate-stimulated starch synthase of starchy maize kernels: purification and properties. *Arch. Biochem. Biophys.* **204**: 578–588.
Ponstein, A. (1990) *Starch synthesis in potato tubers*. Ph.D. thesis, University of Groningen, The Netherlands.
Pozueta-Romero, J., Frehner, M., Viale, A.M. and Akazawa, T. (1991) Direct transport of ADPglucose by an adenylate translocator is linked to starch biosynthesis in amyloplasts. *Proc. Natl Acad. Sci.* **88**: 5769–5773.
Preiss, J. (1984) Bacterial glycogen synthesis and its regulation. *Annu. Rev. Microbiol.* **38**: 419–458
Preiss, J. (1988) Biosynthesis of starch and its regulation. In *The Biochemistry of Plants, Vol. 14: Carbohydrates* (Preiss, J., ed.), Academic Press, San Diego, pp. 181–254.
Preiss, J. (1990) Biology and molecular biology of starch synthesis and its regulation. In *Oxford Surveys of Plant Molecular and Cell Biology Vol. 7* (Miflin, B.J. ed.) Oxford University Press, Oxford.
Preiss, J. and Levi, C. (1979) Metabolism of starch in leaves. In *Encyclopedia of Plant Physiology New Series Vol. 6: Photosynthesis II* (Gibbs, M. and Latzko, E., eds), Springer Verlag, Berlin, pp. 282–312.
Preiss, J. and Levi, C. (1982) Starch biosynthesis and degradation. In *The Biochemistry of Plants Vol. 3: Carbohydrates: structure and function* (Preiss, J., ed.), Academic Press, San Diego, pp. 371–423.
Preiss, J., Cress, D., Hutny, J., Morell, M., Bloom, M., Okita, T. and Anderson, J. (1989) Regulation of starch synthesis. Biochemical and genetic studies. In *Biocatalysis in Agricultural Biotechnology* (Whitaker, J.R. and Sonnet, P.E., eds), American Chemical Society, Washington, pp. 84–92.

Preiss, J., Danner, S., Summers, P.S., Morell, M., Barton, C.R., Yang, L. and Nieder, M. (1990) Molecular characterisation of the *brittle-2* gene effect on maize endosperm ADPglucose pyrophosphorylase subunits. *Plant Physiol.* **93**: 785–790.

Priestley, C.A. (1970) Carbohydrate storage and utilisation. In *Physiology of Tree Crops* (Luckwill, L.C. and Cutting, C.V., eds), Academic Press, London, pp. 113–127.

Rest, J.A. and Vaughan, J.G. (1972) The development of protein and oil bodies in the seed of *Sinapis alba* L. *Planta* **105**: 245–262.

Rijven, A.H.G.C. (1986) Heat inactivation of starch synthase in wheat endosperm. *Plant Physiol.* **81**: 448–453.

Robyt, J. (1984) Enzymes in the synthesis and hydrolysis of starch. In *Starch: Chemistry and Technology* (Whistler, R.L., BeMiller, J.N. and Paschall, E.F., eds), Academic Press, Orlando, pp. 87-123.

Rocha-Sosa, M., Sonnenwald, U., Frommer, W., Stratmann, M., Schell, J. and Willmitzer, L. (1989) Both developmental and metabolic signals activate the promoter of a class I patatin gene. *EMBO J.* **8,** 23–29.

Rohde, W., Becker, D. and Salamini, F. (1988) Structural analysis of the *waxy* locus from *Hordeum vulgare. Nucleic Acids Res.* **16**: 7185–7186.

Saether, N. and Iversen, T.-H. (1991) Gravitropism and starch statoliths in an *Arabidopsis* mutant. *Planta* **184**: 491–497.

Salema, K. and Badenhuizen, N.P. (1967) The production of reserve starch granules in the amyloplasts of *Pellionia daveauana* N. E. Br. *J. Ultrastr. Res.* **30**: 383–399.

Sano, Y. (1984) Differential regulation of waxy gene expression in rice endosperm. *Theor. Appl. Genet.* **68**: 467–473.

Sanwal, G.G., Greenberg, E., Hardie, J., Cameron, E.C and Preiss, J. (1968) Regulation of starch biosynthesis in plant leaves: activation and inhibition of ADPglucose pyrophosphorylase. *Plant Physiol.* **43**: 417–427.

Schiefer, S., Lee, E.Y.C. and Whelan, W.J. (1978) The requirement for a primer in the *in vitro* synthesis of polysaccharide by sweet-corn (1,4)-α- D-glucan synthase. *Carbohydr. Res.* **61**: 239–252.

Schulman, A.H. and Ahokas, H. (1990) A novel shrunken endosperm mutant of barley. *Physiol. Plant.* **78**: 583–589.

Schwartz, D. and Echt, C.S. (1982) The effect of *Ac* dosage on the production of multiple forms of Wx protein by the wx^{m-9} controlling element mutation in maize. *Mol. Gen. Genet.* **187**: 410–413.

Sculthorpe, C.D. (1967) *The Biology of Aquatic Vascular Plants.* William Clowes and Sons Ltd., London.

Shannon, J.C. and Garwood, D.L. (1984) Genetics and physiology of starch development. In *Starch: Chemistry and Technology* (Whistler, R.L., BeMiller, J.N. and Paschall, E.F., eds), Academic Press, Orlando, pp. 25–86.

Shannon, J.C., Creech, R.G. and Loerch, J.D. (1970) Starch synthesis studies in *Zea mays.* II Molecular distribution of radioactivity in starch. *Plant Physiol.* **45**: 163–168.

Shewmaker, C.K., Oakes, J.V., Stalker, D.M. and Boersig, M. (1991) Production of novel carbohydrate in potato tubers. In *Abstracts of 2nd International Potato Molecular Biology Symposium*, St Andrews, Scotland, Potato Marketing Board.

Shure, M., Wessler, S. and Fedoroff, N. (1983) Molecular identification and isolation of the *waxy* locus in maize. *Cell* **35**: 225–233.

Singh, B.K. and Preiss, J. (1985) Starch-branching enzymes from maize. Immunological characterisation using polyclonal and monoclonal antibodies. *Plant Physiol.* **79**: 34–40.

Smith, A.M. (1988) Major differences in isoforms of starch-branching enzyme between developing embryos of round- and wrinkled-seeded peas (*Pisum sativum* L.). *Planta* **175**: 270–279.

Smith A.M., (1990a) Evidence that the waxy protein of pea (*Pisum sativum* L.) is not the major starch-granule-bound starch synthase. *Planta* **182**: 599–604.

Smith, A.M. (1990b) Enzymes of starch synthesis. In *Methods in Plant Biochemistry Vol. 3: Enzymes of primary metabolism* (Lea, P.J., ed), Academic Press, London, pp. 93–102.

Smith, A.M., Bettey, M. and Bedford, I.D. (1989) Evidence that the *rb* locus alters the starch content of developing pea embryos through an effect on ADP glucose pyrophosphorylase. *Plant Physiol.* **89**: 1279–1284.

Smith, A.M., Quinton-Tulloch, J. and Denyer, K. (1990a) Characteristics of plastids responsible for starch synthesis in developing pea embryos. *Planta* **180**: 517–523.

Smith, A.M., Neuhaus, H.E. and Stitt, M. (1990b) The impact of decreased activity of starch-branching enzyme on photosynthetic starch synthesis in leaves of wrinkled-seeded peas. *Planta* **181**: 310–135.

Smyth, D.A. (1988) Some properties of starch-branching enzyme from Indica rice. *Plant Sci.* **57**: 1–8.
Spiltrano, S.R. and Preiss, J. (1987) Regulation of starch synthesis in the bundle sheath and mesophyll of *Zea mays* L. *Plant Physiol.* **83**: 621–627.
Sowokinos, J. (1976) Pyrophosphorylases in *Solanum tuberosum*. I. Changes in ADP-glucose and UDP-glucose pyrophosphorylase activities associated with starch biosynthesis during tuberisation, maturation, and storage of potatoes. *Plant Physiol.* **57**: 63–68.
Sowokinos, J.R. and Preiss, J. (1982) Pyrophosphorylases in *Solanum tuberosum*. III Purification, physical and catalytic properties of ADPglucose pyrophosphorylase in potatoes. *Plant Physiol.* **69**: 1459–1466.
Steup, M. (1988) Starch degradation. In *The Biochemistry of Plants, Vol 14: Carbohydrates* (Preiss, J., ed.), Academic Press, San Diego, pp. 255–296.
Steup, M., Robenek, H. and Melkonian, M. (1983) In vitro degradation of starch granules isolated from chloroplasts. *Planta* **158**: 428–436.
Stitt, M. (1985) Fine control of sucrose synthesis by fructose-2,6-bisphosphate. In *Regulation of Carbon Partitioning in Photosynthetic Tissue* (Heath, R.L. and Preiss, J., eds), Waverley Press, Baltimore, pp. 109–126.
Stitt, M. (1990) Fructose-2,6-bisphosphate as a regulatory molecule in plants. *Annu. Rev. Plant Physiol. Plant Molec. Biol.* **41**: 153–185.
Stitt, M. and Quick, W.P. (1989) Photosynthetic carbon partitioning: its regulation and possibilities for manipulation. *Physiol.Plant.* **77**: 633–641.
Stitt, M. and Steup, M. (1985) Starch and sucrose degradation. In *Encyclopedia of Plant Physiology New Series Vol. 18: Higher plant cell respiration* (Douce, R. and Day, D.A., eds) Springer Verlag, Berlin, pp. 347–390.
Stitt, M., Bulpin, P.V. and ap Rees, T. (1978) Pathway of starch breakdown in photosynthetic tissues of *Pisum sativum*. *Biochim. Biophys. Acta* **544**: 200–214.
Stitt, M., Huber, S. and Kerr, P. (1987) Control of photosynthetic sucrose formation. In *The Biochemistry of Plants Vol. 10:Photosynthesis* (Hatch, M.D. and Boardman, N.K., eds), Academic Press, San Diego, pp. 327–409.
Sumner, J.B. and Somers, G.F. (1944) The water-soluble polysaccharides of sweet corn. *Arch. Biochem.* **4**: 7–9.
Takeda, Y. and Hizukuri, S. (1982) Location of phosphate groups in potato amylopectin. *Carbohydr. Res.* **102**: 312–327.
Torres, N.U., Mateo, F., Melendez-Hevia, E. and Kacser, H. (1986) Kinetics of metabolic pathways. A system in vivo to study the control of flux. *Biochem. J.* **234**: 169–174.
Tsai, C.-Y. (1974) The function of the *waxy* locus in starch synthesis in maize endosperm. *Biochem. Genet.* **11**: 83–96.
Tsai, C.-Y., Salamini, F. and Nelson, O.E. (1970) Enzymes of carbohydrate metabolism in the developing endosperm of maize. *Plant Physiol.* **46**: 299–306.
Tsay, J.S. and Kuo, C.G. (1980) Enzymatic activities of starch synthesis in potato tubers of different sizes. *Physiol. Plant.* **48**: 460–462.
Tsay, C.S., Kuo, W.L. and Kuo, C.G. (1983) Enzymes involved in starch synthesis in the developing mung bean seed. *Phytochem.* **22**: 1573–1576.
Tucker, G.A. and Grierson, D. (1987) Fruit ripening. In *The Biochemistry of Plants, Vol. 12: Physiology of Metabolism* (Davies, D.D., ed.), Academic Press, San Diego, pp. 265–318.
Turner, J.F. (1969) Starch synthesis and changes in uridine diphosphate glucose pyrophosphorylase and adenosine diphosphate glucose pyrophosphorylase in the developing wheat grain. *Aust. J. Biol. Sci.* **22**. 1321–1327.
Turner, S.R., Barratt, D.H.P. and Casey, R. (1990) The effect of different alleles at the *r* locus on the synthesis of seed storage proteins in *Pisum sativum*. *Plant Molec. Biol.* **14**: 793–803.
Tyson, R.H. and ap Rees, T. (1988) Starch synthesis by isolated amyloplasts from wheat endosperm. *Planta* **175**: 33–38.
Viola, R., Davies, H.V. and Chudek, A.R. (1991) Pathways of starch and sucrose biosynthesis in developing tubers of potato (*Solanum tuberosum* L.) and seeds of Faba bean (*Vicia faba* L.). *Planta* **183**: 202–208.
Visser, R.G.F., Hergersberg, M., van der Leij, F.R., Jacobsen, E., Witholt, B. and Feenstra, W.J. (1989) Molecular cloning and partial characterisation of the gene for granule-bound starch synthase from a wildtype and an amylose-free potato (*Solanum tuberosum* L.) *Plant Sci.* **64**: **185–192.**
Visser, R.G.F., Somhorst, I., Kuipers, G.J., Ruys, N.J., Feenstra, W.J. and Jacobsen, E. (1991) Inhibition of the expression of the gene for granule-bound starch synthase in potato by antisense constructs.

Mol. Gen. Genet. **225**: 289–296.

Vos-Scheperkeuter, G.H., de Boer, W., Visser, R.G.F., Feenstra, W.J. and Witholt, B. (1986) Identification of granule-bound starch synthase in potato tubers. *Plant Physiol.* **82**: 411–416.

Vos-Scheperkeuter, G.H., de Wit, J.G., Ponstein, A.S., Feenstra, W.J. and Witholt, B. (1989) Immunological comparison of the starch-branching enzymes from potato tubers and maize kernels. *Plant Physiol.* **90**: 75–84.

Wang, T.L. and Hedley, C. (1991) Seed development in peas: knowing your three 'r' s (or four, or five). *Seed Sci. Res.* **1**: 3–14.

Wang, Z., Wu, Z., Xing, Y., Zheng, F., Guo, X., Zang, W. and Hong, M. (1990) Nucleotide sequence of the rice waxy gene. *Nucleic Acids Res.* **18**: 5898.

Weiner, H., Stitt, M. and Heldt, H.W. (1987) Subcellular compartmentation of pyrophosphate and alkaline pyrophosphatase in leaves. *Biochim. Biophys. Acta* **893**: 13–21.

Wessler, S.R., Baran, B., Varagona, M. and Dellaporta, S.L. (1986) Excision of *Ds* produces *waxy* proteins with a range of enzymatic activities. *EMBO J.* **5**: 2427–2432.

Yazdi-Samadi, B., Rinne, R.W. and Seif, R.D. (1977) Components of developing soybean seeds: oil, protein, sugars, starch, organic acids and amino acids. *Agron. J.* **69**: 481–486.

Yeh, J.Y., Garwood, D.L. and Shannon, J.C. (1981) Characterisation of starch from maize endosperm mutants. *Stärke* **33**: 222–230.

Ziegler, P. (1988) Partial purification and characterisation of the major endoamylase of mature pea leaves. *Plant Physiol.* **86**: 659–666.

2 Cell walls, structure, utilisation and manipulation

G.A TUCKER and J. MITCHELL

2.1 Introduction

Plant cells are surrounded by a rigid wall. The structural components of this wall vary slightly between species but generally they are composed primarily of carbohydrate polymers along with some structural protein. The plant expends a lot of energy in the biosynthesis of cell walls and one of the wall components, cellulose, probably represents the most abundant polysaccharide in the world. The function of the wall is primarily to provide structural support and protection to the plant cells and tissues. Although we know a reasonable amount of detail concerning the molecular structures of the wall polymers, a full description of the three-dimensional structure of the wall itself has not yet been produced.

The cell wall is important industrially for two main reasons. First, by determining the physical properties of the tissue it often dictates the texture of plant products such as fruit and vegetables. The wall polymers are also key elements in determining the quality of processed plant foods such as fruit juices. Second, the wall is a source of commercially important extracts, e.g. cellulose for paper manufacture and carrageen and pectin for gelling agents in the food industry. Cell wall carbohydrates are also significant in nutritional terms not least by providing a large part of the dietary fibre.

Of great importance to the utilisation of these cell wall components in industry is an understanding of their structure–function relationships both in the tissue and following extraction and processing. Factors such as the degree of polymerisation, charge distribution, branching and interaction with other components may all affect these properties. The aim of the plant breeder or industrial chemist is to manipulate the wall structure for optimum function. This can be carried out after extraction either chemically or by the action of added enzymes. In some instances the action of endogenous enzymes during extraction and processing is beneficial, as in the clarification of apple juice by the action of pectinases. In other instances such endogenous enzyme activity can be detrimental, as in the loss of paste viscosity due to pectinase activity during tomato fruit processing.

Using recombinant DNA technology the possibility exists to modify wall components *in situ*. This same technology could also be employed to control endogenous enzymes which may influence processing. Such modification requires detailed information on the biosynthesis and degradation of the various cell wall components. An understanding of the structure–function relationships is also needed in order to be able to 'design' desirable changes. A major limitation, at present, to the intensive application of recombinant DNA technology in this area is a lack of biochemical information on the biosynthesis of wall components. More detailed knowledge of the enzymes involved in the degradation of wall polymers is available, however, particularly in tomato fruit, and this has enabled the manipulation of pectin structure using antisense RNA technology.

2.2 Cell wall structure

The cell walls of higher plants are very complex and are typically composed of around 90% polysaccharide and 10% protein. However, other structural components are present, in particular phenols, and, depending on the type and age of the cell, can constitute a significant proportion of the wall. The structural polysaccharides are very diverse but basically consist of homopolymers or heteropolymers containing one or more of the nine basic monosaccharide building blocks. These building blocks are D-glucose, D-xylose, D-galactose, L-arabinose, L-rhamnose, D -galacturonic acid, D-mannose, L-fucose and D-glucuronic acid (Figure 2.1), and between them they constitute nearly 100% of the cell wall sugars. There are, however, trace amounts of a range of other sugars, notably apiose and also trace

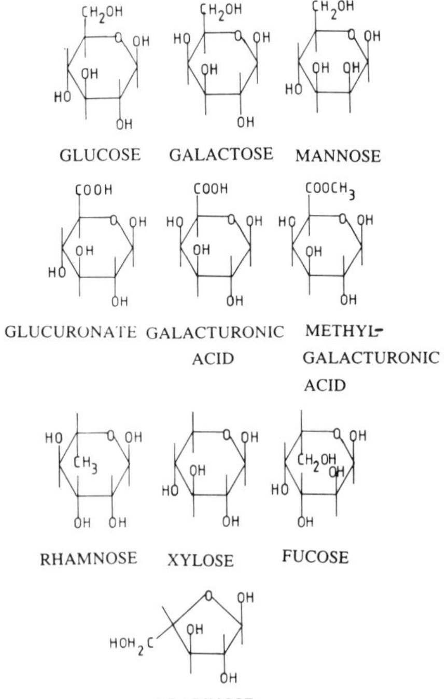

Figure 2.1 The major sugars in plant cell wall polymers.

Table 2.1 Representative list of glycosidic bonds occurring in the major cell wall polymers.

Polymer	Bonds
Arabinans	Ara α-1,2 Ara
Arabinogalactans	Ara α-1,3 Ara
Galactans	Ara α-1,5 Ara
	Gal β-1,4 Gal
	Gal β-1,6 Gal
Cellulose	Glc β-1,4 Glc
Xyloglucan	Fuc 1,2 Gal
	Gal β-1,2 Xyl
	Glc β-1,4 Glc
	Xyl α-1,6 Glc
Glucomannan	Glc β-1,4 Glc
	Glc β-1,4 Man
	Man β-1,4 Glc
	Man β-1,4 Man
Galactoglucomannan	Gal α-1,6 Man
	Glc β-1,4 Glc
	Glc β-1,4 Man
	Man β-1,4 Glc
	Man β-1,4 Man
Mannan	Manβ-1,4 Man
Galactomannan	Gal α-1,6 Man
	Man β-1,4 Man
Heteroxylan	Ara 1,2 Xyl
	Ara α-1,3 Xyl
	Gal β-1,5 Ara
	Gal β-1,4 Xyl
	Xyl β-1,2 Ara
	Xyl 1,4 Gal
	Xyl β-1,4 Xyl
β-D-glucan	Glc β-1,3 Glc
	Glc β-1,4 Glc
Rhamnogalacturonan	Ara 1,4 Rha
	Gal 1,4 Rha
	GalU α-1,4 GalU
	GalU α-1,2 Rha
	Rha 1,4 GalU
	Rha 1,2

amounts of substituted sugars such as 2-*O*-methyl-D-xylose. These sugar building blocks are found linked together in the structural polymers by a wide range of glycosidic bonds (Table 2.1) and, as will be discussed later, may also be involved in non-glycosidic bonding to the protein or phenolic constituents of the wall.

There are several different proteins associated with plant cell walls. Many of these are enzymes involved in wall turnover, others may act as specific markers in cell–cell recognition but some have an obvious structural role within the wall. These proteins obey the normal rules for biosynthesis and amino acid content, but structural proteins in particular seem to have an abnormal distribution of these amino acids leading to their being termed, for instance, hydroxyproline-rich or glycine-rich proteins. These structural proteins usually contain a high proportion of carbohydrate residues.

The structural proteins, although in some cases arising from multigene families, are none the less primary gene products and as such have clearly definable structures. Indeed cDNAs for several such proteins have been identified and their sequence used to infer the structure of the protein, in many cases polymeric chains of repeating amino acid motifs (section 2.2.10). Similarly, individual structural proteins can be readily isolated and purified, thus demonstrating the homogeneity of their structure. In contrast the structural polysaccharides are all secondary gene products, being synthesised by the action of biosynthetic enzymes (the primary gene product). This means that there is a large degree of heterogeneity in their structures. This in turn leads to several problems in defining or working with these polymers. First, it is general practice to group the polysaccharides together not on the basis of a common structure, but on their relative extractability from the cell wall. Thus pectins are generally considered as those polymers which are solubilised in either hot water, chelating agents or dilute acid. Hemicelluloses are polymers extractable under alkaline conditions and α-cellulose is the residue of the wall after such extraction. Attempts are then made to isolate individual polymer types from each of these fractions. Moreover, the heterogeneity of polymers often precludes any definite purification and thus analysis proves difficult. There are, therefore, many problems in attempting to determine the structure of any particular wall polysaccharide. First, in order to solubilise the polymer, severe conditions may be required either to enable solubilisation or to prevent modification during extraction by enzymes in the wall. Second, since any polymer is likely to exhibit heterogeneity, subfractionation is required and may result in the loss of some polymer types. Another key problem is the source of the cell wall since there are marked variations in walls between different cell types, within a cell type with time and between species.

2.2.1 Variation in cell walls

The cell wall is generally between 0.1 and 10 μm thick. The wall laid down during cell growth is usually thin and is called the primary wall. This often differs quite markedly from the secondary wall which is laid down following cessation of cell growth. The primary walls of adjacent cells are normally cemented together by yet a third identifiable structure called the middle lamella. Since wall material is laid down at the surface of the plasmalemma, in fully mature cells it is the secondary wall which is immediately adjacent to the cell surface, with the primary wall and middle lamella on the outside. However, in several cases it may be difficult physically to discern these three regions. In addition many walls undergo lignification following secondary thickening. It follows from the above that cells in different regions, or at different stages of development, will differ markedly in their wall structure and composition. Thus parenchymous tissue from free-growing plants and the fleshy parts of fruit consists principally of primary cell walls and middle lamellae. Secondary thickening occurs in many seed endosperms and woody tissues are highly lignified. In addition to this variation within species there is also a marked difference in the composition of polymers between species. (For reviews see Selvendran, 1985 and Bacic *et al*., 1988.) Due to the nature of the species investigated it is perhaps best to consider these in two groups: the dicots and gymnosperms; and the monocots, including the gramineae and other monocot families.

The primary walls of most dicot species are very similar in polymer composition, although relative amounts of these polymers may vary (Selvendran, 1985). Cellulose is always present and can account for 9–40% of the wall. The major non-cellulosic polymers tend to be pectic substances which include rhamnogalacturonans, arabinans, galactans and the so called type I arabinogalactans. There are also smaller amounts of heteroglucans, in particular xyloglucans. The structural protein of these walls tends to be rich in hydroxyproline. The cambial primary cell walls of dicots are slightly different to the other walls in that in addition to the hemicellulose xyloglucan they tend to contain high (2–10%) levels of xylans or 4-*O*-methylglucuronoxylan. From the few analyses carried out on gymnosperms it would appear that their primary cell walls are very similar to those of the dicots; however, it is possible that the pectic substances in this case may contain relatively more arabinans.

Seeds are a major source of unlignified secondary walls that have been extensively studied because of their commercial importance. In dicots the composition of this wall varies enormously between endospermic and non-endospermic seeds (Selvendran, 1985). The latter (e.g. pea cotyledons) have a wall composition very similar to the primary wall of other dicot tissues. The endospermic cells (e.g. guar endosperm) contain much smaller amounts of pectic substances and cellulose, and larger amounts of hemicelluloses, especially galactomannans. The galactomannans are commercially important for forming gels in the food industry. In many cases xyloglucans are also present in endosperms.

Information on the composition of lignified secondary walls of dicots and gymnosperms comes mainly from the analysis of different types of wood. These represent primarily the walls of xylem fibres and tracheids for dicots and gymnosperms respectively. In both cases cellulose accounts for between 40 and 60% of the walls. In gymnosperms the major non-cellulosic polysaccharides are glucomannans and galactoglucomannans, with 4-*O*-methylglucuronoarabinoxylans present in smaller amounts. In dicots the major non-cellulosic polymers are 4-*O*-methylglucuronoxylans with glucomannans as a minor component. In other lignified tissues, however, e.g. seed hulls from soya bean, levels of pectic substances are higher and the hemicelluloses in this case are mainly acidic xylans and xyloglucan.

The monocots are best considered in two groups, the Gramineae (grasses) have been widely studied due to their commercial importance, and the rest of the monocots can be grouped together. The primary walls of gramineae differ markedly in their composition to those of dicots, pectic substances and heteroglucans forming only a relatively minor part of the wall. The major non-cellulosic components of many gramineae walls are hemicelluloses, especially heteroxylans, including glucuronoarabinoxylans and arabinoxlyans. The walls also contain variable amounts of a so-called mixed linkage β-D-glucan. This polymer contains glucose residues linked by both β-1,3 and β-1,4 glycosidic bonds. The walls contain structural protein but tend to be low in hydroxyproline, unlike the dicots.

The remaining monocot families can be divided into two groups (Bacic *et al.*, 1988) simply on the basis of their relative content of ferulic acid. Some groups contain this phenolic residue covalently attached to the wall polymers, while others do not. While the composition of the primary wall in some of these other monocot families may resemble that in the gramineae, there is a wide range and in many cases the walls appear more like those of primary walls from dicots. Thus, for example, onion walls are high in pectic substances and xyloglucans (Selvendran, 1985), although in many cases the fine structure of these pectic polymers may differ from that of the dicot equivalent.

As with dicots, the non-lignified tissues of monocots tend to be relatively rich in a wide range of hemicellulosic polymers. These include mannans, galactomannans, glucomannans, galactans, heteroxylans and β-1, 3- and 1, 4-glucans.

The lignified secondary walls of monocots are again different between the gramineae and some other monocot families. The gramineae contain between 35 and 40% cellulose with the major non-cellulosic polymers being glucurono arabinoxylans, heteroglucans and minor amounts of β-1, 3- and 1, 4-glucans. The rest of the monocot families are again divided with some having walls more closely resembling the gramineae and others the dicots.

It is obvious from the above brief outline of wall composition in the various taxa that there is a wide range of wall polymers. In addition to these wall polymers there are many other wall-associated polymers which are synthesised in special cells or in response to particular environmental stimuli. In the former group are cutin (a polyester of long chain fatty acids forming the functional component of the cuticle), suberin and waxes. In the latter group is the polysaccharide callose. This is a β-1, 3-linked glucan polymer synthesised by cells in response to wounding. In addition several commercially important polysaccharides are derived from the cell walls of algae, especially sea weeds. These include alginate, agar and carrageen. The structure of algal cell walls differs again from those of higher plants. A description of the algal wall is beyond the scope of this chapter but a good review is that by Percival and McDowell (1978). The structure of the commercially important polymers is dealt with in section 2.3.

In the next section we consider the structure of the major wall components from higher plants and that of the commercially important polymers extractable from algae. However, first a few words of caution. As we have seen, these polymers may be altered during extraction. An obvious example is the possible deesterification and depolymerisation via β-elimination of galacturonic acid polymers extracted under even mildly alkaline conditions. Also, in attempting to purify a homogeneous sample for investigation, only a small proportion of the original total heterogeneous polymer population may be recovered. Much work on structural determination of polymers has been carried out on the walls of cells grown in suspension culture.

Although the polymers obtained in this case are unlikely to be much altered from those in the walls of growing plants, it is obvious that this material represents only the primary wall and that overall composition of cultured cell walls and their free growing equivalents is different (Selvendran, 1985).

The structure of wall polymers has been the subject of intensive investigation over several decades. This topic has been thoroughly reviewed by Aspinall (1980), Darvill *et al.* (1980), McNcil *et al.* (1984), Fry (1986) and Bacic *et al.* (1988).

2.2.2 Cellulose

This constitutes about 20–30% of the dry weight of a typical primary cell wall and occupies about 15% of the wall volume. However, there is a wide range with cellulose accounting for only 2–4% of the wall of cereal endosperm (Fincher and Stone, 1986) and up to 94% of the secondary walls of cotton seed hairs (Meinert and Delmer, 1977). Each cellulose polymer consists of a β-1, 4-linked glucan with each glucan residue oriented at 180° to its neighbour. The molecule takes up the form of an extended ribbon with a two-fold screw axis, cellobiose being the repeat structural unit. This ribbon structure is stabilised by hydrogen bonding (Bacic *et ul.*, 1988). Hydrogen bonding can also occur between adjacent polymers and this leads to the formation of microfibrils of cellulose within the wall (Figure 2.2). These fibrils vary in size but are usually elliptical in cross-section with axes of about 30–300 Å. Most evidence suggests that the cellulose molecules are ordered in a parallel manner within the fibril (Claffey and Blackwell, 1976), although the existence of antiparallel arrangements cannot be excluded (Pizzi and Eaton, 1985). The degree of polymerisation (DP) of the cellulose polymers is also variable. In developing cotton seed hairs the primary wall cellulose polymers are relatively short and show a high degree of heterogeneity ranging from 2000 to 6000 residues. In the secondary walls the molecules tend to be larger and more homogeneous with DP of around 14 000 (Marx-Figini and Shulz, 1966). In one report (Blaschek *et al.*, 1982), cellulose was stated to be biphasic having a DP of either 500 or ranging between 2320 and 4510.

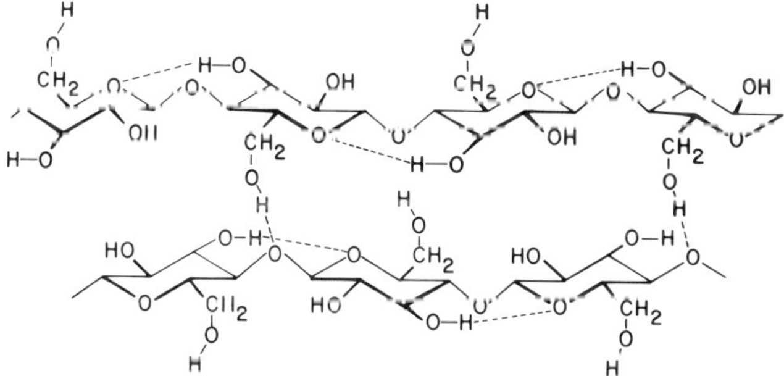

Figure 2.2 Cellulose structure showing how hydrogen bonds act to stabilise the ribbon-like conformation of the polymer and facilitate the innerpolymer attractions within the microfibril.

2.2.3 Xylogucan

Xyloglucans were first identified in tamarind seeds (Kooiman, 1961) in which they disappear during germination, probably because of their use as a carbohydrate reserve. This led to their original name of amyloids. Since then xyloglucans have been identified in a wide range of dicot and monocot cell walls (Hayashi, 1989). However, the structure of this polymer differs between species. All xyloglucans consist of a backbone of β-1, 4-linked glucosyl residues, some of which are substituted at the 6-position with α-linked xylosyl residues. However, the degree of substitution varies between species as does the extent of further substitution with galactose, fucose or arabinose.

In dicots xyloglucan can account for between 20 and 25% of the cell wall. The molar ratio of glucose : xylose : galactose is approximately 4 : 3 : 1 and in addition to these three residues dicot xyloglucans also contain fucose (Hayashi, 1989). An exception to this is found in some solanaceae species in which the xyloglucans have arabinose instead of fucose linked to xylose.

Hydrolysis of pea xyloglucan with an endo β-1, 4-glucanase produces a nonosaccharide and a heptasaccharide in roughly equal proportions (Hayashi and Maclachlan, 1984). The structure of these two are given in Figure 2.3. Partial hydrolysis of xyloglucan yields dimers of these two oligosaccharides thus it is likely that they occur as alternating structural units. In dicots the repeating structural unit of xyloglucan would seem to be the heptasaccharide unit shown in Figure 2.3. The heterogeneity found between species arises from differences in size, which can range from about 7 kDa to 180 kDa (Bacic *et al.*, 1988), the level of substituted xylosyl residues or the distribution of galactosyl and fucosyl–galactosyl residues. Three xyloglucan polymers (180 kDa, 60 kDa and 30 kDa) can be obtained from suspension cultured soybean (Hayashi *et al.*, 1980) and seven types by ion-ex-

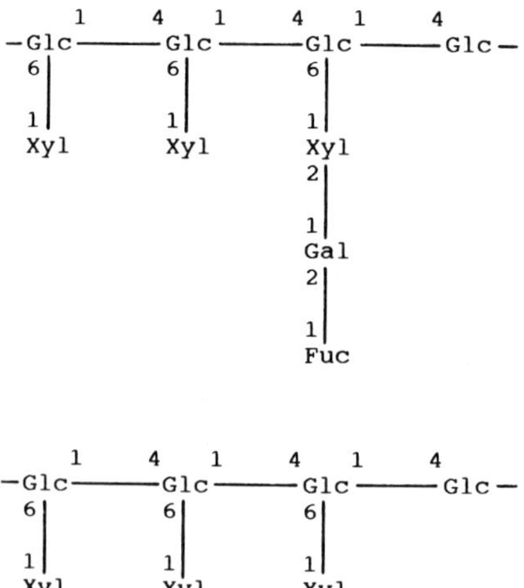

Figure 2.3 Structures of the two main oligosaccharide fragments obtained following the partial digest of cell wall xyloglucan.

change fractionation of polymers from apple cell walls (Selvendran, 1985). These presumably differed in the extent and distribution of side chains. Thus in dicots the xyloglucan is basically a repeating heptasaccharide with varying degrees of substitution, there being three main block structures discernible: (i) regions of highly substituted xylose residues, (ii) regions of low substitution but with various side chains, and (iii) regions of high substitution with di and tri side chains (Selvendran, 1985).

The xyloglucans from monocots, especially those in the graminaceous monocots, account for much less, about 2%, of the cell wall. They are also much less frequently substituted at the xylosyl residues and do not contain fucose (Hayashi, 1989). Hydrolysis of monocot xyloglucans tends to yield a pentasaccharide with glucose : xylose (3 : 2), free glucose and small amounts of a heptasaccharide glucose : xylose (4 : 2), a trisaccharide glucose : xylose (2 : 1) and cellobiose (see references in Hayashi, 1989). Thus it seems that monocot xyloglucans have much of their glucan backbone unsubstituted by xylose. However, it is possible that monocot primary walls have a more highly substituted xyloglucan which is acted upon during growth by α-xylosidases and β-galactosidases to produce a final unsubstituted polymer.

2.2.4 Glucomannans and galactoglucomannans

These polymers constitute a family of polysaccharides with linear extended backbones in which β-1, 4-linked mannosyl and glucosyl residues are arranged randomly. In some cases this backbone may be substituted, again randomly, at the 6-position of the sugars with single galactosyl residues. These galactoglucomannans usually have a gal : glu : man ratio of 1 : 1 : 3 but this can vary (Bacic *et al*., 1988). The presence of galactose side chains increases the solubility of the polymers and is commercially significant for their use as gelling agents in the food industry (see section 2.3).

2.2.5 Mannans and galactomannans

Mannans are linear extended polymers of β-1, 4-linked mannosyl residues. They are a major constituent of some seed endosperms as well as the walls of several green and red algae. In the leguminosae polymers of β-1, 4-linked mannosyl residues have been found which are substituted at the C(*O*)6-position with single α-D-galactosyl residues. This substitution, as for the glucomannans, increases the solubility of the polymer and is commercially significant.

2.2.6 Heteroxylans

This family of molecules constitute the major non-cellulosic polysaccharides in the primary wall of grasses and the secondary walls of angiosperms. They consist of a linear β-1, 4-linked xylanosyl backbone variously substituted by a range of mono-

and oligosaccharide side chains. These include α-arabinosyl and α-glucuronic acid or its 4-*O*-methyl derivative. Again the degree and type of substitution play an important role in determining the solubility of the polymer. In addition there may be substitution with phenolic acids, especially ferulic, coumaric and diferulic.

2.2.7 β(1-3) (1-4)-D-glucans

This is a linear polymer of glucosyl residues linked together by a mixture of β-1, 3 and β-1, 4 glycosidic bonds. Heterogeneity in this instance results from the relative proportion and distribution of these two types of linkage. Generally, these β-glucans, as they are commonly called, contain approximately 30% of the linkages as β-1, 3 and 70% as β-1, 4. However, the distribution of these linkages differs between species. In some instances the β-1, 3 links are never, or rarely, contiguous and the β-1, 4 links occur mainly as runs of 2, 3 or 4 residues with occasional runs of up to 10 glucose residues (Bacic *et al.*, 1988). In other cases, e.g. *Zea mays*, the β-1, 3 links may run in regions of 2, 3 or 4 contiguous linked residues with the β-1, 4 in blocks of more than four (Kato and Nevins, 1984). The long β-1, 4-linked regions are linear ribbon structures with the β-1, 3 links introducing 'kinks' to give an overall irregular shape. These polymers constitute a major non-cellulosic polysaccharide in the gramineous monocots. In addition to the glucan backbone these polymers also have associated with them small amounts of arabinose and xylose. These are probably covalently linked to the β-glucan but could arise from contamination of preparations with other polymers.

2.2.8 Ramnogalacturonans

Dicot cell walls, and in particular the middle lamella region of these walls, contain large amounts of galacturonic acid residues. These are linked together in linear polymers by α-1, 4 glycosidic links. Homogalacturonans are polymers consisting solely or predominantly of these types of linkage. However, the distribution of homogalacturonans as such in cell walls is difficult to assess. It is difficult to isolate homogalacturonans from walls without cleaving covalent bonds, thus their presence in the wall is difficult to determine. However, evidence summarised by McNeil *et al.* (1984) would indicate they do occur in at least some cases.

The majority of the galacturonic acid is, however, found in the rhamnogalacturonans or acidic pectins. In these polymers the α-1, 4-linked galacturonic acid is interspersed with rhamnose residues (Figure 2.4). These again are a diverse range of polymers differing in size, amount and distribution of rhamnose residues, and extent of methyl esterification of the galacturonic acid residues. Polymers range in size from 30 kDa to 300 kDa. The distribution of rhamnose may be random but in several instances there is evidence for at least some organisation. Thus mild acid hydrolysis of citrus fruit pectin yields polygalacturonic acid blocks of about 25 residues indicating a regular distribution of the rhamnose (Powell *et al.*, 1982). More extensive degradation of the primary wall of sycamore cultured cells with a polygalacturonase yields two rhamnogalacturonan fractions, RhaI and RhaII, which have been purified and characterised (McNeil *et al.*, 1984).

Rhamnogalacturonan I (RhaI) is a polymer with a DP of around 2000 composed of galacturonic acid, rhamnose, galactose, arabinose and small amounts of fucose (Figure 2.4). The backbone of this molecule consists of alternating galacturonic acid–rhamnose residues. About half the rhamnose residues are branched and contain glycosyl side chains with an average DP of 7. The nature of these side chains is highly variable with at least 30 different types possible (McNeil *et al.*, 1980, 1982). It is unclear whether RhaI constitutes a homogeneous polymer or a mixture of polymers each with different side chains, since Ishii *et al.* (1989) have reported a family of RhaI-like pectic polysaccharides.

Rhamnogalacturonan II (RhaII) is structurally different from RhaI. It contains about 60 glycosidic residues and includes most of the minor sugars found in the

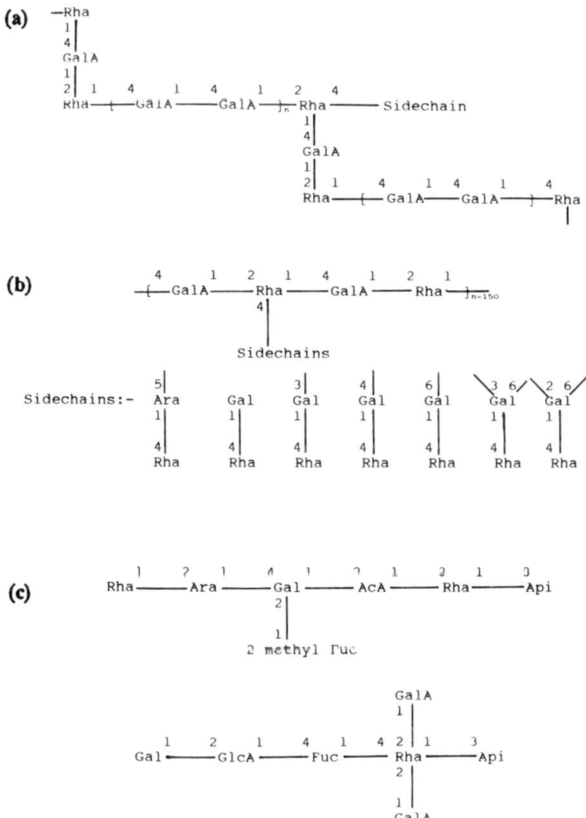

Figure 2.4 Partial structures of acidic pectins from cell walls. (a) Postulated structure of the main backbone of rhamnogalacturonan. Structures of (b) Rhamnogalacturonan I and (c) Rhamnogalacturonan II, two fragments isolated from the walls of suspension cultured sycamore cells.

acidic pectin, these include 2-*O*-methylfucose, 2-*O*-methylxylose, D-apiose, 3-*C*-carboxyl-5-deoxy-L-xylose (aceric acid), and 3-deoxymanno-octaloionic acid. The structure of RhaII is very complex (Melton *et al.*, 1986) and contains at least 20 different glycosidic links. However, it appears to be a repeat of two heptasaccharide units (Figure 2.4).

These highly branched rhamnogalacturonans I and II have been isolated from suspension cultured cells which are predominantly primary cell walls. Similar polymers have been identified from other dicots, gymnosperms and monocots. However, it is likely that these highly branched pectins are restricted to the primary walls of higher plants since analysis of rhamnogalacturonans from the middle lamella of apple fruit cells would indicate a much less branched structure (Knee, 1983).

A further source of variation in the acidic pectins is in the extent and distribution of esterified residues of galacturonic acid. The galacturonic acid can be methyl esterified at the C6 position (Figure 2.1). Highly methylated pectin is termed pectinic acid while relatively deesterified pectin is called pectic acid. In some instances acetylation can also occur. The extent of this esterification is variable but in fruit is often in the range of 50–90%. The distribution of methylesters in the acid pectin may be random or ordered. Recent studies using antibodies against highly esterified or deesterified pectins have suggested that in some tissues at least there seems to be a spatial separation of these two types of polymer (Knox *et al.*, 1990).

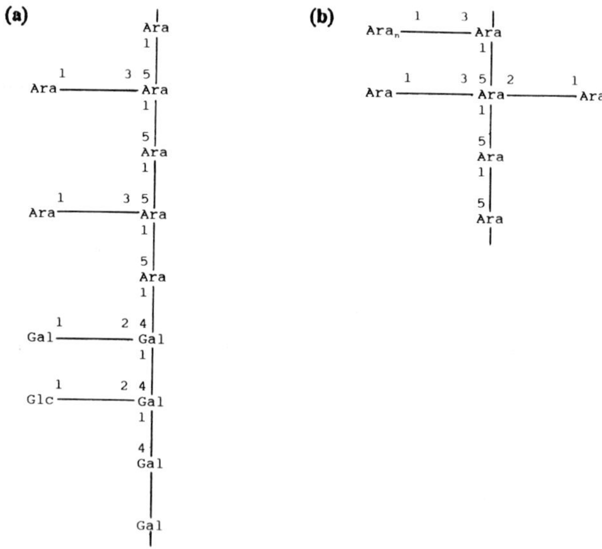

Figure 2.5 Postulated partial structures for (a) arabinogalactan type I and (b) arabinan from cell walls.

2.2.9 Arabinans, galactans and arabinogalactans

These three classes of polymer together constitute the so-called neutral pectic substances. Pure arabinans and galactans do occur in plant cell walls but are relatively rare. The arabinans are typically composed of α-1, 5-linked arabinosyl residues with some degree of branching by linkages at the 2 and/or 3 positions (Figure 2.5). Pure galactans consist of β-1, 4-linked linear polymers, which can assume a helical configuration, with small amounts of 6-linked residues. Two types of arabinogalactans are found associated with extracts from plant cell walls. Type I is characterised by a β-1, 4-linked galactosyl backbone substituted with 5-linked and terminal arabinosyl residues. This polymer is highly variable and can also contain 3, 4-linked galactosyl residues (Aspinall, 1980) (Figure 2.5). Type II is characterised by 3-, 6- and 3, 6-linked galactosyl residues associated with 3- or 5-linked arabinosyl residues. This polymer is again variable but unlike type I does not appear to be a structural component of the cell wall (Fincher *et al.*, 1983). Instead type II arabinogalactans are thought to be associated with extracellular space and possibly associated with the plasmalemma. Type II arabinogalactans are also found in the so-called arabinogalactan proteins (AGPs) which are dealt with in section 2.2.10.

2.2.10 Cell wall proteins, glycoproteins

Several reviews have been published over the last few years relating to cell wall proteins and glycoproteins (Cooper *et al.*, 1987; Cassab and Varner, 1988; Varner and Lin, 1989; Marcus *et al.*, 1991).

The most studied and probably most abundant cell wall protein is the so-called extensin (Wilson and Fry, 1986). This is a group of hydroxyproline rich glycoproteins (HPRG) whose structure, like that of the wall polysaccharides, differs between species and possibly also between tissues. In general, extensins are composed of 50% protein and 50% carbohydrate. The amino acid composition of the protein is relatively unusual, being in general about 33–42 mol% hydroxyproline, with serine, lysine, tyrosine or isodityrosine, valine, proline and histidine being the other abundant amino acids (Cassab and Varner, 1988). This high preponderance of basic amino acids gives the extensin a p*I* of around 10. The actual extensin protein backbone seems to consist of simple repeat units, the basic unit varying between species. Varner and Lin (1989) review several of these repeat sequences. For example, two extensins extractable from suspension cultures of tomato cells (P1 and P2) have the following structure:

P1 consists mainly of two abundant repeat sequences:

a) Ser-Hyp-Hyp-Hyp-Hyp-Thr-Hyp-Val-Thr-Lys
b) Ser-Hyp-Hyp-Hyp-Hyp-Val-Lys-Pro-Tyr-His-Pro-Thr-Hyp-Val-Tyr -Lys

and P2 consists entirely of the repeat sequence:

Ser-Hyp-Hyp-Hyp-Hyp-Val-Tyr-Lys-Tyr-Lys

The carbohydrate content of extensin is generally 97% arabinose and 3% galactose. The arabinose is found attached to hydroxyproline residues predominantly as Hyp-ara3 or Hyp-ara4. The galactose is attached as single units to serine residues. In dicots most of the hydroxyproline residues in the extensin backbone are glycosylated. However, the HPRG from monocots is quite different with a much lower degree of glycosylation. The structure of extensin is an extended helical rod, and this can be seen clearly under the electron microscope. Extensin is synthesised as a soluble unit polypeptide of around 40 kDa but rapidly becomes insoluble in the wall. It is assumed that covalent cross-linking occurs in the wall to form an extended protein network (see section 2.2.13).

Extensin is not the only HPRG found in association with cell walls. There are two other types of glycoprotein in walls which are also rich in hydroxyproline but are distinct from extensin. These are the so-called arabinogalactan proteins (AGPs) and lectins (Marcus *et al.*, 1991). The AGPs are found in all tissues and are composed of around 10% protein and 90% carbohydrate. The protein is rich in hydroxyproline, serine, alanine and glycine and as such tends to be an acidic rather than, as for extensin, a basic protein. The principal carbohydrate residues are arabinose and galactose, with some rhamnose, mannose and galacturonic and/or glucuronic acids. These are present in much longer chain lengths than in extensin and resemble the type II arabinogalactans described in section 2.2.9. A final difference between AGPs and extensin is that these glycoproteins remain freely soluble and as such are not thought to be structural components of the wall but associated with the extracellular space.

The hydroxyproline-rich lectins seem to be restricted to Solanacea species and are thought to be associated with the plasmalemma rather than to be structural components of the wall.

Although extensin is the most thoroughly documented cell wall structural protein, it is unlikely that it represents the only class of proteins in this group. Other candidates that have been suggested to have a structural role in the wall are the so-called repetitive proline-rich protein from soybean (Aeryhart-Fullard *et al.*, 1988), and the glycine-rich protein in petunia (Condit and Meagher, 1987) and rice (Lei and Wu, 1991).

2.2.11 Callose (β-1, 3-glucan)

This is a linear polymer of β-1,3-linked glucosyl residues. It is not common in plant walls but is synthesised transiently in some specialised cells (e.g. cell plate, pollen mother walls) or in response to wounding.

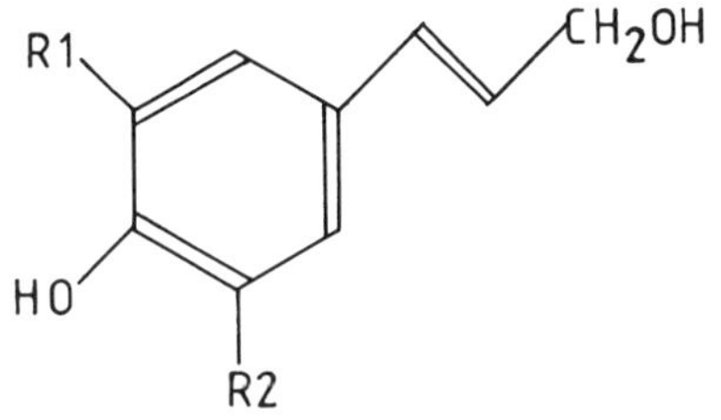

Figure 2.6 Phenolic precursors in the biosynthesis of lignin. R1 = R2 = H, *p*-coumaryl alcohol; R1 = OCH_3, R2 = H, coniferyl alcohol; R1 = R2 = OCH_3 sinapyl alcohol.

2.2.12 Lignin

Although phenolic residues are found associated with several of the above polymers the major phenolic constituent of walls is lignin. Lignin is laid down in the walls of many species once cell growth and expansion has ceased, and tends to replace the aqueous/pectic matrix in previously expanding cells. Lignins are polymers of phenylpropanoid residues almost exclusively derived from *p*-coumaryl, coniferyl or sinapyl alcohols (Figure 2.6). The composition and linkage between monomers differs in species and, because of the wide range of possible C–C or C–O–C bonds and the possible random nature of the polymerisation, no precise structure for lignin can be given.

2.2.13 Structure of algal polysaccharides

2.2.13.1 Alginates. Alginic acid is a 1,4-linked copolymer of α-L-guluronic and β-D-mannuronic acid. It performs a structural role in a variety of brown algae (*Phaeophyceae*). The most important industrial sources of alginates are *Ascophyllum, Laminaria* and *Macrocystis*. A number of Gram-negative bacterial species produce polysaccharides closely resembling alginates but these are not, at least at present, commercially important (Sutherland, 1990) The two uronic acids can be found as blocks (G or M blocks) or as regions where the two types of residues predominantly alternate (Figure 2.7). All three types of structure can be found in

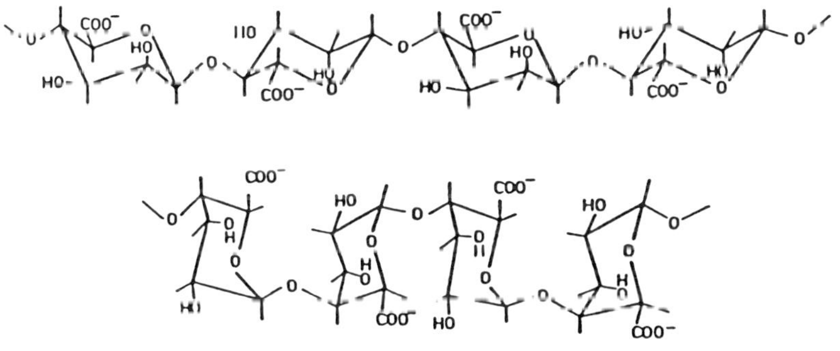

Figure 2.7 Homopolymeric sequences for alginate. Top: poly-D-mannuronate. Bottom: poly-L-guluronate.

the same polymer chain. The sequence of these uronic acids within a chain can be described by second order Markov statistics (Larsen *et al.*, 1969). That is to say a chain is constructed on the basis that the probability of the next residue being mannuronic acid or guluronic acid depends on the preceding two residues in the chain. Although there are eight such probabilities (P_{MMM}, P_{GGG}, P_{MMG},P_{GGM}, P_{MGM}, P_{MGG}, P_{GMG} and P_{GMM}, where, for example, P_{GGM} is the probability of finding a mannuronate residue when the preceding two residues are guluronate) only four of these probabilities are independent. The probabilities can be calculated from knowledge of the triad frequencies as determined by ^{13}C NMR (Grasdalen *et al.*, 1981; Grasdalen, 1983).

2.2.13.2 Carrageenans. Carrageenans are extracted from the Rhodophyceae family of red seaweeds. Important commercial sources are *Eucheuma, Chrondrus* and *Gigartina.* Refined carrageenans are extracted from the seaweed in hot alkali and after filtration the carrageenan is alcohol precipitated, washed and then dried and milled.

The basic repeat unit of the industrially important red seaweed polysaccharides is shown in Figure 2.8. For carrageenans the basic repeat unit is α-1,3-linked D-galactose and β-1,4-linked D-galactose. The galactose units are sulphated to varying extents. The two gelling carrageenans, iota (ι) and kappa (κ), both contain a 3,6-anhydro bridge on the β-1,4-linked galactose unit. These repeat units are compatible with the proposed double helix mechanism for carrageenan gelation (see section 2.3). One of the objectives of the alkali extraction process is to catalyse the conversion of the 6-sulphate group on the 1,4-linked galactose of the biological precursors for kappa and iota carrageenan to the 3,6-anhydro form (Lawson and Rees, 1970). This transformation is induced by an enzyme in the seaweed but further conversion enhances gel strength as it increases the number of helix compatible residues.

Also of commercial significance is lambda carrageenan which has a structure consisting of a disaccharide unit of α-1,3-linked D-galactose-2-sulphate and β-1,4-linked D-galactose-2,6-disulphate (Figure 2.8). Although these are the basic carrageenan repeat structures, there is far more variation between carrageenans than is the case for the alginates, thus a particular sample may contain both kappa and iota character.

2.2.13.3 Agar. Agar is also extracted from red seaweeds particularly the genera *Gelidium* and *Gracilaria.* It is composed of two main polysaccharides, agarose and agaropectin. Agarose is the main gel-forming polysaccharide and its basic structural repeat unit is α-1,4-3,6-anhydro-α-L-galactose-1,3-β-D-galactose-1,4 (Figure 2.8). There will, however, be other structural features present such as the 6-methyl ether, some 6-sulphate derivatives of galactose and a few 3,6 anhydrogalactose-2-sulphate residues. The extent of these will depend on the source and extraction

(a)

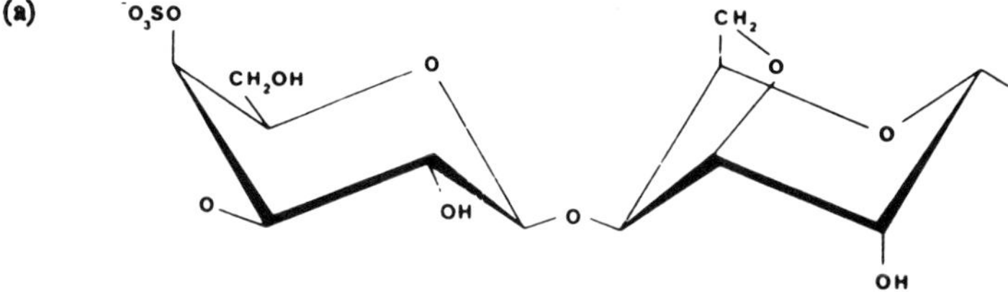

(b)

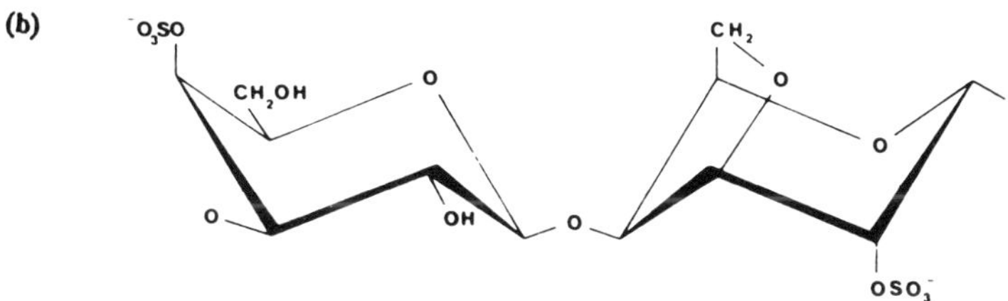

(c)

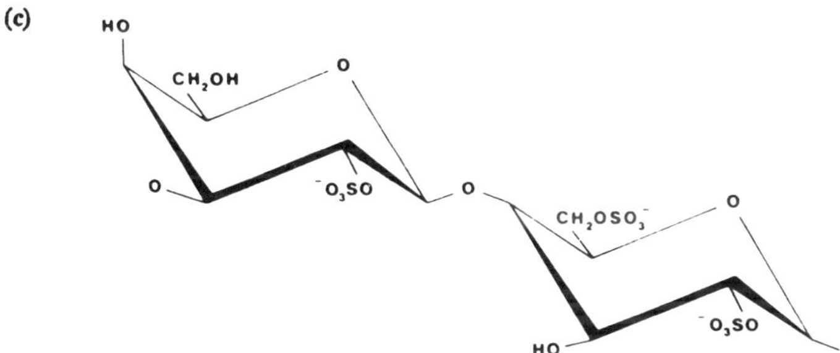

Figure 2.8 Carrageenan structures. The left-hand side is a derivative of a 3-linked galactose unit and the right-hand side is a derivative of a 4-linked galactose unit. (a) kappa carrageenan; (b) iota carrageenan; (c) lambda carrageenan.

method. Although there are some similarities between carrageenans and agarose, commercial preparations of the former have higher molecular weights, i.e. M_W 400 000 for carrageenan (Lecacheux *et al.*, 1985) compared with M_W 100 000 for agarose (Rochas and Lahaye, 1989).

2.2.14 Assembled cell wall structure

Given the wide variation in composition between primary, secondary and middle lamella walls and between tissues and species, especially dicots/monocots, it is not possible to describe a universal wall structure. However, most walls follow a general pattern of structure in which cellulose fibrils are surrounded by hemicellulose and this whole structure is embedded in a matrix of pectin or galactomannans and protein (for an early discussion see Keegstra *et al.*, 1973). The walls may either be highly hydrated or highly lignified. The extent of hydration is important, considering the requirements for an aqueous environment of the enzymes which must be involved in the complex turnover of the wall during growth, development and differentiation. Alternatively the walls of non-growing cells may be highly lignified. The structure of the wall has been reviewed several times in the past ten years (McNeil *et al.*, 1984; Fry, 1986; Bacic *et al.*, 1988; Varner and Lin, 1989).

The general approach to the investigation of wall structure in the past was to employ chemical or enzymic means selectively to degrade the wall polymers and look for cross-links between the polymers. It is beyond the scope of this review to deal with these various techniques but they are covered in Fry (1988).

The types of cross-links that have been found or postulated have been reviewed by Fry (1986) and are summarised in Table 2.2. It has been shown already in section 2.2.2 that cellulose polymers interact via hydrogen bonds to form microfibrilar structures. Hemicellulose can also form non-covalent bonds with cellulose via hydrogen bonding, but this is restricted by the side chain on these polymers. Thus, such interaction can only occur at the surface of the cellulose fibrils and it is thought that the bulk of the hemicellulose fraction of the wall is bound via hydrogen bonds to the surface of the cellulose fibrils. In dicots the major hemicelluloses are the neutral xyloglucans and acidic arabinoxylans while in monocots they are acidic arabinoxylans and neutral β-glucans. Although the bulk of the hemicellulose in the wall is associated with the surface of cellulose fibrils, the length of these polymers,

Table 2.2 Possible cross-links between cell wall polymers. (Adapted from Fry, 1986.)

Nature of link	Polymers involved
Hydrogen bond	Cellulose–cellulose Cellulose–xyloglucan Cellulose–xylan
Ionic bond	Acidic pectin–acidic pectin Acidic pectin–protein
Coupled phenols	Protein–protein Pectin–pectin
Ester bond	Pectin–cellulose
Glycosidic bond	Acidic pectin–neutral pectin
Entanglement	Pectin–protein

and especially the xyloglucans, raises the possibility that they may actually bridge the gap between adjacent microfibrils and hence play an important role in stabilising these fibrils (Hayishi, 1989). It is possible, though evidence is scarce, that some at least of the hemicelluloses may be covalently linked to pectic polymers.

The matrix of dicot cells is made up primarily of cross-linked neutral and acidic pectins. The complexity of these polymers could allow a wide diversity of bonds and proposed links include glycosidic bonds, ionic bonds, and phenolic coupling. It is thought that the neutral pectins are covalently linked via glycosidic bonds to the rhamnose residues in the acid pectin backbone. Thus galactan, arabinan or arabinogalactan polymers are linked via their terminal residues to the 4 position on the rhamnose residue (Figure 2.4). The precise nature of these linkages is unclear. Also, as previously stated, there may be some glycosidic bonding to a small fraction of the hemicellulose polymers.

Pectins are capable of forming several ionic bonds. Perhaps the most important is the ability to form Ca^{2+} bridges (Grant *et al.*, 1973). Regions of deesterified galacturonans on adjacent polymers can interchelate calcium ions in the so-called 'egg box' structure (Figure 2.9). It is estimated that a region of about 25 residues is required to enable the sum of these otherwise weak bonds to allow firm linkage. In addition, such regions of deesterified acidic pectin could link ionically to the basic extensin within the wall. Pectins contain ferulate and *p*-coumarate, ester-

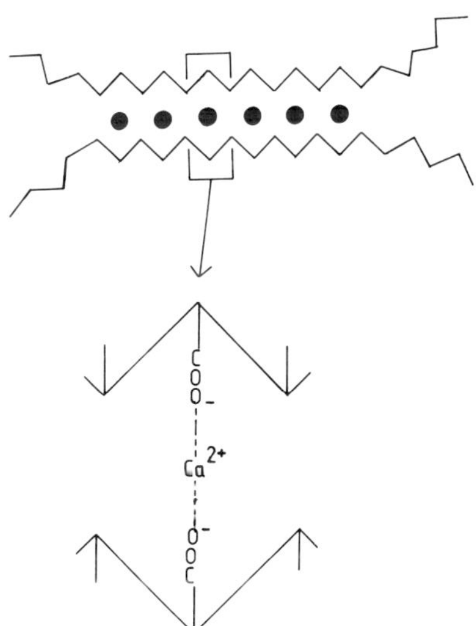

Figure 2.9 Scheme for the interaction of deesterified acid pectin backbones with calcium ions. The so-called 'egg box' structure.

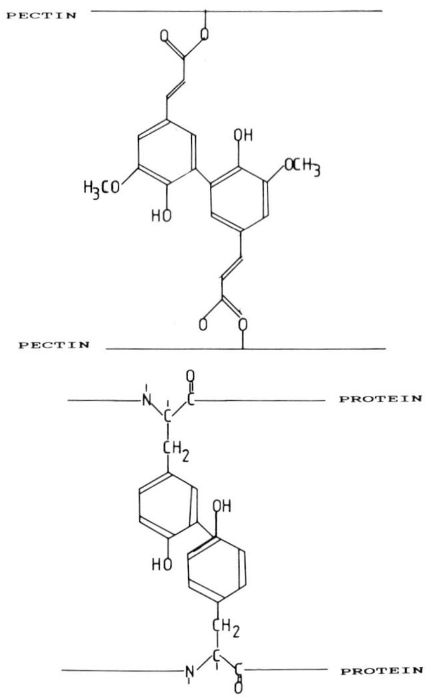

PECTIN
OH
OCH3
H3CO
HO
PECTIN
PROTEIN
CH2
OH
HO
CH2
PROTEIN

Figure 2.10 Coupled phenolic bonds in plant cell walls. Top: diferulic acid coupling is thought to occur between pectin molecules. Bottom: isodityrosine coupling between proteins.

linked through their COOH groups (Fry, 1983). These phenolic constituents could be involved in phenolic coupling. The simplest such coupling would be diferulate (Figure 2.10) and indeed alkaline hydrolysis of walls yields small amounts of this compound (Fry, 1986). Similar phenolic coupling could occur between the ferulate constituents also found in arabinoxylans.

Apart from possible ionic interactions with acidic pectin it is not thought that the cell wall extensin has any other covalent or non-covalent linkages to the other wall components. However, the protein is very insoluble and must therefore be cross-linked to itself. The nature of this cross-link is thought to be via isodityrosine bonds (Figure 2.10) (Fry, 1982). This means that the extensins probably form a mesh. The pores in this mesh are relatively large and could certainly span polysaccharide backbones. It is thus thought that the protein and pectin components of the matrix are enmeshed together, such entanglement acting to further stabilise the wall.

Two fairly recent approaches to the study of assembled wall structure are electron microscopy and immunohistolocalisation. In both cases the need to radically disrupt the wall during investigation is much reduced. The application of sophisticated freeze fracture and electron microscopy allows the direct visualisation of the cell surface, and to some extent the interior of the wall itself (McCann *et al.*, 1990; Roberts, 1990). The other new approach is the use of antibodies, raised against specific wall polymers to investigate the distribution and localisation of these polymers within the wall (Roberts *et al.*, 1985). This approach has been used to study the distribution of esterified and deesterified polyuronides (Knox *et al.*, 1990). It has also been used to localise xyloglucan (Moore *et al.*, 1986), xylan, arabinogalactan and callose (Northcote *et al.*, 1989) and structural proteins (Stafstrom and Staehelm, 1988; Ye *et al.*, 1991) within the wall.

The actual three-dimensional structure of the plant cell wall is obviously complex and is still far from being fully elucidated. It is likely that, just as with the individual wall polymers, variation in the fine structure of the assembled wall will occur between both tissues and more obviously, different species. However, it also seems likely that the general model structure described above for non-lignified walls applies to most if not all cases.

2.3 Utilisation of cell wall polymers

2.3.1 Applications

The food industry is the major user for many cell wall derived materials. Common applications are thickeners or gelling agents. Because starch is substantially cheaper then most of the polysaccharides mentioned, the unique gelling properties of the non-starch polysaccharides are important in determining food uses. To get reliable figures as to the amounts employed in different applications is not easy. Table 2.3 shows data presented by Vincent in 1986. This is now out of date but gives an indication of the distribution of the various polysaccahrides between food application areas. In the following some of the major applications for the commercially important non-starch polysaccharides will be discussed in the context of their properties.

2.3.1.1 Alginate. The most interesting property of alginates is their ability to form heat-stable gels by interaction with divalent ions. In food terms this means that they can be used to produce shapes from waste or comminuted materials. Because the alginate gel is heat-stable, the structures formed can be cooked without losing their shape. There are numerous examples of this type of technology, for example the production of 'artificial' cherries, pomiento strips to stuff cocktail olives, a process for the production of onion rings from comminuted onions which are subsequently frozen (Glickman, 1982) and the reformation of comminuted meat into chunks or fibrous structures which are then used as components of canned petfoods.

Still of substantial importance is the use of alginates as stabilisers in frozen products such as ice cream and ice lollies. Probably the most important of a number of roles that the polysaccharide fulfils in this application is the reduction in the rate of growth of large ice crystals on storage of the product. The mechanism for this appears to be a reduction in crystal growth rate rather than a change in the nucleation process and here the ability of the polysaccharide to gel may be important (Muhr and Blanshard, 1986).

Table 2.3 World production for food use and major applications of non-starch polysaccharides in 1980. (Table adapted from Vincent, 1986.)

Uses (US $ million)	Agar	Alginate	Carrageenan (semi-refined)	Carrageenan (refined-kappa)	Guar gum	Locust bean gum	Cellulose (micro-crystalline)	Pectin
World output (tonnes × 10^3)	0.71	15.7	15.36	4.16	28	13.8	2.8	16.75
Food use in USA and EC (US $ million)								
Bakery	0.4	4.1		2.6			0.08	3.1
Drinks		1.8			0.55		0.23	3.55
Flavours		2.5			4.8		0.06	
Desserts		2.9		29		18	0.15	13.2
Meat products	0.15	6.3	30.25	0.13	1.8			
Health/diet food	0.26						4.32	4.75
Confectionery	0.29	2.45						7.7
Pouches/canned food		2.5	9.5	11.70	2.5	3.7		
Catering	0.92	9.5						
Frozen food					2.7		1.5	
Preserves	0.12							2.0
Baby foods					3.6			
Total	3.19	50.65	48.05	52.43	25.5	25.95	11.64	44.15

Of increasing interest is the use of alginates to immobilise cells by the use of the calcium alginate gelation reaction (Skjaek-Braek *et al.,* 1985) and its role in various medical applications. Currently the most important of the latter is the wound-healing dressings made from alginate fibres but there is a growing recognition that alginates have interesting immunological properties which appear to be related to the mannuronic acid component.

Traditional non-food applications of alginates, such as their use in textile printing and welding electrodes, still consume large quantities of material but have tended to decline in importance compared with food, medical and biotechnological uses.

2.3.1.2 Carrageenans. Two properties of carrageenans determine their major uses: the ability to form gels and their milk-reactivity. The latter involves specific interactions between the carrageenan molecule and the surface of the casein micelle, probably also involving calcium ions. A consequence of this is that the concentration of carrageenan required to form a gel is reduced by the presence of milk. This specific interaction is used in the preparation of milk puddings, the stabilisation of particles in flavoured milks, such as chocolate milk, and in ice cream stabiliser blends where the presence of the carrageenan prevents separation of the mix on ageing. Because of its milk-reactivity, carrageenan can be employed at very low levels in dairy products. For example, only 0.02% would be required in flavoured milks. The non-gelling lambda carrageenan thickens when added to cold milk (Moirano, 1977).

Gelation of carrageenan is used extensively in meat products. The carrageenan locust bean gum mixed gel is used in many canned petfood products. In addition, carrageenan is used to improve water holding in hams and poultry products. Recently it has been used by one of the major hamburger chains in reduced fat products. The texture of carrageenan gels can be controlled by altering the proportion of the individual components (kappa, iota, lambda) in a blend.

Non-food applications of carrageenan have been reviewed by Tye and Stanley (1988) and include use in toothpastes and in air freshner gels.

2.3.1.3 Pectin. The traditional use for pectins as gelling agents in jams and preserves is still by far the most important. The ability to gel in the absence of sugars results in the use of low methoxyl pectins in some diabetic and reduced-sugar products. Pectin also has the ability to stabilise proteins against precipitation by a protective colloid effect. They are therefore found in some low pH milk products, for example drinking yoghurts, and also in some UHT processed milk products.

Degradation by β-elimination means that the heat stability of pectins at neutral pH is very poor. As a result, most applications are confined to low pH systems. Since the high susceptability to β-elimination is due to the presence of the methoxyl group, the neutral pH heat stability can be substantially improved by complete deesterification. The pectate produced has many similarities to a high guluronic acid alginate but there are interesting differences between the two materials. For example, it appears possible to control the gelation of pectates through an autoclaving cycle whereas this has not been achieved for alginates (Mitchell and Taylor, 1985).

2.3.1.4 Guar gum. Guar gum has the advantage of being a very effective thickener. It is non-ionic and is therefore more likely to be compatible with proteins. It is also substantially cold-water soluble. These properties and its low cost tend to make it a general purpose thickener in a variety of applications. For example, in the food industry it is used as a component in many ice cream stabiliser blends, often in combination with carboxymethyl cellulose and carrageenan. It has a wide range of non-food applications particularly in the oil industry as a component in oil drilling muds. Derivatives of guar are available and these can provide substantially enhanced heat stability.

2.3.1.5 Locust bean gum. This polysaccharide is predominantly cold-water insoluble, in contrast to guar gum. In a given preparation there will be material with a range of mannose/galactose ratios; the fraction with the highest degree of substitution is water soluble at ambient temperature (Gaisford *et al.*, 1986). This corresponds to about 30% of the total. The greater cost and lack of cold-water solubility compared with guar gum generally precludes its use as a simple thickener. As mentioned above, locust bean gum is used in gelling applications where its synergy with the red seaweed polysaccharides and xanthan gum can be utilised. In terms of quantity the most important application is undoubtedly in canned petfood products in conjunction with carrageenan. It is a component of many gelling blends of polysaccharides since not only will it enhance gel strength but it will also alter texture, making kappa carrageenan gels more elastic and less prone to syneresis. Locust bean is also used as an ice cream stabiliser. It can be made to gel by the application of freeze–thaw cycles and it is therefore reasonable to speculate that within the freeze concentrated matrix in frozen ice cream it will go some way at least towards promoting a gelled structure which will be effective in controlling ice crystal growth.

2.3.1.6 Glucomannans. There has been a substantial increase in interest in konjak glucomannan which in the past was only employed as an ingredient in some traditional oriental foods. In addition to the synergism with carrageenan, which provides the basis for locust bean gum applications, the heat-irreversible gel formed at high pH in the presence or absence of carrageenan can be used to produce restructured foods or comminuted to give particles that can act as fat replacers (Shelso, 1990).

2.3.2 Structure–function relationships

The industrial importance of cell wall polysaccharides is based on a variety of what in the broadest terms can be described as functional properties. The most general of these are thickening and gelling, and it is appropriate to describe the scientific basis for these in order to provide a framework in which to understand the behaviour of individual materials. An excellent discussion of structure–function relationships is included in the book by Yalpani (1988).

2.3.2.1 Thickening. Polysaccharides in general terms are very effective thickeners because they form high viscosity solutions in water at low concentrations. The rheological properties can be understood to a significant extent in terms of the fundamental parameter, the intrinsic viscosity $[\eta]$.

This is defined by the expression:

$$[\eta] = \frac{1}{c}\left(\frac{\eta}{\eta_0} - 1\right)$$

$$c \rightarrow 0$$

where c is concentration, η is the viscosity of the solution and η_0 is the viscosity of the solvent.

If the conventional view is taken that the polysaccharide behaves hydrodynamically as an equivalent sphere, then the intrinsic viscosity is related to the radius of gyration R_g of the polymer coil and the molecular weight M by the Flory–Fox equation:

$$[\eta] = \phi\, 6^{3/2} R_g^{3} / M$$

where the constant ϕ is the Flory viscosity function.

Because the amount of solvent that has to be considered as part of the equivalent sphere increases by more then the first power of the molecular weight, the intrinsic viscosity depends on the molecular weight, the dependence being described by the Mark–Houwink equation.

$$[\eta] = k M^{\alpha}$$

Some values for the exponent and the constant k are given in Table 2.4 which is taken from Launay *et al.* (1986). This is a good source of a fuller description of these ideas.

Table 2.4 Values for k and α in the Mark–Houwink equation for various polysaccharides

Polysaccharide and solvent	$10^6 k$ ($[\eta]$ in dl/g)	α	Reference
Sodium alginate (0.1 M NaC l)	20	1.0	Smidsrød (1970)
Carboxymethyl cellulose (0.2 M NaC l)	430	0.74	Brown and Henley (1974)
Guar gum (water)	380	0.723	Robinson *et al.* (1982)
Locust bean gum (water)	90	0.79	Sabater de Sabates (1979)
Pectin (DE = 70)	216	0.79	Berth *et al.* (1977)

Although the intrinsic viscosity is a property of the isolated polymer, it is useful in predicting the viscosity of more concentrated solutions since the dimensionless product of the intrinsic viscosity and the concentration ($c[\eta]$) (the coil overlap parameter) determines the extent to which the coils and their hydrodynamically associated solvent occupy the solution. Morris *et al.* (1980) have shown that solution viscosity at zero shear rate can to a large extent be generalised for different polysaccharides in terms of the coil overlap parameter. When $c[\eta]$ is ~ 2, the solution is completely occupied by the coils, and c equals a critical concentration termed c^*. At concentrations above c^* the dependence of the zero shear viscosity on concentration becomes far more severe, increasing from about $c^{1.4}$ to about $c^{3.5}$ (for the galactomannans the dependence is even stronger). For the intrinsic viscosities typically found for industrially important polysaccharides c^* is in the range 0.4 ± 0.2%. Above this concentration polysaccharides solutions are very shear thinning showing a large decrease in viscosity with increasing shear rate. The reason why most polysaccharides give solutions of very high viscosities compared with synthetic polymers is that for the former R_g/M is high because the chain is stiff as a result of the strong steric hindrance to rotation about the glycosidic bond caused by the bulky sugar ring.

2.3.2.2 Gelation. Whereas almost all polysaccharides can function as thickeners in the sense that they give viscous solutions at least at some temperatures, not all will gel. The difference between a thickener and a gelling agent is the lifetime of the associates that form between the chains. Above c^* polysaccharide chains will entangle; however, the lifetime of these entanglements is short. In gelling polysaccharides under certain conditions of solvent composition and temperature, cross-links or junction zones with very long lifetimes are formed. These have specific structures and involve several contiguous sugar residues on each of the participating chains. One of the most interesting aspects of biopolymer science over the last 35 years has been the development of an understanding of the structure of these junction zones. Good, reasonably recent reviews have been given by Morris (1986) and Clark and Ross-Murphy (1987). A challenge which in the authors' opinion still needs to be adequately met is the determination of the relationship between the rheology of the gel and the structure of the network. In terms of applications, both the elastic modulus of the gel (the firmness at low deformations) and the stress required to rupture the gel are important. The former is easier to predict theoretically and probably depends on the number of network chains per unit volume (chains joining adjacent cross-links). However, the use of the rubber elasticity theory is inappropriate because of the stiffness of the network chains and although other recently proposed models (Nishinari, 1988; Higgs and Ball, 1989) have their attractions, their value has yet to be proven.

The other important factor that will determine gelation behaviour in addition to chemical structure is molecular weight. Whereas viscosity will depend strongly on molecular weight, throughput the molecular weight range, as a result of the intrinsic viscosity and hence the coil overlap dependence, the elastic modulus of a gel is independent of molecular weight above a certain threshold value although the break strength appears to increase with molecular weight over the whole range (Mitchell, 1980). Thus the effect of initial molecular weight reduction is to make the gel more brittle. It is worth making the point here that measurement of molecular weight of polysaccharides is far from straightforward. Their tendency to associate and in some cases form microgel structures makes light scattering difficult and although there has been a revival in interest in ultracentrifugation this is not straightforward because of the high degree of non-ideality of polysaccharide solutions. For a recent review see Harding *et al.* (1991).

2.3.3 Structure of junction zones

2.3.3.1 Carrageenan. The gel strength of the gelling polysaccharides extracted from the red seaweeds increases in the order ι carrageenan $<$ κ carrageenan $<$ agarose, i.e. as the sulphate content decreases the gel strength increases. It was the pioneering work of Rees, first at Edinburgh University and then at Unilever Research (Colworth House), which started the development of a structural understanding of polysaccharide gelation (Rees, 1969). Rees and coworkers obtained good quality X-ray diffraction patterns from orientated iota carrageenan gels (Anderson *et al.*, 1969) and this, in combination with computer model building to define the allowed rotation angles about the glycosidic linkage and later some optical rotation studies, interpreted semi-empirically, led to a double helix model for the iota carrageenan junction zone. The individual helices in the coaxial double helix are left-handed and three-fold. The quality of the X-ray data obtained from kappa carrageenan is not so good; nevertheless it seems reasonable to postulate a similar junction zone for it. A structure consisting of a network held together by coaxial double helices is difficult to construct topologically and does not explain in the case of kappa carrageenan why the gel strength is substantially enhanced in the presence of certain ions (K^+, Rb^+ and Cs^+). The domain model (Morris *et al.*, 1980), which retains the double helix but consists of domains containing of the order of six carrageenan chains linked together that subsequently associate through ion mediated bridges, avoids these two problems. Smidsrød and co-workers (Smidsrød *et al.*, 1980; Smidsrød and Grasdalen, 1982) have argued in favour of a single chain to single helix transition and subsequently a cation mediated side-by-side association, a structure originally proposed by Bayley (1955). There has been some evidence also from measurements of the enthalpy change of the transition obtained from isothermal calorimetry (Paoletti *et al.*, 1984, 1985) but it is difficult to discount the X-ray evidence. It has been suggested that single helices or helical dimers can both exist in the particular structure depending on the temperature and the K^+ ion concentration (Rochas and Landry, 1987). It may be that developments in microscopy will resolve some of these questions. Very recently images of the carrageenan double helix have been obtained from scanning tunnelling microscopy (Lee *et al.*, 1992).

Before leaving this topic it should be mentioned that there is evidence from conventional electron microscopy combined with small deformation rheological measurements that carrageenan gelation can be a two-stage process. A fine stranded network, presumably formed by double helices, exists first and then partially collapses to give superstrands which contain far more chains then would be envisaged for the aggregation points of the domains in the domain model (Hermansson, 1989). Finally it should be mentioned that Russian workers have long held the view that the conformational change resulting in helix formation occurs after gelation on cooling (Plaschina *et al.*, 1986). If this is correct, our views of the gelation process would require a complete reappraisal.

2.3.3.2 Alginates. Although monovalent salts of alginic acid are soluble in water, divalent salts, with the exception of magnesium, are insoluble. Gel formation with alginates is generally achieved by the slow introduction of calcium ions into a solution of a monovalent salt of alginic acid. The most frequently used form is sodium alginate. The structural parameter which correlates most strongly with gel strength is the number average length of the G blocks (Skjak-Braek *et al.*, 1985). This will be a good indication of the number of these blocks above the critical length for participation in the ionic cross-link within the alginate gel. The accepted model for this cross-link is the egg box model in which the calcium ions are located in polar cavities between the G block regions of two polymer chains (Grant *et al.*, 1973). Because essentially the unfavourable entropy decrease as the two chains associate is to a first approximation independent of the number of ions bound, whereas the favourable enthalpy decrease increases with ion binding, a certain number of ions has to be involved in the junction for it to be stable. Recent calculations by Stokke *et al.* (1992) have come up with a value for the number of consecutive residues in a single chain of about 10 for Ca^{2+}. The early work of Smidsrd and Haug (1972) showed that the elastic modulus of a calcium alginate gel was independent of molecular weight above a threshold value of a weight average degree of polymersisation of about 150, gelation being possible at values down to 65. As mentioned previously, the rupture properties are dependent on molecular weight over a wider range (Mitchell, 1980).

2.3.3.3 Agar. The most probable structure for the junction zone in agarose gels is composed of associates of three-fold right-handed double helices (Arnott *et al.,* 1974). It has been suggested that many such structures are involved in an individual junction zone.

2.3.3.4 Pectin. Pectins form two types of gel of industrial interest. At high degrees of esterification (> 50%) gels are formed at low pHs in the presence of high levels of sucrose. Other components that lower the water activity, such as glycols or glycerol, may also be employed instead of sugar. It has been suggested that the cosolute disrupts the water structure around the methoxyl groups promoting polymer–polymer association by hydrophobic interactions. This will be enhanced by the charge reduction at low pHs. Commercial high methoxyl pectins have degrees of esterification in the range 55–80%, the setting temperature increasing with degree of esterification. A model for the junction zone obtained from X-ray studies involves a three-fold helix (Walkinshaw and Arnott, 1981a, b), in which several (3–10) chains participate. Acetyl groups interfere with this helix formation and it has been suggested that this is the reason why pectin extracted from sugar beet does not form this type of gel; it has been reported that approximately one in eight of the galacturonic acid residues are acetylated (Pippen *et al.*, 1950; Michel *et al.*, 1984). Neutral sugars will have an effect on gelation properties, for example the kinks introduced by the rhamnose residues in the backbone will not be compatible with a helical junction zone nor will neutral sugar branches.

At degrees of esterification below about 50% pectins will gel with divalent ions. The ion sensitivity of the pectin will depend on the deesterification method. Pecitins deesterified by the action of pectinesterase are more ion-sensitive because the free carboxyl groups have a blockwise distribution due to the sequential mode of action of pectin methylesterase. Industrial low methoxyl pectins are either deesterified by the action of acid or by the action of ammonia. The latter method introduces amide groups into the polysaccharide and results in a material in which the gel strength is less dependent on divalent ion level. There is a near mirror-image relationship between polygulurate and polygalacturonate, and the egg-box model mentioned for alginates is probably also appropriate for low methoxyl pectin gels. Methoxyl groups on the outside face of the two-fold helix forming one side of the junction zone will obviously not interfere with this structure and it may be that it can also accommodate a methoxyl group on the inside face provided there is not one on the opposing chain (Powell *et al.*, 1982).

The complication of the presence of residual methoxyl groups in addition to the neutral sugars mentioned above makes the understanding of pectin gelation more complex than is the case for alginates.

2.3.3.5 Mixed polysaccharide gels. One of the most interesting features of the behaviour of some polysaccharides is their ability to form mixed gels. Predominant in this is the galactomannan locust bean gum. Although it does not gel by itself, its addition to carrageenan or agar allows gelation to take place at lower concentrations of the gelling polysaccharide and indeed at lower molecular weights than would be possible in the absence of the locust bean gum. In addition, mixtures of locust bean gum and the microbial polysaccharide xanthan gum will gel whereas neither polysaccharide forms what can be regarded as a true gel separately. The reasons for this interesting behaviour are still a matter of debate. As mentioned previously, locust bean gum has a mannose to galactose ratio of about 3.5 to 1 whereas the other commercially important galactomannan guar gum, with a mannose to galactose ratio of about 1.8 : 1, does not participate in mixed gels. There is also evidence that within galactomannans of similar mannose to galactose ratios, a more blockwise structure of the galactose substituent results in greater synergism (Dea, 1987). This supports the view originally put forward by Dea *et al.* (1972) that the unsubstituted regions of the mannose backbone can interact with the ordered regions of the second polysaccharide, i.e. the carrageenan or xanthan helices. This picture has been refined by McCleary (1979) who suggested that only a length of one unsubstituted face is required. This means that a structure in which every other mannose residue is galactose-substituted would be highly synergistic. The problem with models that predict specific interactions between the locust bean gum and the ordered form of carrageenan is that no new X-ray diffraction pattern is seen for the mixed gel. The pattern is purely the sum of the carrageenan and the mannan patterns (Cairns *et al.,* 1987). This has led some workers to suggest that phase separation occurs. The exclusion of the carrageenan from the locust bean gum domain enhances the carrageenan concentration in its domain. However, attempts to visualise the phase separated structure using fluorescent labelled polysaccharides (Cairns *et al.*, 1986a, 1987) have not proved successful and we have sympathy for a picture involving specific association between the two macromolecules. The xanthan locust bean gum synergism is interesting. Here there is evidence for a new structure involving the disordered rather then the ordered form of the bacterial polysaccharide (Cairns *et al.*, 1986a, b). However, gelation can also take place under conditions (salt and moderate temperatures) where xanthan is in its ordered form (Mannion *et al.*, 1992). Whether there are two types of interaction or whether locust bean gum stabilises the disordered form of xanthan gum is not clear. The ability of locust bean gum to take part in these mixed gels is one of the main reasons why it commands a higher price than guar gum. A recent development is the production from genetically engineered bacteria of the enzyme α-galactosidase which is used to reduce the galactose content of guar, giving it properties closer to locust bean gum. There is increasing interest in the polysaccharide konjak glucomannan. This can also take part in synergistic interactions with carrageenan and xanthan gum. Indeed, at one stage it was claimed that the xanthan–konjak mannan mixed gel could be formed at a lower concentration than any other polysaccharide system (Shatwell *et al.*, 1990). It is probable that the mechanisms are broadly similar to mixed gels containing locust bean gum. A feature of additional interest is that glucomannan can be made to form heat-irreversible gels on alkaline treatment, sometimes combined with heat, as a result of deacetylation of the glucomannan. When this treatment is carried out in the presence of carrageenan, heat-irreversible mixed gels with interesting properties are obtained (Shelso, 1990). Finally it should be mentioned that mixed gels can be prepared with high methoxyl pectins and alginates at low pH though these systems have not assumed significant industrial importance (Toft *et al.*, 1986).

2.3.4 Modification of polysaccharides to improve gelling performance

It should be apparent from the above considerations that there are many possibilities for modifying polysaccharide structure to improve particularly the gelling performance. The use of alkali in the extraction of carrageenan and agar to catalyse the 6-sulphate to 3,6-anhydro conversion is a good example. This could also be achieved enzymatically either by enzyme action on the extracted polysaccharide or by genetic modification of the plant enzyme systems and subsequent farming of the seaweed. Another example where the use of enzyme systems might be expected to give a commercially useful modification of the structure of a seaweed polysaccharide is the use of the epimerase which converts mannuronic to guluronic acid in alginate. Seaweed sources for high guluronic acid alginates are less readily available than for high mannuronic acid alginates yet the superior gelling performance of the high guluronate alginates means that they command higher prices.

A recent development is the use of α-galactosidase to remove some of the galactose residues from guar gum giving a material with a performance that approaches that of locust bean gum (McCleary and Neukom, 1982). The approach used was to demonstrate the feasibility of the idea using galactosidase extracted from plant sources and obtaining the enzyme in quantity using a biotechnological route. This development has recently been reviewed by Bulpin *et al.* (1990).

As mentioned above, the use of pectin in neutral pH food when heat treatment is involved is limited by the tendency of the polysaccharide to degrade by β-elimination. To utilise pectins in neutral pH foods, a process developed before the war, has been revived. This involves a mild alkali treatment of fruit waste prior to extraction of the pectin. When this method is applied to orange peel, pectinesterase was activated, resulting in an almost complete deesterifaction of the pectin within the cell wall. An attractive feature of this idea is that in some applications the crude material can be used directly. Heat treatment in the presence of phosphate sequestrants will extract the pectate which can be subsequently gelled with calcium ions giving an extremely cheap gelling system (Mitchell and Taylor, 1985).

We have seen major advances in our understanding of structural functional relationships particularly with regard to the gelation process. As a result of this it is often possible to predict the changes required to polysaccharide structure to enhance gelation properties. The recent advances in genetic modification of plants by biotechnological means should allow us to produce materials from which polysaccharides with superior properties can be extracted. An example could be sugar beet which produces a waste material containing a pectin with poor gelling properties because of the presence of acetyl groups and probably also because of a low molecular weight. Suitable manipulation of degrading and deacetylating enzyme systems could produce a material from which useful pectin could be extracted from the waste after sugar extraction.

2.4 Manipulation of cell wall structure

With the advent of recombinant DNA technology the ability to manipulate plant metabolism has been increased dramatically. Obviously one component of plant structure which could be usefully manipulated is the cell wall.

The cell walls of a plant are largely responsible for dictating the size and shape and for providing the rigidity of the plants. Thus, changes in wall chemistry may alter a plant's growth habit, resistance to pests, diseases or environmental stresses. It may also alter a plant's responses to plant growth regulator stimulation. Although alterations to the gross structure of the wall may in future allow us to develop plants with more advantageous agronomic or horticultural characteristics, our knowledge of basic wall chemistry precludes at present any real directed modifications at this level. In particular it is unclear which elements in the complex wall structure are responsible for the load bearing and support properties of the entire wall, and how walls must adapt during growth. This latter aspect of wall metabolism, i.e. control of cell expansion, is worthy of much more study. It is also outside the scope of this review but readers are directed to the review by Labavitch (1981).

Another aspect of wall structure which would be amenable to manipulation is that which may be involved in cell signalling. Small oligosaccharides derived from the breakdown of wall polymers are known to act as hormones and as such elicit a response in the plant cell. Again details of this aspect of wall metabolism are outside the scope of this review but such oligosaccharides have been reviewed recently by Ryan and Farmer (1991).

Perhaps the most imminent prospect for wall manipulation is in the modification of wall polymers useful in industry. For instance cellulose, the most abundant natural polymer, could be used as a feedstock for the production of energy. Thus any wall modification resulting in increased yield, extractability or digestibility would be desirable. Many polymers, as we have seen, are used as gelling agents in the food industry and modification to influence their solubility and gelling properties would also be useful, as would the ability to produce novel polymers tailor-made for a particular purpose.

Thus modification of wall structure, either in the intact plant or of extractable polymers, is likely to find several diverse commercial applications. There are in general two approaches to achieve such modification, either by adapting the biosynthetic routes for the polymers or by manipulating degradative processes either within the plant or following extraction. It should be noted that progress so far in this area is very limited, primarily due to a lack of biochemical information, on biosynthesis in particular. Also, any work in this area must obviously take care not to cause such gross alterations of the wall as to render it unsuitable as a support for the plant in general. However, some indication of the resilience of plant systems to manipulation of their walls comes from experiments with the herbicide 2,6-dichlorobenzonitrile. This compound inhibits cellulose biosynthesis yet it has been shown that cultured tomato cells can adapt to grow in the presence of this herbicide, the resultant wall having very little cellulose–xyloglucan network and being almost entirely composed of pectins (Shedletzku *et al.*, 1990). Whether an intact plant would survive such a gross alteration in its wall structure is unlikely, but this observation does suggest that plants may be able to adapt to less major manipulation of their cell walls.

2.4.1 Cell wall biosynthesis

Several books and reviews are available on the biosynthesis of cell wall polymers. These include those by Ericson and Elbein (1980), Bolwell (1988), Delmer and Stone (1988) and Lewis and Paice (1989). The reader is referred to these for more details on this topic.

The easiest cell wall polymer to manipulate would be the wall protein, in particular the extensin. This is obvious since it is a primary gene product and as such amenable to direct manipulation either by antisense-RNA technology or by over-expression of modified extensin genes. One such modification might be the attempt to reduce tyrosine residues and hence weaken the protein mesh structure of the wall by having less isodityrosine links. To our knowledge, however, no manipulations of extensin in this manner have been reported.

Manipulation of wall polysaccharide synthesis would be more difficult since these are 'secondary' gene products. Manipulation is also restricted by a severe

Table 2.5 Representative list of cell wall polymer synthesising systems using nucleotide diphosphate sugars.

Precursor	Product	Species	Reference
UDP-glucose	β-1,3-glucan	*Brassica oleracea*	Fredriksen *et al.* (1991)
UDP-xylose	1,4-xylan	*Zinnia elegans*	Suzuki *et al.* (1991)
UDP-arabinose	Arabinan	*Phaseolus* vulgaris	Bolwell and Northcote (1981)
UDP-xylose	Xylan	*Phaseolus vulgaris*	Bolwell and Northcote (1981)
UDP-glucose	β-1,4-glucan	*Phaseolus vulgaris*	Benjamin and Callaghan (1985)
UDP-glucose UDP-xylose	Xylogucan	*Pisum sativum*	Ray (1980)
UDP-galacturonic acid	Galacturonan	*Phaseolus aureus*	Fincher and Stone (1981)
UDP-glucose	β-1,3- and 1,4-glucan	*Lolium multiflorum*	Henry and Stone (1982)
UDP-glactose	Galactan	*Phaseolus aureus*	Panayotatos and Villemez (1973)
GDP-glucose	β-1,4-glucan	*Pisum sativum*	Brett (1981)

lack of biochemical detail available on the biochemical mechanisms involved in generating each type of polymer.

Radiolabelling and pulse chase experiments have indicated that for most wall polysaccharides the site of synthesis is the golgi body (see Fry, 1985). Cellulose biosynthesis is the key exception, its synthetic machinery appears to be associated with the plasmalemma membrane (Delmer, 1987). Many cell free systems from plants have been prepared which are capable of synthesising polysaccharides from nucleotide diphosphate activated sugar (NDP-sugar) precursors, and a representative list of these is given in Table 2.5. However, there are several problems with the analysis of this type of experiment. The preparations are inevitably crude membrane extracts and by no means represent pure enzyme systems. It is thus possible that the preparations contain enzymes capable of interconverting the exogenously supplied NDP-sugars. The systems also seem to work by extension of pre-existing polymers, which are then isolated on the basis of radiolabel incorporation and analysed for sugar linkages. Problems arise since it is not possible to differentiate between those links which were already present in the pre-existing primer and those newly synthesised. The products of such *in vitro* synthesis are usually much smaller than any structural polysaccharide and it is thus difficult to identify the putative native equivalent. For example, any β-1, 4-glucan synthesised could represent either a cellulose or hemicellulosic precursor. However, if these problems are taken into consideration several conclusions can be drawn concerning the biosynthesis of wall polysaccharides.

It is evident that the immediate precursors are largely, if not exclusively, NDP-sugars. Table 2.5 shows a representative list of precursors and their putative biosynthetic products. It is apparent that in some cases changes in the concentration of NDP-sugar substrates in these *in vitro* systems influences the end product. For example, in some preparations the level of UDP-glucose supplied alters the ratio of β-1,3 to β-1, 4 links in the synthesised polymers. It is possible that *in vivo* the imprecise synthesis of polysaccharides seen in nature, leading to microheterogeneity, results from minor alterations in substrate supply. It may therefore be possible to manipulate wall polymer structure by manipulating the levels of the various NDP-sugars within particular cells or the plant as a whole. There is a large amount of biochemical data available on the synthesis and transformation of sugar nucleotides in plants (see the review by Feingold and Avigad, 1980). These sugar nucleotides usually account for 10–25% of the total nucleotide pool in the cell and consist of sugars linked primarily to UDP, ADP or GDP. TDP-sugars do occur but are relatively rare. Within this pool UDP-glucose is the predominant species accounting for 60–70% of the total and UDP-galactose is the next most abundant (15–25%). However, the composition of the pool is quite variable between species. The biosynthetic pathways for the sugar nucleotides are fairly well established (Feingold and Avigad, 1980) and those pathways involved in the biosynthesis of the key wall precursors are summarised in Figure 2.11. This pathway chart is by no means complete. There may be other, albeit minor, routes to several of

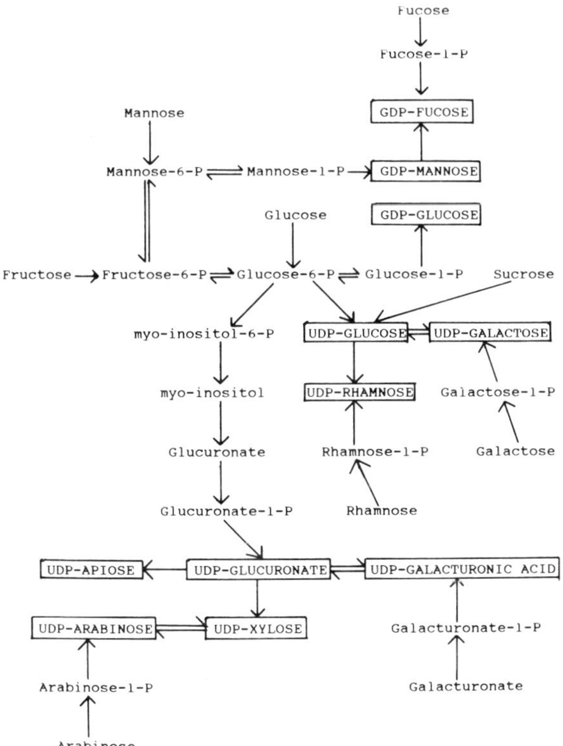

Figure 2.11 Outline of the major pathways for the formation of the nucleotide diphosphate sugars involved in cell wall biosynthesis.

the nucleotide sugars and several other sugar nucleotide products are made within plants, e.g. ADP-glucose which is a key precursor of starch (see chapter 1). However, the pathways shown are useful to illustrate the key groups of enzymes involved.

The first point to realise is that all of the nucleotide sugar precursors can be derived from fructose 6-phosphate, and thus from photosynthesis, without any requirement for carbon input into the cell. It is clear from *in vivo* labelling studies that plants can incorporate sugars directly into their cell walls. Thus exogenously supplied L-arabinose, D-galactose, D-glucuronic acid, D-galacturonic acid and L-fucose have all been shown to be used directly by plants as a source for glycosyl residues in wall synthesis (see Feingold and Avigad (1980) for references). In order for these monosaccharides to be used in this way, they and others must first be activated, the key group of enzymes being the kinases of which several are found in plants. Hexokinase (EC 2.7.7.1) is capable of phosphorylating D-glucose, D-mannose and D-fructose (Saltman, 1933) and a more specific D-fructokinase (EC 2.7.1.4) has also been reported (Medina and Sols, 1956). These enzymes have been widely studied and purified (Feingold and Avigad, 1980) but are unlikely candidates for the fine manipulation of cell walls since their products are key precursors for a wide range of polymers. However, more specialised kinases have been identified in plants, including L-arabinokinase (EC 2.7.1.46), D-galactokinase (EC 2.7.1.6) (Chan and Hassid, 1973), D-glucuronokinase (EC 2.7.1.13) (Leibowitz *et al.*, 1977) and D-galacturonokinase (Neufeld *et al.*, 1961). These kinases may function in the activation of monosaccharides prior to conversion to nucleotide sugars and hence polymer synthesis. However, they may also function in a salvage capacity. This may be especially true for the galacturonokinase.

The key enzymes in the formation of the nucleotide sugars are the pyrophosphorylases and again these have been well documented in plants (Feingold and Avigad, 1980). The most prominant activity in plants usually involves UDP-D-glucose pyrophosphorylase (EC 2.7.7.9) and this enzyme has been intensively studied. UDP-glucose can also be formed from sucrose via sucrose synthase (EC 2.4.1.13) (Feingold and Avigad, 1980). As for hexokinase this pyrophosphorylase plays a key role in the formation of several biosynthetic precursors for products other than cell wall components and would not be a likely candidate for manipulation. However, other glucose pyrophosphorylases, in particular GDP-D-glucose pyrophosphorylase, have been identified (Delmer and Albersheim, 1970; Staver *et al.*, 1978) and since GDP-D-glucose may be involved in wall synthesis (Nikaido and Hassid, 1971) this may be a candidate for manipulation. Other pyrophosphorylases of interest which have been identified include GDP-D-mannose pyrophosphorylase (EC 2.7.7.13), GDP-L-fucose pyrophosphorylase, UDP-D-glucuronate pyrophosphorylase (EC 2.7.7.aa), UDP-D-galacturonate pyrophosphorylase and UDP-L-arabinose pyrophosphorylase.

The other key enzymes in these pathways are those involved with the transformation of sugars. Important examples here are the epimerases such as UDP-D-glucuronate-4-epimerase, which catalyses the conversion of UDP-glucuronic acid to UDP-galacturonic acid, and the UDP-D-glucose-4-epimerase which converts UDP-glucose to UDP-galactose. Similar enzymes are also involved in the transformation of xylose to arabinose. Finally, the production of the uronic acids and pentoses could be manipulated via alterations in the myo-inositol pathway or of the enzyme UDP-D-glucose dehydrogenase (EC 1.1.1.22) which catalyses the NAD-linked conversion of UDP-glucose to UDP-glucuronic acid. Thus it can be seen that manipulation of substrate supply is possible. However, little is known about the molecular biology of these enzymes and problems could arise if a pathway is central to several biosynthetic routes or if more than one route exists for the production of the substrate.

Direct manipulation of the synthetic enzymes for cell wall polymers is possible but here much less is known about the biochemistry, let alone the molecular biology. Table 2.5 gives a representative list of systems which have been identified. It is beyond the scope of this review to cover all these in detail but we shall consider a few cases in particular, namely the synthesis of cellulose, xyloglucan and pectin molecules.

Cellulose synthesis has been extensively reviewed by Delmer (1983, 1987) and Roberts (1984). It has been demonstrated by radiolabelling that cellulose synthesis occurs at the plasma membrane. The system in bacteria is relatively well understood (Delmer and Stone, 1988) while that in plants is not so clear. In bacteria a lipid–sugar biosynthetic intermediate has been identified. Such a mechanism for synthesis may also operate in plants (Brett and Northcote 1975; Datema *et al.*, 1985) for cellulose and other wall polysaccharides. Freeze-fracture studies of plant membranes often reveal structures associated with cellulose microfibrils, these being either linear, rosette or globular (Delmer and Stone, 1988), and such structures are associated with cellulose synthesis in bacteria. The enzymes involved, however, are poorly characterised. Several groups have been able to demonstrate the incorporation of UDP-glucose into β-glucan chains. However, in most cases this has been shown to be either incorporation into β-1, 3 links, as in callose, or the incorporation into β-1, 4 is very low and could represent hemicellulose synthesis. Benjamin and Callaghan (1985) and Jacobs and Northcote (1985) have both demonstrated good synthesis of β-1, 4-glucans from UDP-glucose. The synthesis of β-1,3-linked callose is a phenomenon common to most plant preparations. Since callose is not normally a constituent of the plant cell wall, it has been postulated that this callose synthase activity arises from disruption of the plasma membrane during extraction and could represent an alternative biosynthetic mechanism of the *in situ* cellulose synthase. Work to isolate callose synthase (EC 2.4.1.34) is well advanced and Fredrickson *et al.* (1991) recently reported a soluble preparation capable of synthesising callose from UDP-glucose. This preparation was highly purified and contained only five polypeptides, suggesting a multimeric structure for this enzyme.

The biosynthesis of the cell wall heteropolymers is more complex and these systems are found in the golgi body. Thus xyloglucan synthesis involves the cooperation of glucan synthetase and UDP-xylose xylosyltransferase (Ray, 1980). In the widely studied pea system, UDP-glucose and UDP-xylose are incorporated into a xyloglucan-like polymer; however, the synthesis of the glucan backbone appears to be able to proceed independent of any UDP-xylose, suggesting that backbone and side chain synthesis are independent (Ray, 1980). However, preincubation with UDP-glucose stimulates subsequent UDP-xylose incorporation. In the case of soybean, however, there seems to be an absolute requirement for both UDP-glucose and UDP-xylose to be available all the time and UDP-glucose will only be incorporated in the presence of UDP-xylose (Hayashi and Matsuda, 1981). The mechanism of addition of the other sugars, e.g. galactose and fucose, to the polymer is unknown. It should be noted that UDP-xylose has also been shown to be incorporated into xylan polymers (Bolwell and Northcote, 1981; Suzuki *et al.*, 1991) and that this activity is also associated with the golgi bodies (Bolwell and Northcote, 1983).

Similar cooperation must also exist in the biosynthesis of the heteropolymeric pectin molecules. Thus UDP-galactose and UDP-arabinose are incorporated, via an as yet undefined system, into arabinogalactans (Ericson and Elbein, 1980) and the rhamnogalacturonan is similarly synthesised probably from UDP-galacturonic acid and presumably an NDP-rhamnose although no definitive information on any enzyme systems are available in this case. One aspect of pectin synthesis that is relatively well understood is that of methylation of the galacturonic acid residues. It is clear that this methylation occurs within a membrane compartment, presumably the golgi bodies, and that it is the polymer, rather than the UDP-galacturonic acid precursor, which is methylated. The enzyme involved, pectin methyl transferase, uses *S*-adenosyl methionine as the methyl donor (Kauss *et al.*, 1969). It is not clear, however, to what extent this methylation occurs or how it may be regulated. In fact the whole question of regulation is complex and again outside the scope of this review (but see Bolwell, 1988).

The final aspect of biosynthesis which could be amenable to manipulation would be the reactions occurring within the wall to construct the final three-dimensional structure. However, very little is known about these enzyme systems. Manipulation of isodityrosine residues between wall proteins has already been suggested but manipulation of the peroxidase activities thought responsible for the diferulic acid links between wall polymers, or indeed the synthesis of lignin, might be a useful approach to altering wall properties. There is, however, little biochemical information available on these specific enzymes, although one report has shown that the addition of ferulic acid to pectin polymers occurs in the golgi bodies (Fry, 1987).

Another target enzyme may be pectinesterase. By generating deesterified regions on the pectin molecules, this enzyme may be implicated in the formation of non-ionic bonds in the cell wall. This enzyme will, however, be dealt with in more detail in section 2.4.2 since it may also be involved in wall degradation.

2.4.2 Cell wall degradation

Controlling degradation of the wall may also prove a useful mechanism for affecting wall structure and function. Altering degradative enzymes in tissues could also be advantageous in manipulating changes in wall polymers following extraction.

There are many enzymes found in plant tissues capable of digesting or metabolising cell wall polysaccharides. A representative list is given in Table 2.6. Most studies on wall degrading enzymes have centred on fruit tissue where the relatively large wall degradation accompanying ripening is thought to be

Table 2.6. Representative list of plant enzymes capable of degrading wall polymers.

Arabinosidase	
α-galactosidase	(EC 3.2.1.22)
β-galactosidase	(EC 3.2.1.23)
β-1,3-glucanase	(EC 3.2.1.6)
β-1,4-glucanase	(EC 3.2.1.4)
α-glucosidase	(EC 3.2.1.20)
α-mannosidase	(EC 3.2.1.24)
β-mannosidase	(EC 3.2.1.25)
Pectinesterase	(EC 3.1.1.11)
Polygalacturonase endo acting	(EC 3.2.1.15)
Polygalacturonase exo acting	(EC 3.2.1.67)
β-xylosidase	(EC 3.2.1.37)

responsible, at least in part, for the softening of the fruit. The role of wall hydrolases in ripening is beyond the scope of this chapter, but readers are directed to Huber (1983), Tucker and Grierson (1981), Brady (1987) and Fischer and Bennett (1991). It is clear that cell wall degradation must also occur in all cells during wall turnover accompanying expansion and elongation (Labavitch, 1981), as well as in more specialised cases such as sieve cell formation, abscission and pollen tube formation. These other processes are less well studied than fruit ripening. It is likely, however, that the enzymes involved in the wall degradation occurring in all these diverse physiological phases are similar, and we shall deal with a few of these key enzymes in a little more detail.

The four enzymes most studied in fruit are polygalacturonase (EC 3.2.1.15), pectinesterase (EC 3.1.1.11), β-1, 4-glucanase or, to give it its more trivial name, cellulase (EC 3.2.1.4) and β-galactosidase (EC 3.2.1.23). A consideration of these four enzymes will highlight factors probably important for the future studies on the less well documented enzymes in Table 2.6. Polygalacturonase (PG) exists as either an endo-acting or exo-acting (EC 3.2.1.67) form, in a wide range of fruit. Its mode of action is to cleave the α-1, 4 link between adjacent deesterified galacturonic acid residues in the acidic pectin backbone. Similar enzymes which cleave methylated galacturonic acid backbones are known from fungi and bacteria, but if present in plant tissue must be extremely rare. The activity of PG is not restricted to fruit tissue, for instance this enzyme is found in tomato abscission zones (Tucker *et al.*, 1984) and pollen (Pressey, 1991). The major endo-PG from tomato fruit has been purified by several groups and found to be a polypeptide with a molecular weight of 46 000 (Tucker *et al.*, 1980; Mohd-Ali and Brady, 1982). cDNA (Grierson *et al.*, 1986; Sheehy *et al.*, 1987) and genomic clones (Bird *et al.*, 1988) have also been isolated for tomato PG and, as we shall see later in this chapter, antisense RNA technology has been employed to manipulate the activity of this enzyme in fruit tissue (Smith *et al.*, 1988; Sheehy *et al.*, 1988).

It should be noted that endo-PG in tomato fruit occurs in at least two isoforms (Tucker *et al.*, 1980; Mohd-Ali and Brady, 1982), PG1 and PG2. Both isoforms have been purified (Moshrefi and Luh, 1984). The major isoform, PG2, consists of a single polypeptide with a molecular weight of 46 000 while PG1 appears to consist of this same polypeptide in association with a second subunit with a molecular weight of about 38 000. The nature and function of this second subunit is being investigated (Pogson *et al.*, 1991). It has been found that the level of PG activity in tomato leaf abscission zones is not affected by antisense manipulation which results in a marked reduction in the fruit enzyme (Taylor *et al.*, 1990). Since the antisense construct was under the control of a constituative CaMV 35S promoter, this observation would suggest that the fruit and abscission related PG enzymes are the products of separate genes. This hypothesis is further supported by a lack of cross hybridisation between fruit PG cDNA and abscission zone mRNA and also by a lack of cross reactivity between antibodies to the fruit PG and the PG isolated from abscission zones (Taylor *et al.*, 1990).

Pectin esterase (PE) occurs as a wide range of isozymes. For instance, in tomato fruit at least four isoforms have been identified (Pressey and Avants, 1972) and different forms seem to be present in different tissues (Harriman *et al.*, 1991). The ratio of isoforms changes during ripening (Tucker *et al.*, 1982). The mode of action of PE is the deesterification of galacturonic acid methyl esters to generate galacturonic acid residues with a free carboxyl group at the C6 position. As such it is possible that the action of PE might generate reaction sites for the action of PG. However, such a synergistic mode of action for these two enzymes remains to be demonstrated *in situ*. As we shall see later, such synergistic reactions do occur *in vitro*. PE activity is found associated with a wide range of tissues other than fruit, e.g. leaves (Markovic *et al.*, 1989). The major isoform in ripe tomato fruit has been purified (Tucker *et al.*, 1982) and fully sequenced (Markovic and Jornvall, 1986). A cDNA (Ray *et al.*, 1988) and genomic clones (Harriman and Handa, 1990) for PE have also been identified from tomato fruit and again these have been used to manipulate the activity of PE in the fruit (Tucker *et al.*, 1992), a topic to be returned to in section 2.4.3.

β-1, 4-glucanase is often referred to as cellulase and again this enzyme occurs in several isoforms. However, it is rare for this enzyme to be capable of degrading native cellulose. Instead it is often assayed using artificial substrates such as carboxymethylcellulose (CMC-cellulase). The identification of the *in situ* substrate for this enzyme is unclear but could actually be the backbones of hemicelluloses such as xyloglucan. There are, however, more specific, if less widespread, enzymes capable of the specific degradation of xyloglucans (Fanutti *et al.*, 1991). Cellulase is again found in a wide range of tissues other than fruit and is especially associated with cell wall turnover during cell expansion (Labavitch, 1981) and with abscission (Horton and Osborne, 1967). Again cDNAs for this enzyme have been identified (Tucker and Milligan, 1991) but as yet no reports on the manipulation of activity by recombinant DNA technology have been published.

β-Galactosidase activity is determined by use of the artificial substrate *p*-nitrophenyl galactoside. It is not uncommon for an enzyme identified using this substrate to be unable to cleave the β links between adjacent galactose residues in a cell wall polymer. If, however, it can be demonstrated that the enzyme is active against a polymeric substrate, then it is more correct to call it a galactanase. Indeed several β-galactanases have been identified in plants. Like other hydrolases they tend to occur in isoforms and are common to a wide range of tissues, and along with β-glucanase are often associated with turnover during cell expansion (Labavitch, 1981).

It is evident that a very wide range of cell wall degrading enzymes are found in plant tissues and that these are involved in several aspects of plant growth and development. There is, therefore, much scope for the manipulation of these enzymes at a genetic level. Our knowledge of the molecular biology of several of these enzymes, namely PG, PE and cellulase, has reached a stage where these can be altered by using recombinant DNA techniques. This will be dealt with in section 2.4.3. It is likely that in the near future similar manipulation of other wall degrading enzymes will be attempted. There are two aspects to consider in this type of work. Firstly, while it may be desirable to manipulate an enzyme in a particular tissue, e.g fruit, consideration must be given to whether the enzyme occurs in other tissues in which the result of any manipulation may prove deleterious. Secondly, how wall degrading enzymes are controlled *in situ* is poorly understood. It is evident from studies on the pecteolytic enzymes in tomato fruit that such control exists and must be quite rigorous. There is sufficient PG and PE activity in the fruit to fully degrade the pectin in a matter of hours, if not minutes. Yet, while such degradation occurs freely *in vitro,* it is severely restricted *in vivo* (Seymour *et al.*, 1987). How such restrictions on enzyme action *in situ* operate is unclear but this may have a profound effect for any enzyme manipulations envisaged.

2.4.3 Manipulation of wall polymers: a case history

Having considered the general features of wall synthesis/degradation, how could these be manipulated to alter wall structure/function or polymer structure? There are several possible means. First, biosynthetic processes could be specifically enhanced to favour the synthesis of particular polymers, either by an increase in NDP-sugar substrates or by activation of specific synthases, the latter by removal of enzyme constraints or introduction of extra genetic material to get over-expression of particular enzymes. In a similar vein, new polymers may be made *in situ* by the incorporation of genes for new biosynthetic enzymes. However, such manipulation requires a massive advance in our understanding of the biochemistry of polymer synthesis.

An alternative approach would be to manipulate the degradative enzymes. Again, over-expression may prove useful in some cases but in most instances suppression of activity by antisense RNA technology is probably more appropriate. This technique has been employed to manipulate PG and PE levels in tomato fruit (Grierson *et al.*, 1990; Smith *et al.*, 1988, 1990; Sheehy *et al.*, 1988; Schuch *et al.*, 1989; Tucker, 1990; Tucker *et al.*, 1992).

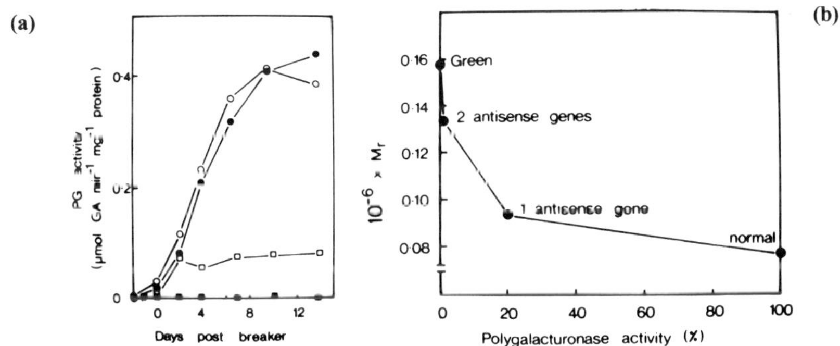

Figure 2.12 Effect of antisense polygalacturonase genes on enzyme activity and pectin depolymerisation in transgenic tomato fruit. (a) Enzyme activity in normal (● and ○) fruit and transgenic fruit containing one antisense gene (□) or two antisense genes (■). (b) Pectin depolymerisation. Soluble pectin was extracted from normal or transgenic fruit 10 days after the onset of ripening and the weight average molecular weight determined by equilibrium ultracentrifugation.

Tomato plants transformed with an antisense DNA construct based on PG cDNA show much reduced expression of PG activity in the ripening fruit (Figure 2.12). Levels of activity can be reduced to below 1% of normal (Smith *et al.*, 1988, 1990). This reduction in activity prevents the depolymerisation of polyuronides which accompanies normal ripening (Figure 2.12) and is entirely consistent with the mode of action of this enzyme (Schuch *et al.*, 1989; Tucker *et al.*, 1990). The effect on fruit softening is only slight: transgenic fruit still soften but are normally more resistant to cracking and store much better, making them easier to harvest and transport (Schuch *et al.*, 1991). A significant advantage is also seen, however, when these fruit are processed. PG action in normal fruit, combined with PE, acts to depolymerise pectins in processed pastes thus reducing the viscosity. This reduction in viscosity is commercially undesirable and so it is normal practice for processors to heat tomatoes to a temperature in excess of 80°C, an expensive process, in order to inactivate these enzymes. Transgenic fruit with less than 1% PG, however, produce pastes without heating, with rheological properties which are better than those from normally heat-treated fruit (Figure 2.13) (Schuch *et al.*, 1991).

Similar experiments have been carried out for PE in tomato (Tucker *et al.*, 1992). In this case a cDNA for one of the two major isoforms of this enzyme found in Ailsa Craig tomatoes has been isolated and used to generate antisense plants. In this instance PE activity is reduced to around 10% of normal (Figure 2.14), the residual PE activity presumably resulting from the continued production of the second PE isoform, which appears to be relatively unaffected by the antisense transformation. In the transgenic fruit the degree of pectin esterification is, as expected, increased (Figure 2.14), but again this has no significant effect on the softening of the fruit. In

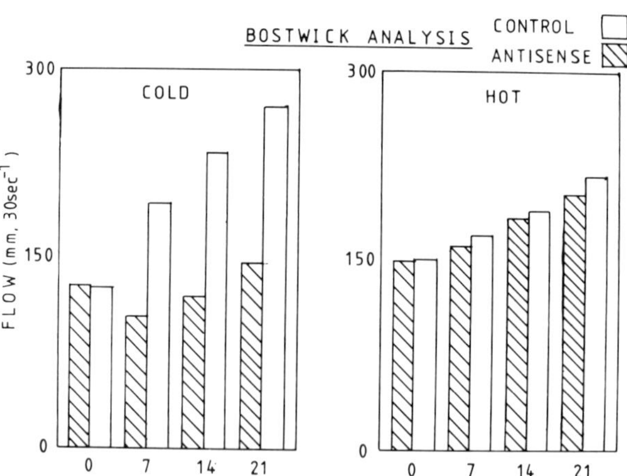

Figure 2.13 Effect on fruit processing of the reduction in polygalacturonase activity in transgenic tomato fruit. Fruit were harvested at various ripening stages and processed to a paste either directly (cold) or following a heat treatment in a microwave oven (hot). Resultant pastes were then analysed for viscosity by a standard Bostwick test. Good quality pastes, with high viscosity, record low values in this analysis.

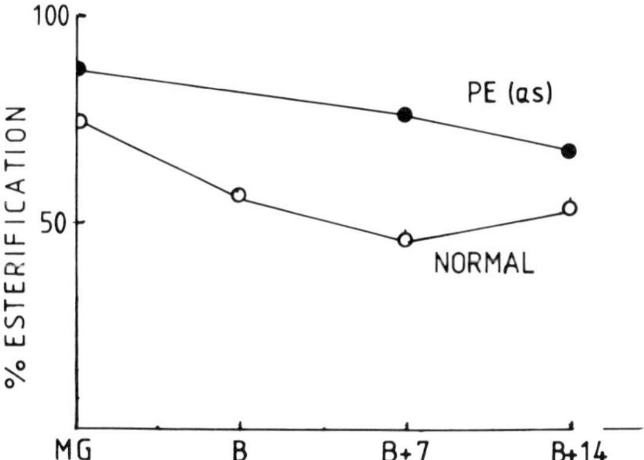

Figure 2.14 Effect of reduced pectinesterase (PE) activity on pectin esterification in tomato fruit. The degree of esterification of total cell wall pectin was monitored throughout the ripening of normal (○) or antisense PE exhibiting 10% normal enzyme activity (●) of tomato fruit.

this case serum viscosity, another quality attribute of processed pastes, is increased and again this could be commercially desirable (Robertson, unpublished).

The cell wall is a complex structure with varied and important functions for the plant. Wall products are also extracted and used commercially. There is, therefore, immense scope for the manipulation of wall polymers. The complex nature is, however, both a blessing and a problem. The blessing is that there are a wide range of possible 'targets' for manipulation as suggested in this chapter. The problem is that of predicting the consequences of such manipulation in a complex situation. The future commercial application of genetic engineering to cell wall manipulation, especially in the intact plant, will rely on a better biochemical understanding of the wall and its turnover.

References

Anderson, N.S., Cambell, J.W., Harding, N.M., Rees, D.A and Samuel, J.W.B. (1969) X-ray diffraction studies of polysaccharide sulphates. Double helix models for κ- and ι-carrageenans. *J. Molec. Biol.* **45**: 85–99.

Arnott, S., Fulmer, A., Scott, W.E.,Dea,I.C.M.,Moorhouse,R. and Rees,D.A. (1974) The agarose double helix and its function in agarose gel structure. *J. Molec. Biol.* **90**.269–284.

Aspinall, G.O.(1980) Chemistry of cell wall polysaccharides. In *The Biochemistry of Plants, Vol. 3* (Preiss, J., ed.), Academic Press, San Diego, pp.473–500.

Averyhart-Fullard, V., Datt, K. and Marcus, A. (1988) A new hydroxyproline-rich protein in the soyabean cell wall. *Proc. Nat Acad Sci. USA*. **85**: 1082–1085.

Bacic, A., Harris, P.J. and Stone, B.A. (1988) Structure and function of plant cell walls. In *The Biochemistry of Plants, Vol 14* (Preiss,J., ed.), Academic Press, San Diego, pp. 297–371.

Bauer, W.D. (1977) Plant cell walls. In *The Molecular Biology of Plant Cells* (Smith, H., ed.) Blackwell Scientific, pp. 6–23.

Bayley, W.D.(1988) X ray and infrared studies on carrageenan. *J. Molec. Biol.* **17**:194–205.

Benjamin, M. and Callaghan, T. (1985) Biosynthesis of (1,4)-β-D-glucan in cell free preparations from mung beans. In *Biochemistry of Plant Cell Walls* (Brett, C.T. and Hillman, J.R. eds.), Society for Experimental Biology Seminar Series 28, pp. 243–258.

Berth, G., Anger, J. and Linow, F. (1977) Light scattering and viscosimetric studies for molecular weight determination on pectins in aqueous solutions. *Die Nahrung* **21**: 939.

Bird, C.R., Smith, C.J.S., Ray, J.A., Moreau, P., Bevan, M.W., Bird, A.S., Hughes, S., Morris, P.C., Grierson, D. and Schuch, W. (1988) The tomato polygalacturonase gene and ripening-specific expression in transgenic plants. *Plant Molec. Biol.* **11**: 651–662.

Blaschek, W., Koehler, H., Semler, U. and Franz, G. (1982) Molecular weight distribution of cellulose in primary cell walls. *Planta* **154**: 550–555.

Bolwell, G.P. (1988) Synthesis of cell wall components: aspects of control. *Phytochemistry* **27**: 1235–1253.

Bolwell, G.P. and Northcote, D.H. (1981) Control of hemicellulose and pectin synthesis during differentiation of vascular tissue in bean (*Phaseolus vulgaris*) callus and in bean hypocotyl. *Planta* **152:** 225–331.

Bolwell, G.P. and Northcote, D.H. (1983) Arabinan synthase and xylan synthase activities of *Phaseolus vulgaris*. Subcellular localisation and possible mechanism of acton. *Biochem J.* **210**:497–507.

Brady, C.J. (1987) Fruit ripening. *Ann. Rev. Plant Physiol.* **38**: 155–178.

Brett, C.T. (1981) Polysaccharide synthesis from GDP-glucose in pea-epicotyl slices. *J. Exp. Biol.* **32**: 1067–1077.

Brett, C.T. and Northcote, D.H. (1975) The formation of oligoglucans linked to lipid during synthesis of β-glucan by characterised membrane fractions isolated from peas. *Biochem. J.* **148**: 107–117.

Brown, C.T. and Henley, D. (1964) Studies on cellulose derivatives. Part IV. The configuration of the polyelectrolyte sodium carboxymethyl cellulose in aqueous sodium chloride solutions. *Makromol. Chem.* **79**: 68–88.

Bulpin, P.V., Gidley, M.J., Jeffcoat, R. and Underwood, D. R. (1990) Development of a biotechnological process for the modification of galactomannan polymers with plant α-galactosidase. *Carbohyd. Polymers* **12** : 155–168.

Cairns, P., Miles, M.J. and Morris, V.J. (1986a) X-ray diffraction studies of kappa carrageenan tara gum mixed gels. *Int. J. Biol. Macromol.* **8**: 124–127.

Cairns, P., Miles, M.J. and Morris, V.J. (1986b) Intermolecular binding of xanthan gum and carab gum. *Nature* **322**: 89–90.

Cairns, P., Miles, M.J., Morris, V.J. and Brownsey, G.J. (1987) X-ray fibre-diffraction studies of synergistic, binary polysaccharide gels. *Carbohydr. Res.* **160**: 411–423.

Cassab, G.I. and Varner, J.E. (1988) Cell wall proteins. *Ann. Rev. Plant Physiol.* **39**: 321–353.

Chan, P.H. and Hassid, W.Z. (1975) One step purification of D-galactose and L-arabinose kinases from *Phaseolus aureus* seedlings by ATP-sepharose affinity chromatography. *Anal. Biochem.* **64**: 372–379.

Claffey, W. and Blackwell, J. (1976) Electron diffraction of valonia cellulose. A quantitative interpretation. *Biopolymers* **15**: 1903–1915.

Clark, A. and Ross-Murphy, S.M. (1987) Structural and mechanical properties of biopolymer gels. *Adv. Polymer Sci.* **83**: 57.

Condit, C. and Meagher, R.B. (1987) Expression of a gene encoding a glycine-rich protein in petunias. *Molec. Cell Biol.* **7**: 4273–4279.

Cooper, J.B., Chen, J.A., van Holst, G.-H. and Varner, J.E. (1987) Hydroxyproline-rich glycoproteins of plant cell walls. *Trends Biochem. Sci.* **12**: 24–27.

Darvill, A.G., McNeil, M., Albersheim, P. and Delmer, D.P. (1980) The primary cell walls of flowering plants. In *The Biochemistry of Plants, Vol 1* (Tolbert, N.E., ed.), Academic Press, pp. 91–162.

Datema, R., Schwarz, R.T., Rivas, L.A. and Pont, L.R. (1983) Inhibition of β-(1,4)-glucan biosynthesis by deoxyglucose. The effect on the glucosylation of lipid intermediates. *Plant Physiol.* **71**: 76–81.

Dea, I.C.M. (1987) The role of structural modification in controlling polysaccharide functionality. *Indust. Polysacch.* **3**: 207–216.

Dea, I.C.M. McKinnon, A.A. and Rees, D.A. (1972) Tertiary and quarternary structure in aqueous polysaccharide systems which model cell wall cohesion: reversible changes in conformation and association of agarose, carrageenans and galactomannans. *J. Molec. Biol.* **68**: 153–172.

Delmer, D.P. (1983) Biosynthesis of cellulose. *Adv Carbohyd. Chem. Biochem.* **41**: 105–115.

Delmer, D.P. (1987) Cellulose biosynthesis. *Ann. Rev. Plant Physiol.* **38**: 259–290.

Delmer, D.P. and Albersheim, P. (1970) The biosynthesis of sucrose and nucleoside diphosphate glucoses in *Phaseolus aureus. Plant Physiol.* **45**: 782–786.

Delmer, D.P. and Stone, B.A. (1988) Biosynthesis of plant cell walls. In *The Biochemistry of Plants, Vol. 14* (Preiss, J., ed.), Academic Press, pp.373–420.

Ericson, M.C. and Elbein, A.D. (1980) Biosynthesis of cell wall polysaccharides and glycoproteins. In *The Biochemistry of Plants, Vol. 3* (Preiss, J., ed.), Academic Press, pp.589–616.

Fanutti, C., Gidley, M.J. and Reid, J.S.G. (1991) A xyloglucanoligosaccharide specific α-D-xylosidase or exooligoglucan-xylohydrolase from germinated nasterium (*Tropaeolum majus* L.) seeds. Purification, properties and its interaction with a xyloglucan specific endo (1,4)β-D-glucanase and other hydrolases during storage xyloglucan mobilization. *Planta* **184**: 137–147.

Feingold, D.S. and Avigad, G. (1980) Sugar nucleotide transformations in plants. In *The Biochemistry of Plants, Vol. 3* (Preiss, J., ed.), Academic Press, pp. 101–170.

Fincher, G.B. and Stone, B.A. (1981) Metabolism of non-cellulosic polysaccharides. In *The Encyclopaedia of Plant Physiology, Vol. 13B* (Tanner, W. and Loewus, F.A, eds.) Springer-Verlag, pp.68–132.

Fincher, G.B. and Stone, B.A. (1986) Cell walls and their components in cereal grain technology. *Adv. Cereal Chem.* **8** : 207–295.

Fincher, G.B. Stone, B.A. and Clarke, A.E. (1983) Arabinogalactan- proteins. Structure, biosynthesis and function. *Ann. Rev. Plant Physiol.* **34**: 47–70.

Fischer, R.L. and Bennett, A.B. (1991) Role of cell wall hydrolases in fruit ripening. *Ann. Rev. Plant Physiol. Molec. Biol.* **42**: 675–703.

Fredrikson, K., Kjellbom, P. and Larsson, C. (1991) Isolation and polypeptide composition of (1, 3)-β-glucan synthase from plasma membranes of *Brassica oleracea*. *Physiol. Plant.* **81**: 289–294.

Fry, S.C. (1982) Isodityrosine, a new cross-linking amino acid from plant cell wall glycoprotein. *Biochem. J.* **204**: 449–455.

Fry, S.C. (1983) Feruloylated pectins from the primary cell wall: their structure and possible functions. *Planta* **157**: 111–123.

Fry, S.C. (1986) Cross-linking of matrix polymers in the growing cell walls of angiosperms. *Ann. Rev. Plant Physiol.* **37**: 165–186.

Fry, S.C. (1987) Intracellular feruloylation of pectic polysaccharides. *Planta* **171**: 205–211.

Fry, S.C. (1988) *The Growing Plant Cell Wall: Chemical and Metabolic Analysis*. Longman Scientific, chapter 5 and references therein.

Gaisford, S.E., Harding, S.E., Mitchell, J.R. and Bradley, T.D. (1986) A comparison between the hot and cold water soluble fractions of 2 locust bean gum samples. *Carbohydr. Polymers* **6**: 423–442.

Glickman, M. (1982) Food applications of gums. In *Food Carbohydrates* (Lineback and Inglett, eds.) AVI, pp.270–295

Grant, G.T., Morris, E.R., Rees, D.A., Smith, P.J.C. and Thorn, D. (1973) Biological interactions between polysaccharides and divalent cations: the egg-box model. *FEBS Lett.* **32**: 195–198.

Grasdelen, H. (1983) High-field, one H-nmr spectroscopy of alginate: sequential structure and linkage conformations. *Carbohyd. Res.* **118**: 255–260.

Grasdelen, H., Larsen, B. and Smidsrød, O. (1981) ^{13}C nmr studies of monomeric composition and sequence in alginate. *Carbohyd. Res.* **89**: 179–191.

Grierson, D., Tucker, G.A., Keen, J., Ray, J., Bird, C.R. and Schuch, W. (1986) Sequencing and identification of a cDNA clone for tomato polygalacturonase. *Nucl. Acids Res.* **14**: 8595–8603.

Grierson, D., Smith, C.J.S., Watson, C.F., Morris, P.C., Gray, J.E., Davies, K., Picton, S.J., Tucker, G.A., Seymour, G.B., Schuch, W., Bird, C.R. and Ray, J. (1990) Regulation of gene expression in transgenic tomato plants by antisense RNA and ripening-specific promoters. In *Genetic Engineering of Crop Plants* (Lycett, G.W. and Grierson, D., eds), Butterworths, London, pp.115–125.

Harding, S.E., Varum, K.M., Stokke, B.T. and Smidsrød, O. (1991) Molecular weight determination of polysaccharides. *Adv. Carbohyd. Anal.* **1**: 63–144.

Harriman, R.W. and Handa, A.K. (1990) Identification and characterisation of three pectinmethylesterase genes in tomato. *Plant Physiol.* **93** (supplement): 44.

Harriman, R.W., Tieman, D.M. and Handa, A.K. (1991) Molecular cloning of tomato pectinmethylesterase gene and its expression in rutgers, ripening inhibitor, non-ripening and never-ripe tomato fruit. *Plant Physiol.* **97**: 80–87.

Hayashi, T. (1989) Xyloglucan in the primary cell wall. *Ann. Rev. Plant Physiol. Molec. Biol.* **40**: 139–168.

Hayashi, T. and Maclachlan, G. (1984) Pea xyloglucan and cellulose: I Macromolecular organisation. *Plant Physiol.* **75**: 596–604.

Hayashi, T. and Matsuda, K. (1981) Biosynthesis of xyloglucan in suspension cultured soybean cells. Evidence that the enzyme system of xyloglucan biosynthesis does not contain β-(1,4)- glucan-4-β-D-glucosyl transferase activity. *Plant Cell Physiol.* **22**: 1571–1584.

Hayashi, T., Kato, Y. and Matsuda, K. (1980) Xyloglucan from suspension cultured soybean cells. *Plant Cell Physiol.* **21**: 1405–1418.

Henry, R.J. and Stone, B.A. (1982) Factors influencing β-glucan synthesis by particulate enzymes from suspension cultured *Lolium multiflorum* endosperm cells. *Plant Physiol.* **69**.

Herrmansson, A.-M. (1989) Rheological and microstructural evidence for transient states during gelation of kappa carrageenan in the presence of potassium. *Carbohyd. Polymers* **10**: 163–182.

Higgs, P.G. and Ball, R.C. (1989) Some ideas concerning the elasticity of bio-polymer networks. *Macromolecules* **22**: 2432–2437.

Horton, R.F. and Osborne, D.J. (1967) Senescense, abscission and cellulase activity in *Phaseolus vulgaris*. *Nature* **214**: 1086–1088.

Huber, D.J. (1983) The role of cell wall hydrolases in fruit softening. *Horticult. Rev.* **5**: 169–219.

Ishii, T., Thomas, J., Darvill, A. and Albersheim, P. (1989) Structure of plant cell walls: XXVI The walls of suspension cultured sycamore cells contain a family of rhamnogalacturonan I like pectic polysaccharides. *Plant Physiol* **89**: 421–428.

Jacob, S.R. and Northcote, D.H. (1985) *In vitro* glucan synthesis by membranes of celery petioles: the role of the membranes in determining the type of linkage formed. *J. Cell Sci.* (supplement) 2: 1–11.

Kato, Y. and Nevins, D.J. (1984) Enzymic dissociation of Zea shoot cell wall polysaccharides: II Dissociation of (1,3),(1,4)-β-D-glucan by purified (1,3), (1,4)-β-D-glucan-4-glucanohydrolase from *Bacillus subtilis*. *Plant Physiol.* **75**: 745–752.

Kauss, H., Swanson, A.C., Arnold, R. and Odzuck, W. (1969) Biosynthesis of pectic substances. Localisation of enzymes and products in a lipid-membrane complex. *Biochem. Biophys. Acta* **192**: 55–61.

Keegstra, K., Talmadge, K.W., Bauer, W.D. and Albersheim, P. (1973) The structure of plant cell walls: III A model of the walls of suspension cultured sycamore cells based on the interconnections of the macromolecular components. *Plant Physiol.* **51**: 188–196.

Kitamura, S., Takeo, K., Kuge, T. and Stokke, B.T. (1991) Thermally induced conformational transition of double stranded xanthan in aqeous salt solutions. *Biopolymers* **31**: 1243–1255.

Knee, M. (1973) Polysaccharides and glycoproteins of apple fruit cell walls. *Phytochem.* **12**: 637–653.

Knox, J.P., Linstead, P.J., King, J., Cooper, C. and Roberts, K. (1990) Pectin esterification is spacially regulated both within cell walls and between developing tissues of root apices. *Planta* **181**: 512–521.

Kooiman, P. (1961) The constitution of Tamarindus amyloid *Rec. Trav. Chim. Pays. Bas.* **80**: 849–865.

Labavitch, J.M. (1981) Cell wall turnover in plant development. *Ann. Rev. Plant Physiol.* **32**: 385–406.

Larsen, B., Painter, T., Haug, A. and Smidsrød, O (1969) A statistical description of alginate molecules in terms of a penultimate-unit copolymer. *Acta. Chem. Scand.* **23**: 355–370.

Launay, B., Doublier, J.L. and Cuvelier, O. (1986) Flow properties of aqueous solutions and dispersions of polysaccharides. In *Functional Properties of Food Macromolecules* (Mitchell, J.R. and Ledward, D.A. eds.), Elsevier.

Lawson, C.J. and Rees, D.A. (1970) An enzyme for the metabolic control of polysaccharide conformation and function. *Nature* **227**: 392–393.

Lecacheux, P., Panarmo, R., Brigand, G. and Martin, G. (1985) Molecular weight distribution of carrageenans by size exclusion chromatography and low angle laser light scattering. *Carbohyd. Polymers* **5**:423–440.

Lee, I., Atkins, E.D.T. and Miles, M.J. (1992) Visualisation of the carrageenan double helix using STM. *Ultra Microscopy*: in press.

Lei, M. and Wu, R. (1991) A novel glycine-rich cell wall protein in rice. *Plant Molec. Biol.* **16**: 187–198.

Leibowitz, M.D., Dickinson, M.D.B., Loewus, F.A. and Loewus, M. (1977) Partial purification and study of pollen glucuronokinase. *Arch. Biochem. Biophys.* **179**: 559–564.

Lewis, N.G. and Paice, M.G. (1989) Plant cell wall polymers. Biogenesis and biodegradation. *Am. Chem. Soc. Symp. Ser.* 399.

Mannion, R.O., Mitchell, J.R., Launay, B., Cuvelier, G., Hill, S.E., Harding, S.E. and Melici, C.D. (1992) Xanthan\locust bean gum interactions at room temperature. *Carbohyd. Polymers*: in press.

Marcus, A., Greenberg, J. and Averyhart-Fullard, V. (1991) Repetitive proline-rich proteins in the extracellular matrix of the plant cell. *Physiol. Plant.* **81**: 273–279.

Markovic, O. and Jornvall, H. (1986) Pectinesterase: The primary structure of the tomato enzyme. *Eur. J. Biochem.* **158**: 455–462.

Markovic, O., Machova, E. and Erdelska, O. (1989) Localisation of pectinesterase in leaves and flower parts of *Medicago sativum* (L.) and *Nicotiana tabacum* (L.) *Inst. Chem. Cchr. Slov. Acad. Sci.* **44**: 401–406.

Marx-Figini, M. and Shutz, G.V. (1966) Uber die kinetik und den mechanimis der biosynthese der cellulose in den hohren pflanzen. *Biochem. Biophys. Acta* **112**: 81–101.

McCann, M.C., Wells, B. and Roberts, K. (1990) Direct visualisation of cross-links in the primary cell wall. *J. Cell. Sci.* **96**: 323–324.

McCleary, B.V. (1979) Enzymic hydrolysis, fine structure and gelling interaction of legume-seed D-galacto-D-mannans. *Carbohyd. Res.* **71**: 205–230.

McCleary, B.V. and Neukom, H. (1982) Effect of enzymic modification on the solution and interaction properties of galactomannans. *Prog Food Nutr. Sci.* **6**: 109–118.

McNeil, M., Darvill, A.G. and Albersheim, P. (1980) Structure of plant cell walls: X Rhamnogalacturonan I. A structurally complex pectic polysaccharide in the walls of suspension cultured sycamore cells. *Plant Physiol.* **66**: 1128–1134.

McNeil, M., Darvill, A.G. and Albersheim, P. (1982) Structure of plant cell walls: XII Identification of seven differently linked glycosyl residues attached to *O*-4 of the (2, 4)-linked L-rhamnosyl residues of rhamnogalacturonan I. *Plant Physiol.* **70**: 1586–1591.

McNeil, M., Darvill, A.G., Fry, S. and Albersheim, P. (1984) Structure and function of the primary cell wall of plants. *Ann. Rev. Biochem.* **53**: 625–663.

Medina, A. and Sols, A. (1956) *Biochim. Biophys. Acta* **19**: 378–379.

Meinert, M.C. and Delmer, D.P. (1977) Changes in biochemical composition of the cell wall of the cotton fibre during development. *Plant Physiol.* **59**: 1088–1097.

Melton, I.D., McNeil, M., Darvill, A.G. and Albersheim, P. (1986) Structural characteristics of oligosaccharides isolated from the pectic polysaccharide rhamnogalacturonan II. *Carbohyd. Res.* **146**: 279–305.

Michel, F., Thibault, J.-F. and Doublier, J.-L. (1984) Viscometric and potentiometric study of high methoxy pectins in the presence of sucrose. *Carbohyd. Res.* **4**: 283–297.

Mitcham, E.J. Gross, K.C. and Ng T.J. (1989) Tomato fruit cell wall synthesis during development and senescence. *Plant Physiol.* **89;** 477–481.

Mitcham, E.J., Gross, K.C. and Ng, T.J. (1991) Ripening and cell wall synthesis in normal and mutant tomato fruit. *Phytochem.* **30**: 1777–1780.

Mitchell, J.R. (1980) Rheology of gels. *J. Text. Stud.* **11**: 315–337.

Mitchell, J.R. and Taylor, A.J. (1983) Crude pectate gelling agents in heat processed foods. In *Upgrading Wastes for Feeds and Foods* (Ledward, D.A., Taylor, A.J. and Lawrie, R.A., eds.), Butterworths, pp. 247–268.

Mohd-Ali, Z. and Brady, C.J. (1982) Purification and characterisation of the polygalacturonase of tomato fruit. *Aust J. Plant Physiol.* **9**: 155–159.

Moirano, A.L. (1977) Sulphated seaweed polysaccharides. In *Food Colloids* (Graham, H. ed.), AVI Publishing Company, Westpoint, Connecticut, pp. 347–381.

Moore, P.J., Darvill, A.G., Albersheim, P. and Staedkelin, L.A. (1986) Immunological localisation of xyloglucan and rhamnogalacturonan I in the cell walls of suspension cultured sycamore cells. *Plant Physiol.* **82**: 787–794.

Morris, V.J. (1986) Gelation of polysaccharides. In *Functional Properties of Food Macromolecules* (Mitchell, J.R. and Ledward, D.A., eds.), Elsevier Applied Science, pp. 171.

Morris, V.J. and Miles, M.J. (1986) Effect of natural modifications on the functional properties of extracellular bacterial polysaccharides. *Int. J. Biol. Macromol.* **8**: 342–348.

Morris, E.R., Rees, D.A. and Robinson, G. (1980) Cation specific aggregation of carrageenan helices—the domaim model of biopolymer gelation. *J. Molec. Biol.* **137**: 349–362.

Moshrefi, M. and Luh, B.S. (1984) Purification and characterisation of two tomato polygalacturonase isoenzymes. *J. Food Biochem* **8**: 39–54.

Muhr, A.H. and Blanshard, J.M.V. (1986) Effect of polysaccharide stabilizers on the rate of growth of ice. *J Food Technol.* **21**: 683–710.

Neufeld, E.F., Feingold, D.S., Ilves, S.M., Kessler, G. and Hassid, W.Z. (1961) Phosphorylation of D-galacturonic acid by extracts from germinating seeds of *Phaseolus aureus*. *J. Biol. Chem.* **236**: 3102–3105.

Nikaido, H. and Hassid, W.Z. (1971) Biosynthesis of saccharides from glycopyranosyl esters of nucleoside pyrophosphates (sugar nucleotides). *Adv. Carbohyd. Chem. Biochem.* **26**: 351–483.

Nishinari, K., Koide, S. and Ogino, K. (1985) On the temperature dependence of elasticity of thermo-reversible gels. *J. Phys.* (France) **46**: 793–797.

Northcote, D.H., Davey, R. and Lay, J. (1989) Use of antisera to localise callose, xylan and arabinogalactan in the cell plate primary and secondary walls of plant cells. *Planta* **178**: 353– 366.

Panayotates, N. and Villemez, C.L. (1973) The formation of a β-(1,4)-galactan chain catalysed by a *Phaseolus aureus* enzyme. *Biochem. J.* **133**: 263–271.

Paoletti, S., Smidsrd, O. and Grasdalen, H. (1984) Thermodynamic stability of the ordered conformations of carrageenan polyelectrolytes. *Biopolymers* **23**: 1771–1794.

Paoletti, S., Delben, F., Cesaro, A. and Grasdalen, H. (1985) Conformational transition of kappa carregeenan in aqueous solution. *Macromol.* **18**: 1834–1841.

Percival, E. and McDowell, R.H. (1981) Algal walls-composition and biosynthesis. In *Encyclopaedia of Plant Physiology Vol 13B* (Tanner, W. and Loewus, F.A., eds.), Springer-Verlag, pp. 277–316.

Pippen, E.L., McCready, R.M. and Owens, H.S. (1950) Gelation properties of partially acetylated pectins. *J. Am. Chem. Soc.* **72**: 813–816.

Pizzi, A. and Eaton, A. (1985) The structure of cellulose by conformational Analysis-2. The cellulose polymer chain. *Macromol. Chem. Sci* **A22**: 105–168.

Plashchina, I.G., Muratalieva, I.R., Braudo, E.E. and Tolstoguzov, V.B. (1986) Studies of the gel formation of κ-carrageenan bove the coil-helix transition temperature range. *Carbohyd. Polymers* **6**: 15–34.

Pogson, B.J., Brady, C.J. and Orr, G.R. (1991) On the occurrence and structure of subunits of endo-polygalacturonase isoforms in mature-green and ripening tomato fruit. *Aust. J. Plant Physiol.* **18**: 65–79.

Powell, D.A., Morris, E.R., Gidley, M.J. and Rees, D.A. (1982) Conformation and interactions of pectins: II Influence of residue sequence on chain association in calcium pectate gels. *J. Molec. Biol.* **155**: 517–531.

Pressey, R. (1991) Polygalacturonase in tree pollen. *Phytochemistry* **30**: 1753–1755.

Pressey, R. and Avants, J.K. (1972) Multiple forms of pectinesterases in tomatoes. *Phytochemistry* **11**: 3139–3142.

Ray, P.M. (1980) Cooperative action of the glucan synthetase and UDP-xylose xylosyltransferase of golgi membranes in the synthesis of a xyloglucan like polysaccharide. *Biochem. Biophys. Acta.* **629**: 431–434.

Ray, J., Knapp, J., Grierson, D., Bird, C. and Schuch, W. (1988) Identification and sequence determination of a cDNA clone for tomato pectinesterase. *Eur. J. Biochem.* **174**: 119–124.

Rees, D.A. (1969) Structure, conformation and mechanism in the formation of polysaccharide gels and networks. *Adv. Carbohyd. Chem. Biochem.* **24**: 267–332.

Roberts, K. (1984) Cellulose synthesis—another brick in the wall. *Nature* **311**: 105–106.

Roberts, K. (1990) Structures at the plant cell surface. *Curr. Opin. Cell. Biol.* **2**: 920–928.

Roberts, K., Phillips, J., Shaw, P., Grief, C. and Smith, E. (1985) An immunological approach to the plant cell wall. In *Biochemistry of Plant Cell Walls* (Brett, C.T. and Hillman, J.R., eds.), Society of Experimental Biology Seminar Series 28, pp. 177–198.

Robinson, G., Ross-Murphy, S.B. and Morris, E.R. (1982) Viscosity–molecular weight relationship, intrinsic chain flexibility and dynamic solution properties of guar galactomannan. *Carbohyd. Res.* **107**: 17–32.

Rochas, C. and Lahaye, M. (1987) Molecular organisation of κ-carrageenan in aqueous solution. *Carbohyd. Polymers* **7** : 435–448

Rochas, C. and Lahaye, M. (1989) Average molecular weight distribution of agarose and agarose-type polysaccharides. *Carbohyd. Polymers* **10**: 289–298.

Ryan, C.A. and Farmer, E.F. (1991) Oligosaccharide signals in plants: A current assessment. *Ann. Rev. Plant Physiol. Molec. Biol.* **42**: 631–674.

Sabater de Sabates, A. (1979) Contribution l'etude des relations entre characteristiques macromolecules et proprits rhologiques en solution aqueuse concentre d'un paississant alimentarie: la gomme de croube. Ph.D. thesis, University of Paris XI ENSIA.

Saltman, P. (1953) Hexokinases in higher plants. *J. Biol. Chem.* **200**: 145–154.

Schuch, W., Bird, C.R., Ray, J., Smith, C.J.S., Watson, C.F., Morris, P.C., Ray, J.E., Arnold, C., Seymour, G.B., Tucker, G.A. and Grierson, D. (1989) Control and manipulation of gene expression during tomato fruit ripening. *Plant Mol. Biol.* **13**: 303–311.

Schuch, W., Kanczler, J., Robertson, D., Hobson, G.E., Tucker, G.A., Grierson, D., Bright, S. and Bird, C. (1991) Fruit quality characteristics of transgenic tomato fruit with altered polygalacturonase activity. *Horticult. Sci.* **26**: 1517–1520.

Selvendran, R.R. (1985) Development in the chemistry and biochemistry of pectic and hemicellulosic polymers. *J. Cell Sci.* (supplement) 2:51–88.

Seymour, G.B., Lasslett, Y. and Tucker, G.A. (1987) Differential effects of pectolytic enzymes on tomato polyuronides *in vivo* and *in vitro. Phytochemistry* **26**: 3137–3139.

Shatwell, K.R., Sutherland, I.W., Ross-Murphy, S.B. and Dea, I.C.M. (1990) Influence of the acetyl substituent on the interactions of xanthan with plant polysaccharides 3. xanthan-konjac mannan systems. *Carbohyd. Polymers* **14**: 131–147.

Shedletsky, E., Shmuel, M., Delmer, D.P. and Lamport, D.T.A. (1990) Adaption and growth of tomato cells on the herbicide 2,6-dichlorobenzonitrile leads to production of unique cell walls virtually lacking a cellulose-xyloglucan network. *Plant Physiol.* **94**: 980–987.

Sheehy, R.E., Pearson, J., Brady, C.J. and Hiatt, W.R. (1987) Molecular characterisation of tomato fruit polygalacturonase. *Molec. Gen. Genet.* **208**: 30–36.

Sheehy, R.E., Kramer, M. and Hiatt, W.R. (1988) Reduction of polygalacturonase activity in tomato fruit by antisense RNA. *Proc. Nat. Acad. Sci. USA* **85**: 8805–8809.

Shelso, G. J. (1990) Commercialisation of new synergistic application of carrageenan. *In Gums and Stabilisers for the Food Industry, Vol. 5* (Phillips, G. O., Wedlock, D. J. and Williams, P. A., eds.), IRL Press, Oxford, pp. 563–569.

Skajak-Braek, G., Grasdalen, H., Draget, K, and Smidsrød, O. (1989) Inhomogenous calcium alginate bonds. In *Biomedical and Biotechnological Advances in Industrial Polysaccharides* (Crescenzi, V., Dea, I.C.M., Paoletti, S., Stivata, S.-S. and Sutherland, I., eds.), Combon and Bleach, New York, pp. 345–362.

Smidsrød, O. (1970) Solution properties of alginates. *Carbohyd. Res.* **13**: 359–372.

Smidsrød, O. and Grasdalen, H. (1982) Some physical properties of carrageenan in solution and gel state. *Carbohyd. Polymers* **2**: 270–272.

Smidsrød, O. and Haug, A. (1972) Properties of poly (1,4)-hexuronates in the gel state: II Comparison of gels of different chemical composition. *Acta. Chem. Scand.* **26**: 71–78.

Smidsrød, O., Anderson, I.I., Grasdalen, H., Larsen, B. and Paynter, T. (1980) Evidence for a salt-promoted freeze-out of linkage. Conformations in carrageenans as a pre-requisite for gel formation. *Carbohyd. Res.* **80**: C11–C16.

Smith, C.J.S., Watson, C.F., Ray, J., Bird, C.J., Morris, P.C., Schuch, W. and Grierson, D. (1988) Antisense RNA inhibition of polygalacturonase activity in transgenic tomatoes. *Nature* 334: 724–726.

Smith, C.J.S., Watson, C.F., Morris, P.C., Bird, C.R., Seymour, G.B., Gray, J.E., Arnold, C., Tucker,G.A., Schuch, W., Harding, S.E. and Grierson, D. (1990) Inheritance and effects on ripening of antisense polygalacturonase genes in transgenic tomatoes. *Plant Molec. Biol.* **14**: 369–379.

Stafstrom, J.P. and Staehelm, L.A. (1988) Antibody localisation or extensin in cell walls of carrot storage roots. *Planta.* 174: 321–332.

Staver, M.J., Glick, K. and Baisted, D.J. (1978) Uridine diphosphate glucose-sterol glucosyl-transferases and nucleoside diphosphatase activities in etiolated pea seedlings. *Biochem. J.* **169**: 297–303.

Sutherland, I.W. (1990) *Biotechnology of Microbial Exopolysaccharides*. Cambridge University Press, Cambridge.

Suzuki, K., Ingold, E., Sugiyama, M. and Komamine, A. (1991) Xylan synthase activity in isolated mesophyll cells of *Zinnia elegans* during differentiation to tracheary elements. *Plant Cell Physiol.* **32**: 303–306.

Taylor, J.E., Tucker, G.A., Lasslett, Y., Smith, C.J.S., Arnold, C.M., Watson, C.F., Schuch, W., Grierson, D. and Roberts, J.A. (1990) Polygalacturonase expression during leaf abscission of normal and transgenic tomato plants. *Planta* **183**: 133–138.

Toft, K., Grasdalen, H. and Smidsrød, O. (1986) In *Chemistry and Function of Pectins* (Fishman, M.L. and Jen, J.J., eds.), ACS Symposium Series, **310**: 117–132.

Tucker, G.A. (1990) Genetic manipulation of fruit ripening. *Biotechnol. Genet. Engng. Reviews* 8. 133–159.

Tucker, G.A. and Grierson, D. (1987) Fruit ripening. In *The Biochemistry of Plants, Vol. 12* (Davies, D.D., ed.), Academic Press, pp. 265–318.

Tucker, M.L. and Milligan, S.B. (1991) Sequence analysis and comparison of avocado fruit and bean abscission cellulases. *Plant Physiol.* **95**: 928–933.

Tucker, G.A., Robertson, N.G. and Grierson, D. (1980) Changes in polygalacturonase isoenzymes during the ripening of normal and mutant tomato fruit. *Eur. J. Biochem.* **112**: 119–124.

Tucker, G.A., Robertson, N.G. and Grierson, D. (1982) Purification and changes in activities of tomato pectinesterase isoenzymes. *J. Sci. Food Agric* **33**: 396–400.

Tucker, G.A., Schindler, C.B. and Roberts, J.A. (1984) Flower abscission in mutant tomato plants. *Planta* **160**: 164–167.

Tucker, G.A., Seymour, G.B., Bundick, Y., Robertson, D., Smith, C.J.S. and Grierson, D. (1992) Use of antisense RNA technology to manipulate pectin degradation in tomato fruit. *New Zeal. J. Horticult. Crop Bot.* **20**. 119–124.

Tye, R.J. and Stanley, N.F. (1988) The carrageenan industry. *Abs Am. Chem. Soc.* **196**: 32.

Varner, J.E. and Lin, L.-S. (1989) Plant cell wall architecture. *Cell* **56**:231–239.

Vincent, A. (1986) Food texture additives. *IFST Proceedings* **19**: 107–111.

Walkinshaw, M.D. and Arnott, S. (1981a) Conformation and interactions of pectins I. X-ray diffraction analysis of sodium pectate in neutral and acidified forms. *J. Molec. Biol* **153**: 1055 1073.

Walkinshaw, M.D. and Arnott, S. (1981b) Conformation and interactions of pectins II. Models for junction zones in pectinic acid and calcium pectate gels. *J. Molec. Biol.* **153**: 1075–

Wilson, L.G. and Fry, J.C. (1986) Extension—a major cell wall glycoprotein. *Plant Cell Environ.* **9**: 239–260.

Yalpani, M. (1988) *Polysaccharides: Synthesis, Modification and Structure\Property Relationships*. Elsevier.

Ye, Z.-H., Song, Y.-R., Marcus, A. and Varner, J.E. (1991) Comparative localisation of three classes of cell wall proteins. *Plant J.* **1**: 175–183.

3 Biochemistry and molecular biology of lipid biosynthesis in plants: potential for genetic manipulation

A. R. SLABAS, T. FAWCETT, G. GRIFFITHS and
K. STOBARD

3.1 The nature of lipids

Lipids are a heterogeneous class of chemicals which can be broadly defined as insoluble in water and highly soluble in non-polar organic solvents. Within this definition are included a wide range of chemical compounds including terpenes and steroids, together with a large variety of chemical compounds still waiting to have their exact structure defined. Early interest in these compounds was initiated by chemists who wished to apply their skills of separation and structural analysis to natural products. As a result there is a large volume of literature on the lipid composition of natural oils (for example Hilditch and Williams, 1964). In this review we will limit ourselves to considerations of fatty acids and esterified fatty acids.

3.1.1 Nomenclature of fatty acids

The nomenclature of fatty acids is normally based on the hydrocarbon chain having the largest number of carbon atoms with substituents being defined *from the carboxyl end* as 1 or from the carbon atom at 2 as the α-position. In comparing the structure of unsaturated fatty acids it may be convenient to define the double bond *from the terminal methyl group*, so that acids derived from one another, by chain elongation, can be readily identified. For example linoleic acid $CH_3(CH_2)_4CH = CH.CH_2\ CH = CH.(CH_2)_7.CO_2H$ is (Δ)-9, 12- octadecadienoic acid or C18:2 (Δ9, 12) when numbering from the carboxyl group, or (*n*-6, *n*-9)-octadecadienoic acid when numbering from the methyl group.

The study of fatty acids has a long chemical history and hence many of the fatty acids have trivial names. To assist those new to the area Table 3.1 gives the structure and trivial names of several of the more common saturated and unsaturated fatty acids. Apart from these linear molecules, substituted fatty acids also occur in nature. Examples of these are hydroxy fatty acids including ricinoleic acid (12-hydroxy-*cis*-9-octadecanoic) which is a dominant storage product of castor bean (*Ricinus communis*) and cyclic fatty acids, such as dihydrosterculic acid (8-[2-octyl *cyclo* propanyl]octanoic).

Table 3.1 Structure, systematic name and trivial name of common fatty acids.

Type	Systematic name	Common name
Saturated		
C2:0	*n*-Ethanoic	Acetic
C4:0	*n*-Butanoic	Butyric
C8:0	*n*-Octanoic	Caprylic
C10:0	*n*-Decanoic	Capric
C12:0	*n*-Dodecanoic	Lauric
C14:0	*n*-Tetradecanoic	Myristic
C16:0	*n*-Hexadecanoic	Palmitic
C18:0	*n*-Octadecanoic	Stearic
C20:0	*n*-Eicosanoic	Arachidic
C22:0	*n*-Docosanoic	Behenic
C24:0	*n*-Tetracosanoic	Lignoceric
C26:0	*n*-Hexacosanoic	Cerotic
C28:0	*n*-Octacosanoic	Montanic
Monounsaturated		
C16:1 [Δ 9c]	*cis*-9-hexadecenoic	Palmitoleic
C16:1 [Δ 11c]	*cis*-11-hexadecenoic	Palmitvaccenic
C18:1 [Δ 9c]	*cis*-9-octadecenoic	Oleic
C18:1 [Δ 11c]	*cis*-11-octadecenoic	Vaccenic
C22:1 [Δ 13c]	*cis*-13-docosenoic	Erucic
Diunsaturated		
C18:2 [Δ 9, 12c]	*cis, cis*-9, 12-octadecadienoic	Linoleic
Triunsaturated		
C18:3 [Δ 6, 9, 12c]	all *cis*-6, 9, 12-octadecatrienoic	γ-Linolenic
C18:3 [Δ 9, 12, 15c]	all *cis*-9, 12, 15-octadecatrienoic	α-Linolenic

3.1.2 Distribution and function of lipids

Fatty acids are found in all cells. They are an important structural component of the lipid bilayer of membranes in which they are present in the form of various phospholipid classes. Analyses of the chemical structure of fatty acids in membranes, which traditionally has been performed by gas-liquid chromatography, reveals little variation in the acyl side chains. The lipids are mainly long chain C16–C18 with little variation in the degree of saturation. Apart from providing the physical limits to cells and the boundaries to subcellular organelles, these lipids also provide the correct environment for a number of important proteins, such as acyltransferases and translocators, to function in. Indeed certain membrane enzymes, such as UDP-galactose : diacylglycerol galatosyltransferase (Coves *et al*., 1988), have been shown to have a direct lipid requirement for catalytic activity.

The epicuticular waxes are a type of lipid (cutins) long chain hydroxy fatty acid polymers, and have an important structural and protective role. This has been elegantly demonstrated by transforming a non-virulent fungus into a virulent one by expression of a cutinase gene in the recombinant organism so allowing it to penetrate the plant cutin 'coat' and commence the invasive process (Dickman *et al*., 1989).

Table 3.2 Some properties of membrane and storage lipids. The mass of storage lipids is very high in certain tissues and the acyl component shows wide interspecific variation.

Lipid property	Membrane lipid	Storage lipid
Mass	Small	Up to 45% dry weight of seed
Acyl group	C16–C18	C2–C24

Although generally conserved, with respect to chain length of fatty acid, there are well documented cases of alteration in the degree of unsaturation of membrane lipids brought about by environmental stress (Lynch and Thompson, 1984). In certain plants and microorganisms the degree of unsaturation of lipids increases as the temperature is lowered (Murata, 1983). The physical chemical reason for this is well understood; at higher temperature membranes containing saturated fatty acids remain fluid. However, due to melting point considerations, on lowering the temperature such membranes become less fluid. If in temperature acclimatisation more unsaturated fatty acids are present in membranes then such adapted membranes can retain the fluidicity present at higher temperatures. Thus at lower temperatures biochemical function is not impeded.

Apart from structural roles, it is now clear that a number of lipid molecules play a key role in the host range specificity of nodulating bacteria. A sulphated and acylated glucosamine oligosaccharide has recently been shown to act as a signal of *Rhizobium meliloti* in eliciting root nodule organogenesis in alfalfa (Truchet *et al.*, 1991). It has been shown that the nature of the alkyl side chain is important in determining the host range specificity.

From a commercial perspective, apart from the potential of engineering frost resistance into plants, the major interest is in storage lipids. The reasons for this are apparent when considering the potential for manipulating storage lipids over membrane lipids (Table 3.2). A wide variety of plant storage lipids are used for industrial purposes in the detergent, food and chemical industries and the ability to be able to customise or upgrade them represents one of the main challenges in the area of plant lipids.

3.1.3 Is the genetic manipulation of oil seeds on a commercial scale a reality?

Many of the early targets set for genetic manipulation of plants were concerned with improving the nutritional quality of proteins in seeds. Such approaches were mainly based on increasing the lysine or methionine content, in which such materials were nutritionally deficient. These modifications could only make minor variations to the total protein content, however, as the degree of expression of the foreign gene was low (in the order of 1%) in comparison to the mass of seed storage protein present. The scenario is different for proteins which can fundamentally alter the rheological or processing properties of the plant material, such as those affecting pasta or bread-making quality (Payne *et al.*, 1984).

With the genetic manipulation of plant lipids the situation is somewhat different from the marginal benefits which can be obtained from altering storage proteins. This is due to three main considerations.

(1) The storage product to be altered is a lipid, hence expression of an enzyme at a relatively low level which could alter the biochemical pathway will have major effects on the end product. An example of this is the effect of acylthioesterase II on the quality of oil in lactation in rat mammary gland (Smith and Libertini, 1979).
(2) The potential for product alteration is very high. As many different varieties of storage triglycerides occur naturally in different species, changing the storage lipid profile should be a permissible event.
(3) Single species of fatty acid can dominate the profile of the triglyceride of seeds, hence it should be possible to drastically reduce fractionation costs for oils derived from plants by genetically engineering them to produce one dominant product.

Storage lipids represent a large concentrated reserve of lipids which can be readily harvested and represent a good feedstock for both the chemical and food industries. Since the seeds can be harvested in a dry mature state they have advantages in storage and transportation. It is to this end and because of demands on substainable agriculture, renewable raw resources and environmentally friendly chemical synthesis that considerable interest has been expressed in the potential for genetic manipulation of oils. Several industrial concerns such as Monsanto, Du Pont, Unilever and Calgene have interests in this area. In addition research programmes within agrochemical concerns are being pursued because enzymes of lipid biosynthesis are targets for herbicide action (Parker *et al.*, 1990).

From a chemical point of view the end product of such genetic manipulation would be acceptable since recombinant protein and extraneous DNA is normally removed and destroyed during the post harvest processing of the extracted oils. There can be little doubt then that modification of oils has enormous commercial potential. It is the authors' view that attitudes prevalent among some commercial concerns will, however, have to alter to enable them to exploit this market niche. This is likely to be brought about by socioeconomic pressures, which will increase during the next decade or so. At present there is not an enormous potential of high premium costing for seeds from an agronomic point of view. The market is mainly driven by two simple aims: to produce more and to pay less for the production process and hence increase the profitability margin. The greatest opportunity for a commercial concern is to take a share of a world market which it does not have at the moment and so increase its profitability. Advantage will be with those industrial concerns which know which specific modification they need to make for the manufacturing process.

3.1.4 What information is required for the successful genetic manipulation of oil seeds?

Most of the requirements for genetic manipulation of plant lipids can be discerned from a study of the restrictions which nature imposes on such lipids in plants. Since the composition of membrane lipids shows little variation in plants and their occurrence is ubiquitous, any alteration to such a fundamentally conserved structure is likely to be lethal. This focuses attention on storage triglyceride synthesis which

(1) occurs in specific tissues of the seed;
(2) involves deposition of lipid occurs within a particular time frame;
(3) involves basic biosynthesis in specific subcellular compartments;
(4) involves specific enzymes in biosynthesis.

From these considerations it is apparent that the following basic information will be required for a successful genetic engineering programme:

(1) a detailed understanding of the biochemistry of lipid biosynthesis in plants, to allow informed decisions to be made about potential target genes and identification and purification of the enzymes to be carried out so that the genes encoding them can be cloned
(2) a knowledge of tissue specific and temporal specific regulatory elements of gene expression, so that any introduced gene is expressed in the correct place at the correct time
(3) an understanding of protein targeting mechanisms so that extraneously introduced enzymes or proteins are targeted to the correct compartment, be it plastid, microsome or oil body.

The growth of molecular biology has given us tools with which to isolate and manipulate genes almost at will yet, sadly, the emphasis on fundamental biochemical and protein chemistry skills required to identify key enzymes is diminishing. There can be little doubt as to the importance of enzymology and biochemistry and for those who have read Arthur Kornberg's *For the love of enzymes* (Kornberg, 1989) it will be clear why this discipline must be perpetuated. A major increase in plant protein biochemistry will be needed to exploit the opportunities offered by molecular biology. Most of the advances in this area require experimental material which is rich in lipid synthesising enzymes. For a long period of time leaf material has been used as a convenient experimental source. This is rather unfortunate and consequently progress has not been as rapid as it might have been if more work had concentrated on seeds.

In the remainder of this chapter we will assess the state of knowledge of lipid biosynthesis in plants, concentrating on that required to alter the composition of storage lipids by genetic engineering. We would like to point the reader to two recent reviews in this general area by Ohlorogge *et al.* (1991) and Somerville and Browse (1991), plus a recent review from our own laboratory which concentrates

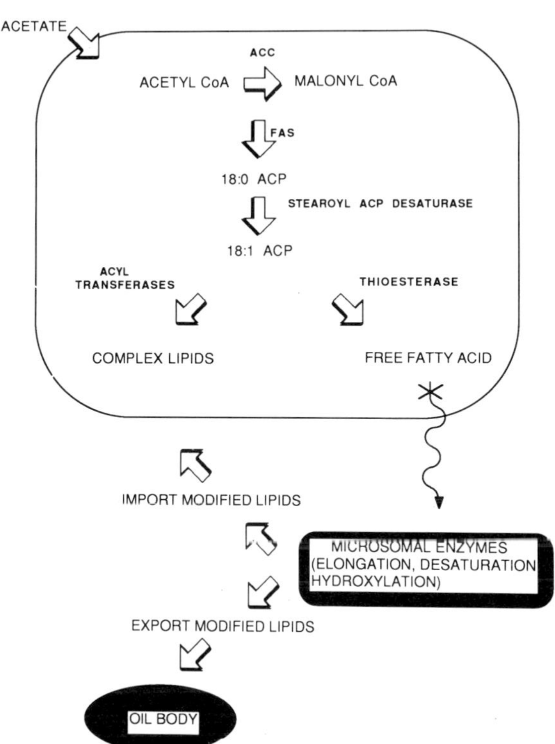

Figure 3.1 The biosynthesis of fatty acids in plants. Acetate enters the plastid and is converted to acetyl-CoA by a mechanism still under debate, acetyl-CoA is converted to malonyl-CoA by the action of acetyl-CoA carboxylase (ACC) and these two substrates are used by fatty acid synthetase (FAS) to make long chain fatty acyl-ACP derivatives, which may be subjected to desaturation by a soluble (Δ9)-desaturase. ACP-derivatives can then be cleaved to free fatty acid by the action of thioesterase or incorporated into plastid lipids by acyltransferases. Free fatty acids can be transported by an unknown mechanism to the endoplasmic reticulum for modification or incorporation into storage lipids. Fatty acids modified outside the plastid can be imported back into that organelle for incorporation into complex lipids.

on advances in the biochemistry and molecular biology of plant lipid metabolism over the last 10 years (Slabas and Fawcett, 1992).

3.2 Biosynthesis of lipids

3.2.1 What is the precursor of lipid biosynthesis and how it is imported into the plastid?

The biosynthesis of fatty acids was first investigated using metabolic labelling experiments involving the feeding of acetate. These clearly demonstrated that acetate can be incorporated into fatty acids. Using other labelling experiments it has been shown that in plant homogenates lipids are synthesized from acetyl-CoA and malonyl-CoA. Acetyl-CoA provides the carbon atoms for C1 and C2 of the final lipid produced. The remainder of the chain is provided by malonyl-CoA (for reviews see Stumpf, 1980, 1987). The key biochemical steps that are involved are:

(1) conversion of acetate to acetyl CoA by a mechanism still under debate
(2) conversion of acetyl-CoA to malonyl-CoA by acetyl-CoA carboxylase (ACC)
(3) a conversion of acetyl-CoA plus malonyl-CoA to fatty acids by the fatty acid synthetase complement of enzymes (FAS).

These steps are schematically shown in Figure 3.1. Despite the fact that acetate is incorporated into fatty acids by isolated chloroplasts this is not evidence that it is the direct precursor of lipid synthesis in plastids. Recent experiments from the laboratory of Thomas (Masterson *et al.*, 1990) strongly support the contention that L-acetyl carnitine is probably the precursor of acetyl-CoA in plastid lipid biosyn-

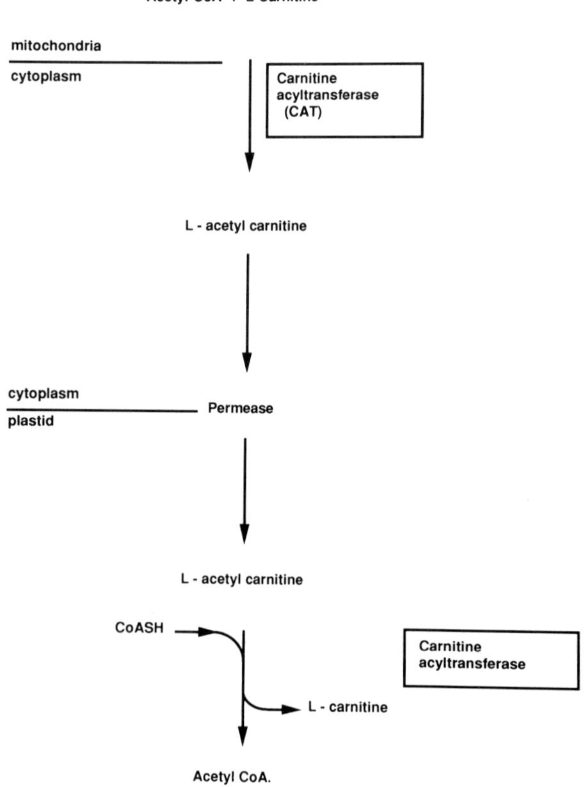

Figure 3.2 Import of acetate into plastids. Acetyl groups are probably transported to the plastid as acetyl carnitine and the conversion mediated by carnitine acyltransferase.

thesis. In labelling experiments it acts as a superior substrate to acetate both in terms of the rate and extent of incorporation into fatty acids. The import mechanism of the precursor thus probably involves the following steps: conversion of acetyl-CoA to acetyl carnitine in the mitochondria by the action of carnitine acyltransferase; transport of acetyl carnitine to the plastid; entry into the plastid by a permease and subsequent conversion back to acetyl-CoA by carnitine acyltransferase (Figure 3.2.)

Research needs to be performed in this area to examine in detail this hypothesis. Of equal importance is the definition of the nature of the fatty acid exported from the plastid.

3.2.2 What is the exported product of lipid biosynthesis from the plastid?

The species of lipid exported from the plastid is not clear. Fatty acids are potent detergents and hence could be expected to disrupt membranes. The exported species has not as yet been determined; however, the possibility does exist that like the import of acetate, export of fatty acids could be mediated by carnitine acyltransferase. This hypothesis is open to experimental verification by incubating *in vivo* radiolabelled plastids with L-carnitine and seeing if L-oleolycarnitine is produced as a reaction product. Current dogma maintains that following the synthesis of long chain acyl-ACPs (acyl carrier proteins) within the plastid, the product is hydrolysed by acyl-ACP thioesterase. The free fatty acid liberated is believed to be activated to an acyl-CoA by acyl-CoA synthase which is then exported to the cytoplasm. Definitive experiments in this area still remain to be performed.

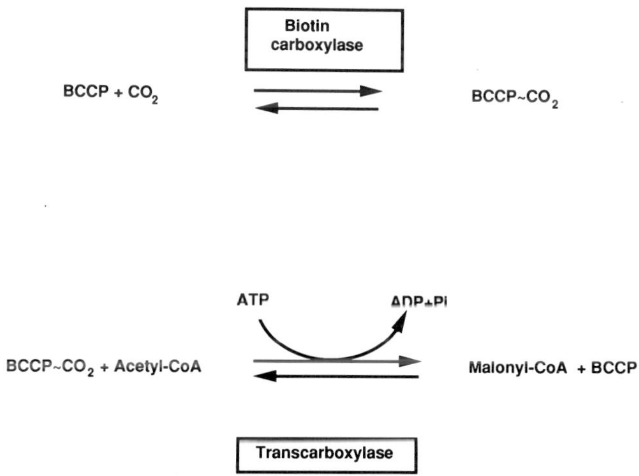

Figure 3.3 Mechanism of bacterial acetyl-CoA carboxylase. Biotin carboxyl carrier protein (BCCP) is carboxylated by the action of biotin carboxylase, the carboxylated intermediate then acts as a CO_2 donor in a reaction with acetyl-CoA mediated by transcarboxylase.

3.2.3 The role and regulation of acetyl-CoA carboxylase in lipid biosynthesis

Acetyl-CoA carboxylase (ACC) represents the first committed step in lipid biosynthesis in plants. It represents one of a class of biotin-containing enzymes of which there are several in both plants and animals (Wurtele and Nikolau, 1990; Chandler and Ballard, 1986). The enzyme in bacteria is composed of three separate polypeptide chains : biotin carboxyl carrier protein (BCCP), biotin carboxylase and transcarboxylase. The two component reactions involved are shown in Figure 3.3.

For many years it has been misconceived that in plants chloroplast ACC has a different subunit molecular weight to seed forms. The discrepancies in M_W are enormous (summarised by Hellyer *et al.*, 1986). Direct isolation of wheat germ and rape leaf ACC (Slabas *et al.*, 1986b), and immunoblotting (Slabas *et al.*, 1988) have demonstrated the presence of a high molecular weight form (basic subunit size of over 200 kDa) in both of these sources. Recently, a 50 kDa protein from embryogenic carrot cells has been purified. Analysis of a cDNA clone, obtained by screening with antibodies specific to this protein, has indicated that it may be a cytoplasmic form of ACC (Nikolau and Wurtele, 1991).

ACC plays a key role in the regulation of triglyceride biosynthesis in rape seed. Experiments by Turnham and Northcote (1983) have demonstrated that ACC activity is rapidly reduced once maximal lipid accumulation has been achieved in rape seed, suggesting that it is the loss of ACC activity during seed development which limits the extent of lipid synthesis.

While the cDNA for animal ACC has been cloned (Dai *et al.*, 1986), such studies are not very advanced in plants. The limited availability of the plant protein has also restricted studies on its regulation. In animal systems it is known that ACC is highly regulated, in response to hormonal balance, by phosphorylation. The sites of the covalent modification have been clearly identified (Ottey *et al.*, 1989). There is no evidence that phosphorylation of plant ACC occurs and no indication that such regulation would be required. Light/dark transitions dramatically alter pH, magnesium and adenylate concentration in the plastid, which could bring about a rapid regulation of the enzymes' activity and hence total lipid synthesis.

3.2.4 The structure and components of plant fatty acid synthetase

The component reactions of fatty acid synthetase (FAS) are shown in Figure 3.4. It is not intended here to present detailed information on each of the enzymes as this is available elsewhere (Harwood *et al.*, 1990). However, certain key points need to be made.

The biosynthesis of fatty acids occurs in plastids; in leaves these are the chloroplasts while in seeds they are proplastids. Plant fatty acid synthetase consists of a freely dissociable set of enzymes and contains more than six catalytic components plus ACP. Biochemical studies have been performed measuring the activity of FAS enzymes during lipid deposition; these enzyme activities are induced prior to the onset of lipid deposition and their level remains high even after

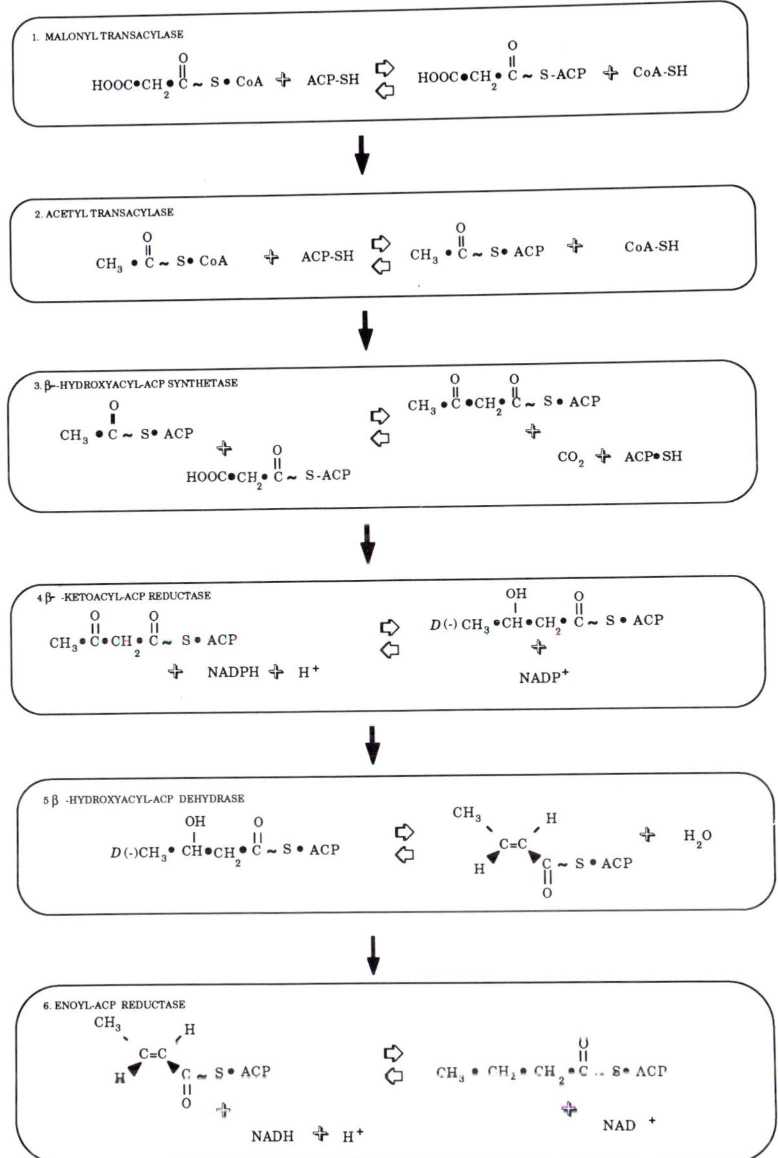

Figure 3.4 Component reactions of plant fatty acid synthetase (FAS).

the attainment of total lipid content (Slabas *et al.*, 1987b).

The most detailed study of a plant FAS component has been on ACP. This protein is of low M_W (approximately 10 kDa) and resembles coenzyme-A in structure, containing a 4-phosphopantetheine moiety which is covalently attached to a highly conserved serine residue (see Figure 3.5.).

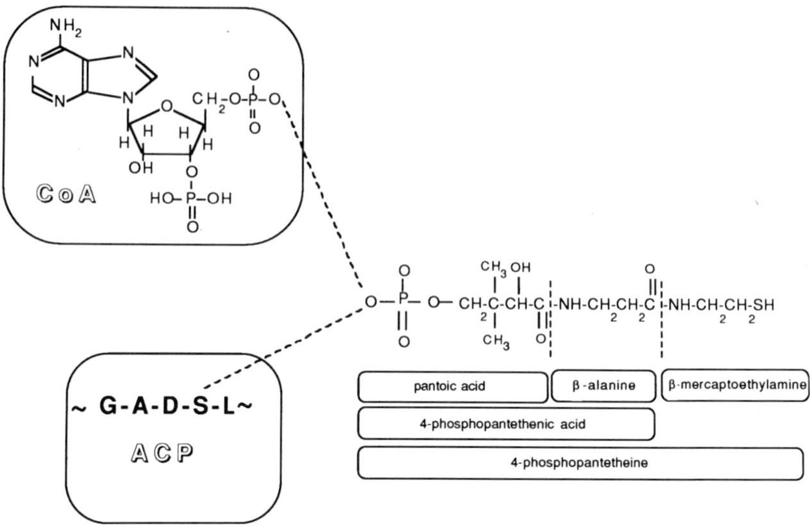

Figure 3.5 Comparison of coenzyme-A (CoA) and acyl carrier protein (ACP). Both molecules contain a 4′-phosphopantetheine moiety; in the case of ACP it is covalently attached to a conserved serine residue.

ACP has a central role in fatty acid biosynthesis and as acyl-ACP is the substrate for acyltransferases responsible for the synthesis of plastid lipids (Frentzen *et al.*, 1983) and the stearoyl-ACP desaturase in chloroplasts which catalyses the first lipid desaturation reaction (see Ohlrogge, 1987). While early studies on acyltransferases from chloroplasts were undertaken using coenzyme-A substrates, for convenience of substrate availability, subsequent experimentation involving competition between acyl-CoA and acyl-ACP showed dominant selectivity for acyl-ACPs for these enzymes (Frentzen *et al.*, 1983). In seeds ACP, like a number of other FAS components, requires detergent solubilisation, although it is not an integral membrane protein (Slabas *et al.*, 1987a).

It is now becoming clear that like bacterial fatty acid synthetase, the plant model is far more complex than originally perceived. Isoforms exist for a number of components, for example three different ACPs have been identified in barley. In addition certain other enzymes have limited substrate specificities. A selection of examples is given in the following sections.

3.2.5 The condensation reaction

Early experiments on the condensation reaction of fatty acid synthetase, catalysed by β-ketoacyl ACP synthase (KAS), have led to the conclusion that two condensing enzymes are present (Shimakata and Stumpf, 1982a). KAS I is involved in synthesis of fatty acids up to C16:0 and KAS II is involved in the conversion of

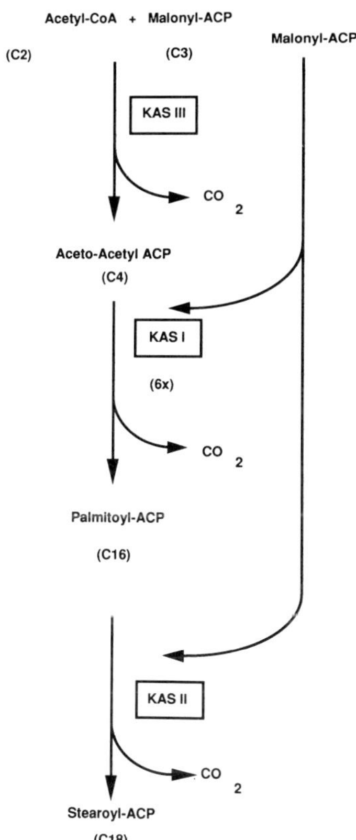

Figure 3.6 The role of β-ketoacyl synthase (KAS; condensing enzymes) isoforms in fatty acid synthesis. KAS III utilises acetyl-CoA and malonyl-ACP to form the four carbon acetoacetyl-ACP; KAS I is involved in the extension of the fatty acid to C16 and KAS II in the conversion of C16 to C18. KAS I and KAS II use ACP derivatives only.

C16:0 to C18:0. The former enzyme is substantially more sensitive to inhibition by the antibiotic cerulenin and has been separated from the latter by column chromatography. The role of these two enzymes has been confirmed by reconstitution experiments (Shimakata and Stumpf, 1982a). Both KAS I and KAS II are inhibited by cerulenin, which covalently modifies the enzyme (Kauppinen *et al.*, 1988). The discovery of a third condensing enzyme (KAS III) in bacteria, acetoacetyl-acyl carrier protein synthetase (Jackowski *et al.*, 1987), has stimulated plant biochemists to look for and subsequently find this isoform in plants (Jaworski *et al.*, 1989). Both KAS I and KAS II use acyl-ACPs as substrate. KAS III uses acetyl-CoA and malonyl-ACP as substrates and thus is mechanistically distinct from the other two condensing enzymes. Condensing enzymes represent a group of enzymes which are important components in the synthesis of a number of chemical compounds such as waxes, chalcone and resveratrol (Hahlbrock and

Grisbach, 1979; Schroder *et al.*, 1988) hence there is much interest in them experimentally from a structure/function perspective. The role of condensing enzymes in basic fatty acid synthesis is shown schematically in Figure 3.6.

The discovery of KAS III in plants and bacteria raises an important point. To date a number of studies have implicated acetyl-CoA:ACP transacylase (ACAT) as a rate limiting enzyme in fatty acid synthetase (Stumpf and Shimakata, 1983). However, with the discovery of KAS III, there does not seem to be an absolute requirement for ACAT in fatty acid synthesis. One possibility to explain the detection of ACAT activity in plants is that KAS III also possesses ACAT activity as a partial reaction in its catalytic mechanism. Similar observations have been made with acyl-ACP synthetase in *E.coli* of a partial reaction of an enzyme involved in phosphatidylethanolamine metabolism (Cooper *et al.*, 1989). This is shown schematically in Figure 3.7.

3.2.6 The end product of fatty acid synthetase—thioesterases and acyltransferases

The end product of fatty acid synthetase is (C18:0)-ACP. This fatty acid as its ACP is readily desaturated to form C18:1 (Δ9)-ACP by stearoyl-ACP desaturase. Stearoyl-ACP desaturase has been purified and cloned independently by two groups (Shanklin and Somerville, 1991; Thompson *et al.*, 1991). The product of this reaction (C18:1)-ACP has then two opportunities for further metabolism. It can be converted to a free fatty acid by the action of an acyl-ACP hydrolase (Hellyer and Slabas, 1990) and be subsequently exported from the plastid, possibly as a carnitine derivative. Alternatively it can be used as a substrate for complex lipid biosynthesis (monogalactosyldiacylglycerol, MGDG and digalactosyldiacylglycerol, DGDG) in the plastid, using glycerol phosphate acyl transferases present in the plastid. Figure 3.1 shows the metabolic fate of newly synthesised lipids.

Roughan has elegantly formulated a hypothesis to rationalise the diversity of

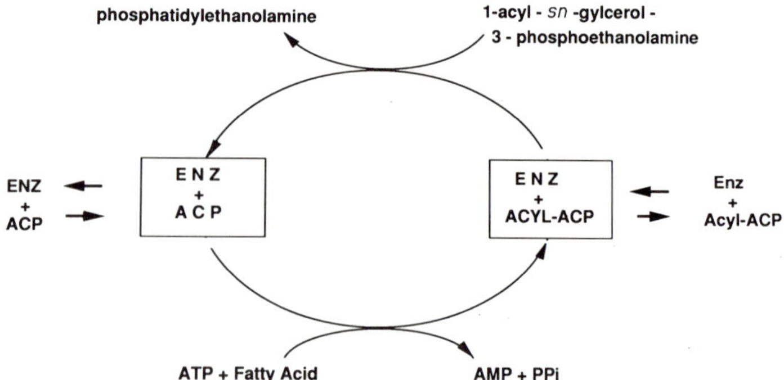

Figure 3.7 Acyl-ACP synthase as a partial reaction of 1-acyl-*sn*-glycero-3-phosphoethanolamine acyltransferase. The first step in the acyltransferase reaction is the ATP-catalysed ligation of a fatty acid to the enzyme-bound ACP subunit (acyl-ACP synthetase activity). The acyl-moiety is then transferred to 1-acyl-*sn*-glycero-3-phosphoethanolamine to produce phosphatidylethanolamine (1-acyl-*sn*-glycero-3-phosphoethanolamine acyltransferase activity). Redrawn and adapted from Cooper *et al.* (1989).

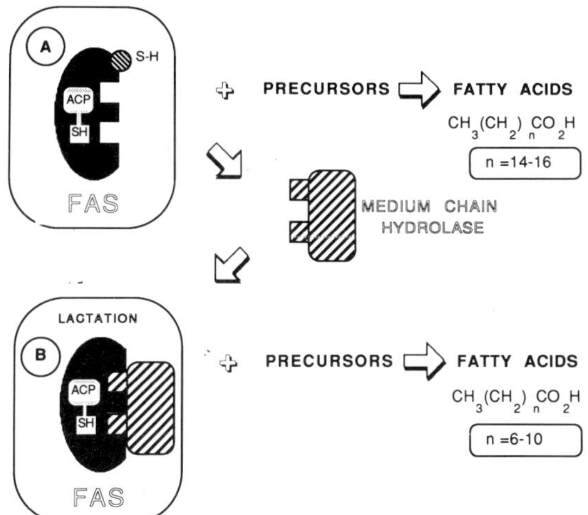

Figure 3.8 Action of medium-chain hydrolase during lactation in rat mammary gland. (A) Fatty acid synthesis is normally terminated by acyl-ACP thioesterase I, giving acyl chain lengths of C16. (B) During lactation a second thioesterase is induced which produces medium-chain (C6–C10) fatty acids.

glycerolipid types found in plants (Roughan, 1987). His two-pathway model for the synthesis of glyceroplipids has 'prokaryotic' and 'eukaryotic' elements. The synthesis of glycerolipid within the chloroplast was termed the prokaryotic pathway because the fatty acid composition resembles that of some cyanobacteria. The extra-chloroplast pathway, confined largely to the endoplasmic reticulum, was termed eukaryotic. These pathways can be distinguished as C16 fatty acids are strictly excluded from the *sn*-2 position in the eukaryotic pathway whereas C18 fatty acids are excluded from the *sn*-2 position in the prokaryotic pathway (Roughan, 1987). Central to this hypothesis is the balance of acyltransferase to thioesterase activity in the plastid. If in an organism the ratio of acyltransferase to thioesterase activity is high, the dominant glycerolipids will be of a prokaryotic type, produced within the plastid. If, on the other hand, the ratio is in favour of thioesterase, there will be high metabolic activity in the microsomal system and the lipids will correspondingly be of a eukaryotic type.

3.2.7 Acyl-ACP thioesterases—how many types are there and could they have more than one role?

The role of acylthioesterases in determining the product formed in fatty acid synthesis is well documented in animals (Smith and Libertini, 1979; Poulose *et al.*, 1985). Animal fatty acid synthetase has all six catalytic domains plus the ACP on one polypeptide chain. In addition it has a terminal acyl-ACP thioesterase I domain which cleaves off the covalently bound fatty acid once it has reached C16:0 in chain length. The product is a long-chain free fatty acid. During lactation in rats a second protein, medium-chain hydrolase (acyl-ACP thioesterase II), interacts with fatty acid synthetase so causing premature chain termination (Safford *et al*., 1987). This is shown schematically in Figure 3.8.

In plants acyl-ACP thioesterase has been purified to homogeneity from rape seed. This enzyme has a subunit M_W of 38 kDa and has a high specificity for oleoyl-ACP (Hellyer and Slabas, 1990). Such substrate specificity is consistent with a thioesterase which would fit into the termination step after the soluble (Δ9)-desaturase has acted.

Are there other acyl-ACP thioesterases which could be present in plants involved in the biosynthesis of medium-chain fatty acids in an analogous fashion to the animal enzyme? Recently such a short-chain specific hydrolase has been isolated and cloned from Californian bay laurel. This enzyme, when expressed in a *fad D* mutant of *E.coli*, causes premature chain termination of fatty acid synthesis and a concomitant increase in the level of medium-chain fatty acids (T. Volker, personnal communication). While the enzyme is involved in the synthesis of medium-chain fatty acids in Californian bay laurel, this does not mean that it is necessarily the sole mechanism of such synthesis in plants. In goats, the mechanism of medium-chain fatty acid biosynthesis is known not to involve an acylthioesterase II, but appears to be an endogenous property of the synthetase when incubated with microsomes (Knudsen and Grunnet, 1982).

3.2.8 The structure of triglycerides

Triglycerides (TAG) are the major storage lipids in plants. The basic structure of TAG (Figure 3.9) consists of a glycerol backbone to which are esterified three fatty acids. When glycerol is represented by a Fischer projection the three carbon atoms are numbered 1 to 3 from top to bottom and the prefix *sn*-(stereospecific numbering) indicates that this system is being used. Such a nomenclature defines the position of a particular fatty acid in the lipid molecule.

A combination of acyl quality (especially of unsaturation) and stereospecific distribution determines the qualitative and physical properties of the TAG. With the development of techniques for the stereospecific analysis of TAG, however, it is apparent that far from being randomly distributed, the fatty acids are often highly enriched at given positions on the glycerol backbone. Triacylglycerols extracted

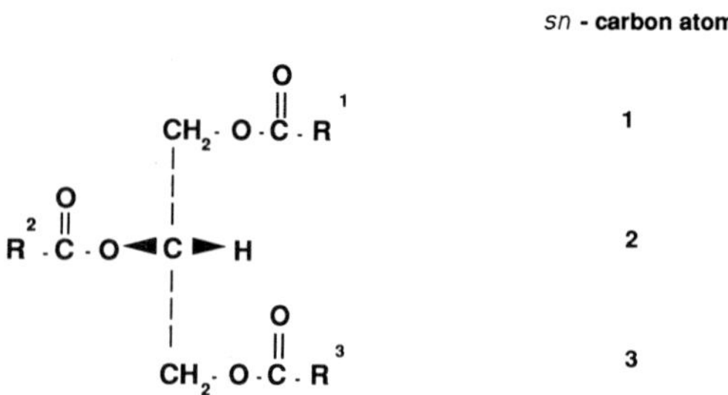

Figure 3.9 Fischer projection of a triacylglycerol molecule. The R-groups represent the acyl substituents.

from tissues, therefore, are generally abundant in particular configurations. This is most probably a reflection of the specificity/selectivity characteristics of the acyltransferases involved in their synthesis. It is these properties which are difficult to mimic (in a cost-effective manner) in chemical synthetic formulations.

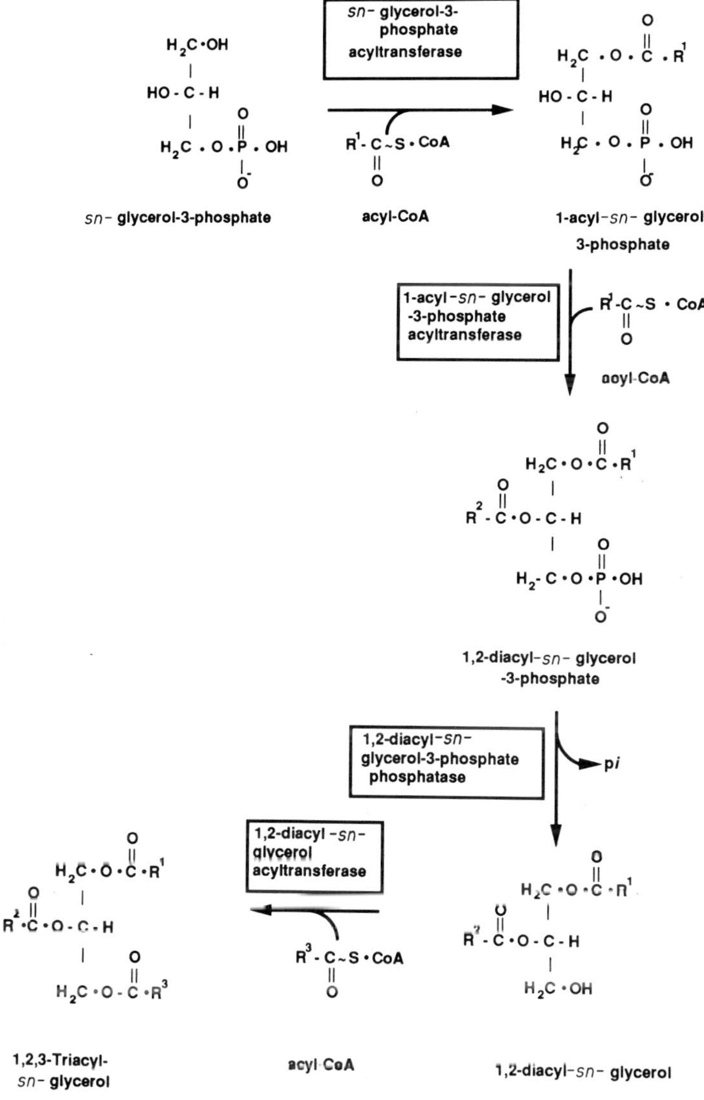

Figure 3.10 Synthesis of triacylglycerol. The pathway involves the two-step acylation of glycerol to form 1, 2-diacylglycerol-3-phosphate, followed by a specific dephosphorylation and a third acyl transfer reaction.

3.2.9 Assembly of fats and oils

A number of recent review articles have been written on the biosynthesis of triacylglycerols in plants. For a comprehensive treatise, the reader is referred to the work of Gurr (1980) which summarises the earlier literature and to Stymne and Stobart (1987) and Griffiths and Harwood (1991) for more recent coverage. Studies in the early sixties with radiolabelled glycerol (Barron and Stumpf, 1962) showed that the pathway for fat synthesis in plants was essentially similar to that found in animals by Kennedy (1961). These studies were further extended by the work of Gurr, Slack and their co-workers in the 1970s and early 1980s. The pathway involves the two step acylation of glycerol 3-phosphate to form 1,2-diacylglycerol-3-phosphate (phosphatidic acid) via 1-acylglycerol-3-phosphate (lysophosphatidic acid; LPA). This is then followed by a dephosphorylation to yield diacylglycerol which is then acylated at the *sn*-3 position to form triacylglycerol (see Figure 3.10).

In seeds which accumulate a polyunsaturated type of oil additional reactions are present which involve phosphatidylcholine. This phospholipid is the substrate for the (Δ12)-desaturase converting oleate to linoleate (Stymne and Appelqvist, 1978; Slack *et al.* 1979) and can utilise substrate at both the *sn*-1 and *sn*-2 positions. The linoleate at position *sn*-2 can re-enter the acyl-CoA pool and be exchanged with the other eighteen carbon fatty acids, principally oleate. Such a freely reversible exchange mechanism can generate polyunsaturated fatty acids available for acylation of the glycerol backbone in the Kennedy pathway (see Figure 3.11.). In

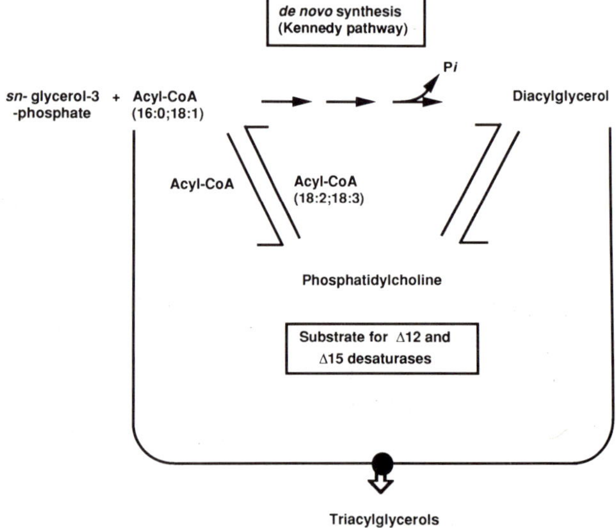

Figure 3.11 Desaturation reactions involving phosphatidylcholine. Phosphatidylcholine is the substrate for (Δ12)-desaturation converting oleate to linoleate. The linoleate at position *sn*-2 can re-enter the acyl-CoA pool to provide polyunsaturated fatty acids for triglyceride synthesis. In addition cholinephosphotransferase catalyses the reversible reaction between diacylglycerol and phosphatidylcholine, providing another route for the incorporation of polyunsaturates into triglycerides.

addition, there is also a reversible reaction between diacylglycerol and phosphatidylcholine (considered to be catalysed by cholinephosphotransferase) which would also allow oleate groups to enter phosphatidylcholine for desaturation (Slack *et al.*, 1985; Stobart and Stymne, 1985). Both the acyl exchange and the equilibration of the diacylglycerol and phosphatidylcholine pools appear to act in concert as an enrichment cycle for generating polyunsaturated fatty acids. The acyltransferases involved in the acylation of glycerol 3-phosphate generally show a high degree of selectivity and usually direct saturated fatty acids to the *sn*-1 position and unsaturates to the *sn*-2 position (Griffiths *et al.*, 1985; Ichihara *et al.*, 1987; Frentzen, 1986). The acylation of position *sn*-3 (catalysed by diacylglycerol acyltransferase) was initially considered to be of low selectivity but there is growing evidence that it may exert a strong influence on the nature of the acyl quality of this position (Bafor *et al.*, 1990a; Griffiths and Harwood, 1991).

3.2.10 Synthesis of α- and γ- linolenic acids

So far we have covered the assembly of linoleate-rich oils in seed tissues; however, many crops (rape, linseed and soy for example) contain quantities of α-linolenate (18:3Δ9,12,15) which is undesirable for oxidative reasons. In leaves the (Δ15) desaturase which synthesises α-linolenate from linoleate largely utilises acyl substrate esterified in MGDG (Jaworski, 1987). Unfortunately, the enzyme appears to be highly labile and to date few *in vitro* studies have been achieved with it. In oil seeds the current evidence suggests that phosphatidylcholine plays a major role in Δ15 desaturation (Stymne and Appelqvist, 1980) which is supported by recent studies using linseed mutants (Stymne *et al.* 1990). In some plants (e.g. evening primrose, borage and blackcurrent) the oil synthesised is rich in an alternative type of 18:3 fatty acid, referred to as γ-linolenic acid (18:3; Δ6,9,12) which has some interesting medicinal properties (Horrobin and Manku, 1983).

In animals the (Δ6)-desaturase (as is general for animal desaturases) generates a further double bond (at the Δ6 position) between an already existing double bond (at the Δ9 position as in linoleate) and the carboxyl (proximal) end of the fatty acid (Holloway, 1983). The substrate for this is linoleoyl-CoA. Desaturases in plants, however, tend to operate differently and insert double bonds between an established double bond and the methyl (distal) end of the fatty acid. The substrates for these are complex lipids and in oil seeds, particularly the phosphatidylcholine of the endoplasmic reticulum (ER). Evidence suggests that γ-linolenic acid in plants (developing borage seeds) is synthesised sequentially from oleate (C18:1; Δ9) via linoleate (C18:2; Δ9,12) using microsomal phosphatidylcholine as substrate (Stymne and Stobart, 1986). The (Δ6)-desaturase largely utilises linoleate substrate at position *sn*-2 of phosphatidylcholine whereas the (Δ12)-desaturase enzyme is active with oleate at both positions *sn*-1 and *sn*-2 (Griffiths *et al.*, 1988a). Much more work is required to understand how (Δ6)-desaturase activity is regulated in plants and how γ-linoleic acid formation is related to the assembly of the storage triacylglycerols.

3.2.11 Synthesis of a highly unsaturated fatty acid, octadecatetraenoic acid (C18:4)

With respect to (Δ6)-desaturase activity, it is interesting that some members of the *Boraginaceae* possess a novel C18-polyunsaturated fatty acid, octadecatetraenoic acid (C18:4; Δ6,9,12,15) in their leaf lipids (Jameison and Reid, 1969). The seed oils of *Cynoglossum* and *Anchusa* also contain α- and γ- linolenates and in addition have the C18-tetraenoic acid (Griffiths *et al.*, 1989) although the cotyledons lack chloroplasts. It would appear, therefore, that tissue which has (Δ15)-and (Δ6)-desaturase activity will accumulate to some extent the octadecatetraenoic acid. The highly polyunsaturated fatty acid, however, appears to be synthesised in a most precise fashion through the action of the (Δ6)-desaturase on α-linolenate and cannot be formed through (Δ15)-desaturase activity via γ-linolenate (Griffiths *et al.*, 1989). These observations on the synthesis of octadecatetraenoic acid raise questions regarding acyl substrate recognition for desaturases. Studies on the (Δ12)-desaturase indicate that substrate specificity in complex lipids may be conferred by the distance of the established double bond from the methyl end of the fatty acid chain (Howling *et al.*, 1972). The (Δ6)-desaturase, however, appears to operate from a double bond located toward the carboxyl end. From the limited information available, the (Δ15)-desaturase may align against all the established double bonds in the molecule irrespective of their position. Hence, in the case of the γ-linolenate substrate, the primary cue for the (Δ15)-desaturase may be the double bond at the Δ6 position and therefore no octadecanoic acid will arise, i.e. the (Δ15)-desaturase will be acting at a position which already possesses a double bond (the Δ12 position).

These observations would have implications if it becomes possible to transform plants with the (Δ6)-desaturase gene. If the recipient organism already possesses (Δ15)- desaturase activity (e.g. oil-seed rape) then the resulting fatty acid product(s) might accumulate significant quantities of the tetraenoic acid.

3.2.12 How are triacylglycerols with non-membrane fatty acids assembled?

Large numbers of plant species accumulate triacylglycerols with high amounts of uncommon (or non-membranous) fatty acid constituents (see Badami and Patil, 1982). In order to maintain membrane integrity and function, mechanisms must exist to ensure that these potentially deleterious components are excluded from the phospholipids. Our understanding of how this is acheived is embryonic but detailed information in this area may be of importance if we are successfully to transform plants to produce such components without altering the membrane properties. A number of important points concerning the biosynthesis of specific lipids have emerged by investigating plant species which have a distinct fatty acid composition in their storage lipid.

Medium-chain length fatty acid. In *Cuphea lanceolata*, for example, which accumulates mainly caproate (C10:0), the 1-acylglycerol-3-phosphate acyltransferase has acyl specificities that divert the assembly of acyl groups to either di-medium-

or di-long-chain diacylglycerols (Bafor *et al.*, 1990a). The diacylglycerol acyltransferase selectivity utilises the medium-chain diacylglycerols for TAG synthesis, thereby enriching the diacylglycerol pool with long-chain fatty acid for membrane lipid synthesis.

Ricinoleate biosynthesis. Recent work with microsomes from developing castor bean endosperm has demonstrated that the *sn*-2 position of phosphatidylcholine is the site of the Δ12 hydroxylation of oleate to ricinoleate (Bafor *et al.*, 1991). The product is, however, rapidly released from the membrane, in preference to the common acyl constituents, through the action of a phospholipase A_2-type activity. A similar mechanism was also found to operate in other oil seed tissues, namely rape and safflower supplied with ricinoleoylphosphatidylcholine. It is interesting that such a selective hydrolysis of oxidative fatty acid products by phospholipase A has also been recently implicated in membrane repair mechanisms in animal tissues (Kuijk *et al.*, 1987).

Stearate-rich fats (cocoa). In cocoa the bulk of stearate production appears after the synthesis of the membrane lipids. Again, the diacylglycerol acyltransferase appears to show a relatively high degree of selectivity for this fatty acid generating triacylglycerol and thereby precluding its entry into phospholipids (Griffiths and Harwood, 1991). The mechanism by which high levels of stearate are released from the plastids has received little attention, although in a recent study it was concluded that in contrast to some other tissue types (see Pollard *et al.*, 1991; Sambanthamurthi and Oo, 1990) there was no induction of an acyl-ACP thioesterase during development with an increased selectivity towards stearoyl-ACP (Griffiths and Harwood, unpublished results).

Erucate-rich oils. Erucate (22:1), like stearate above, is synthesised after the bulk of the membrane lipid synthesis has ceased. This fatty acid is located at the primary ends of the triacylglycerol (*sn*-1 and *sn*-3 positions). Rape microsomes are incapable of directly acylating glycerol 3-phosphate with erucoyl-CoA yet can readily construct oil with the common fatty acids in such preparations (Griffiths *et al.*, 1988b). Similarly, *de novo* synthesis of erucate from oleate and malonyl-CoA in rape microsome homogenates was unsuccessful in generating erucoyl-triacylglycerol (Taylor *et al.*, 1990) which is contrary to the work of Fehling *et al.* (1990) in *Lunaria*. Thus, the mechanism by which erucate is incorporated into the *sn*-1 position of triacylglycerol is still unclear. By comparison, species of *Limnanthes* (in which erucate is found at all three positions of the glycerol backbone) readily utilise erucoyl-CoA in the acylation of position *sn*-2 in microsomes supplied with 1-acylglycerol 3-phosphate acceptor (Cao *et al.*, 1990, Lohden *et al.*, 1990).

Petroselinic acid. Petroselinic acid (octadeca-6-enoic acid), a major constituent of many seed oils from the *Umbelliferae*, is of great interest for its industrial application (Kleiman and Spencer, 1982). Little is known about the biosynthesis of this fatty acid, but preliminary results indicate that it is linked to *de novo* fatty acid synthesis, perhaps via an ACP associated reaction (Cahoon and Ohlrogge, 1991). Recently it was shown that the LPA acyltransferase from rape and safflower seeds could not, unlike corresponding carrot seed enzyme, acylate petroselinic acid to *sn*-2 if the same acid were present at the *sn*-1 position of the *sn*-LPA substrate (Dutta *et al.*, 1991). Thus, if genes governing the synthesis of petroselinic acid were introduced into rape, petroselinic would probably be incorporated into positions *sn*-1 and *sn*-3 of the triacylglycerols.

3.2.13 How are oils packaged in plant cells?

Oil reserves in plant cells appear as electron dense spherical bodies with an average diameter of 1–2 μm. There has been much controversy over the origin of these oil bodies. Principally there are two schools of thought: (i) the oil is deposited in the interior of the bilayer of the endomembranes which then buds off leaving the oil surrounded by a half unit membrane (Qu *et al.*, 1986); (ii) the oil is initially synthesised without any direct membrane involvement but is stabilised or sequestered by proteins (and some lipid) at the oil–cytoplasm interface (Smith, 1974; Stobart *et al.*, 1986). Current evidence suggests that these oil body proteins are quite distinct from proteins of the endoplasmic reticulum profiles and are highly hydrophobic in nature. Electrophoretic analyses of these proteins from a taxonomically diverse range of oil seeds show that each consists of a small number of major components, with more numerous minor components (Qu *et al.*, 1986; Murphy *et al.*, 1989; Tzen *et al.*, 1990). These vary in M_W from about 15 000 to 90 000 with few or none of the major components in common between different species (Qu *et al.*, 1986).

Extensive amino acid sequences of the maize oil body proteins have been determined from a partial cDNA for the M_r 16 000 component and from a cDNA and gene for the M_r 18 000 component (Vance and Huang, 1987; Qu and Huang, 1990). The two proteins have a degree of homology especially in the central region and are predicted to form four segments of an α-helix. The first of these is hydrophilic and is possibly located in the cytosol while the fourth is amphipathic and may sit at the surface of the lipid body. These are separated by a pair of hydrophobic antiparallel helices separated by a hairpin bend which are proposed to span the membrane and project into the lipid. This latter structure is proposed to act as an internal signal for intracellular trafficking as well as an anchor for attachment to the oil body (Qu and Huang, 1990). For a recent comprehensive review on this topic see Murphy (1990).

3.3 Molecular aspects of plant lipid biosynthesis

3.3.1 Current knowledge

There have been great advances in the molecular biology of plant lipid biosynthesis. Details of these are outside the scope of this chapter and interested readers are referred to Slabas and Fawcett (1992). Here the properties of proteins which have been purified to homogeneity are summarised, those which have been subjected to amino acid sequence analysis are indicated and those for which cDNA or genes have been cloned and sequenced are listed (Table 3.3). The main findings of these detailed molecular studies are as follows:

(1) The majority of these enzymes are nuclear coded.
(2) A leader sequence is present which serves to import the protein into the appropriate subcellular organelle. Studies have been performed fusing these leader sequences to foreign proteins and this has resulted in correct subcellular targeting.
(3) The ACP gene from rape seed has been extensively studied and an appropriate 5′ gene control region has been identified which allows both temporal specific and tissue specific expression using β-glucuronidase (GUS) reporter assays in transgenic plants.
(4) Recent experiments in our own laboratory have identified specific oligonucleotide sequences in the 5′ region of the ACP gene which bind proteins in gel retardation assays. These studies are being extended to identify the *trans*-acting factors responsible for regulating transcription of the gene, with a view to cloning them.

In short, much is now known about the components of plant fatty acid synthetase and with a concerted effort a full description should be available in the future. The desired information to genetically manipulate oil composition is beginning to become available. This is not only important from a commercial aspect but also from a scientific one. With the ability to overproduce these proteins in appropriate biological hosts, it will be possible to look directly at the subtle protein–protein interactions which occur with these components and which may be important for the mechanism and regulation of lipid biosynthesis.

3.4 Potential manipulations of plant lipid biosynthesis

At the present time there is an excess production of plant oils in Europe. It therefore follows that there is agricultural capacity available for the production of crops containing novel lipids or lipids which are present in a form more readily amenable to chemical processing. As a feedstock for the chemical industry, activated fatty acids, i.e. those containing chemically derivitisable groups such as OH, CO_2H, or epoxy, are highly desirable. The production of more useful oils could reduce the current excess of vegetable oil production in Europe.

3.4.1 Conversion of plant oils

Conversion of oil made by plants into a product for which there is a greater demand, by using biotransformation or the action of industrial enzymes, is a means of increasing the volume of vegetable oils for commercial use. The manipulation of

Table 3.3 Details of components involved in plant fatty acid synthesis. Only proteins which have been purified to homogeneity, been directly sequenced at the amino acid level or for which the corresponding cDNA or gene has been cloned are documented. The numbers in parentheses refer to the references given at the foot of the table.

Protein	Source	Purified to homogeneity	Amino acid sequence	Subunit M_w (Da)	Native form	cDNA cloned	Gene cloned
ACP	Spinach leaf ACP I	+(1)	+(2,3)	10 300(1)		+(4)	
	Avocado mesocarp	+(1)		11 400(1)			
	Barley leaf						
	ACP I	+(5)	+(5)	8 800(5)		+(6)	+(7)
	ACP II	+(8)	+(8)			+(9)	
	ACP III					+(6)	+(7)
	Arabidopsis leaf						+(10,11)
	Brassica campestris seed					+(12)	
	Brassica napus seed	+(13)	+(13)	9 200(13)		+(14)	+(15)
ACC	*Brassica napus* seed	+(16)		220 000(16)			
	Soybean seed	+(17)		240 000(17)			
	Parsley cell suspension	+(18)		220 000(18)	α_2(18)		
	Carrot	+(19)		50 000(19)		+(19)	+(19)
	Brassica napus leaf	+(20)		220 000(20)			
Malonyl CoA: ACP transacylase	Avocado mesocarp	+(21)		42 500(21)			
	Barley	+(5)		34 500(5)	α_1(5)		

Protein	Source	Purified to homogeneity	Amino acid sequence	Subunit M_w(Da)	Native form	cDNA cloned	Gene cloned
β-Ketoacyl-ACP synthetase	*Brassica napus* seed KAS I	+[22]		43 000[22]	α_2[22]		
	Barley leaf KAS I	+[23]	α-subunit[23] β-subunit[23]	45 000[23] 46 000[23	90 000 as αα,αβ,ββ[23]	+[23]	
	Castor seed	+[24]		46 000[24] 50 000[24]		+[24] +[24]	
β-Hydroxyacyl-ACP dehydrase	Spinach leaf	+[25]		19 000[25]	α_4[25]		
Enoyl-ACP reductase (NADH)	*Brassica napus* seed	+[26,27]	+[28]	35 000[27]	α_4[27]	+[29]	
	Arabidopsis leaf					+[30]	
	Spinach leaf	+[25]		32 500[25]	α_4[25]		
β-Ketoacyl-ACP reductase (NADPH)	Spinach leaf	+(25)		24 ,200(25)	a4(25)		
	Avocado mesocarp	+[31]	+[31]	28 000[31]	α_4[32]		
	Brassica napus seed	+[32]	+[32]	30 000[32]	α_4[32]		
	Arabidopsis leaf					+[33]	
Soluble (Δ9) desaturase	Avocado mesocarp	+[34]	+[34]	35 000[34]	α_2[34]		
	Castor seed					+[34]	
	Cucumber					+[35]	
	Safflower seed	+[36]	+[36]	40 000[36]		+[36]	

Table 3.3 (Continued)

Protein	Source	Purified to homogeneity	Amino acid sequence	Subunit M_w(Da)	Native form	cDNA cloned	Gene cloned
Acyl-ACP thioestarese	*Brassica napus* seed	+[37]	+[38]	38 000[37]			
(Medium-chain specific)	Californian bay laurel				+[39]		
Glycerol-3-P acyltransferase	Pea leaf	+[40,41]	+[41]	40 500[40] 40 000[41]		+[41]	
	Squash leaf	AT 2+[42] AT 3+[42]	+[43] +[43]	39 000[42] 39 000[42]	α_1[42] α_1[42]	+[43]	
	Arabidopsis leaf					+[44]	+[44]
LTPs	Spinach leaf	+[45]	+[46]	9 000[45]	α_1[45]	+[47]	
	Maize seedlings	+[48]		9 000[48]	α_2[48]	+[49]	
	Castor	+[50]	+[51]	9 000[50,51]	α_1[50]		
	Finger millet	+[52]	+[52,53]	9 300[52]			
	Barley	+[54]	+[53,54]	9 300[54]	+[55]		

[1] Simoni *et al.* (1967); [2]Kuo and Ohlrogge (1984); [3]Matsumara and Stumpf (1968); [4]Scherer and Knauf (1987); [5]Hoj and Svedson (1983); [6]Hansen (1987); [7]Hansen and von Wettstein-Knowles (1991); [8]Hoj and Svedson (1984); [9]Hansen and Kauppinen (1991); [10]Lampa and Jacks (1991); [11]Post-Beittenmiller *et al.* (1989); [12]Rose *et al.* (1987); [13]Slabas *et al.* (1987a); [14]Safford *et al.* (1988); [15]de Silva *et al.* (1990); [16]Slabas and Hellyer (1985); [17]Charles and Cherry (1986); [18]Egin-Buhler and Ebel (1983); [19]Nikolau and Wurtele (1991); [20]Slabas *et al.* (1986b); [21]Hilt (1984); [22]Mackintosh *et al.* (1989); [23]Siggaard-Anderson *et al.* (1991); [24]Genez *et al.* (1991); [25]Shimakata and Stumpf (1982b); [26]Slabas *et al.* (1986a); [27]Slabas *et al.* (1990); [28]Slabas *et al.* (1991); [29]Kater *et al.* (1991); [30]Fawcett and Slabas, unpublished; [31]Sheldon *et al.* (1990); [32]Sheldon *et al.* (1991); [33]Slabas *et al.* (1992); [34]Shanklin and Somerville (1991); [35]Shanklin *et al.* (1991); [36]Thompson *et al.* (1991); [37]Hellyer and Slabas (1990); [38]Hellyer and Slabas (1987); [39]Volker, personnal communication; [40]Douady and Dubacq (1987); [41]Weber *et al.* (1991); [42]Nishida *et al.* (1987); [43]Ishitaki *et al.* (1988); [44]Nishida *et al.* (1990); [45]Kader *et al.* (1984); [46]Bouillon *et al.* (1987); [47]Bernhard *et al.* (1991); [48]Douady *et al.* (1985); [49]Tchang *et al.* (1988); [50]Watanabe and Yamada (1986); [51]Takashima *et al.* (1986); [52]Campos and Richardson (1984); [53]Bernhard and Somerville (1989); [54]Svennson *et al.* (1986); [55]Mundy and Rogers (1986).

oil quality, post-harvesting, can be achieved by the use of immobilised microbial enzymes on inert supports or by use of entire organisms. Conversions which rely on one enzyme, such as lipases, can be carried out in a continuous mode utilising immobilised enzyme technology, while those reactions which require cofactors or involve more than one step are carried out in suspension culture. The microbial enzymes required for these biotransformations can be induced under appropriate growth conditions. Extracellular microbial lipases can remain catalytically active in a medium composed mainly of organic solvents. Under these conditions the enzymes can effectively catalyse a wide range of esterification and transesterification reactions providing the normal hydrolytic action of the enzymes is restricted due to the limited availability of water. These properties are being used to produce valuable confectionery fats. For example, when palm-oil fraction and stearic acid were treated under appropriate conditions with the regiospecific *Mucor miethei* lipase, specific interesterification resulted in a fat with the same physical properties as cocoa butter, which is of much greater cost commercially (MacRae, 1989). It is possible to select for lipases with different specificities which will have use in catalysing defined chemical reactions. Such microbial enzymes can be over-expressed with relative ease using established microbial gene technology.

The use of microbes in converting plant oils is becoming more important. Cocoa butter substitutes can be readily synthesised by yeast cultures inhibited with sterculic acid, an inhibitor of ($\Delta9$)-desaturase (Moreton, 1988), and in mutants of *Candida curvata* (Ykema *et al.*, 1989) γ-linolenic acid can also be produced, on an industrial scale, by fungi in fermentors. Several mutants exist in β-oxidation which will convert inexpensive fatty acids to dicarboxylates (Casey, 1990). Optimisation of these mutants of *Candida cloacae* by genetic engineering could give rise to large-scale substrates for the chemical industry. The reason there are so few processes in existence is mainly economic, i.e. there is a lack of suitably priced products that can be made by this technology and only speciality products with prices in excess of $5000/ton may stand a chance of commercial success.

3.4.2 Genetic manipulation of plant lipids

Alteration of the type of storage oil made by plants would appear to be permissible, as there are a wide range of naturally occurring storage lipids in plants. These alterations can be made by either traditional plant breeding programmes or the use of genetic engineering techniques. These are not mutually exclusive as any genetic engineering programme will require a certain amount of traditional breeding to bring transgenic plants into the market place. The genetic engineering approach, although attractive, demands an intimate knowledge of the biochemistry and genetic organisation of the target plant. Potential problems arise due to tissue and compartment-specific expression of genes within plants and the apparent plastic metabolism which they possess. However, two examples of altered plant lipid profiles by genetic engineering have very recently been reported showing the potential in this area:

(1) Two groups have independently purified the soluble (Δ9) desaturase and cloned the corresponding cDNAs (Shanklin and Somerville, 1991; Thompson *et al.*, 1991). The group at Calgene used the cDNA to modulate the stearoyl-ACP desaturase level, using antisense technology, in transgenic rape seed. This resulted in a marked decrease in the level of oleic acid with an increase in the level of stearic acid (Kridle *et al.*, 1991).

(2) The second example involved the transformation of different tobacco plants with constructs containing cDNAs encoding two soluble glycerol 3-phosphate acyltransferase enzymes with different substrate selectivities. The resulting phophatidylglycerol molecules of the transformants had an altered acyl composition at the *sn*-1 position (Murata, 1991). The type of acyl moiety at this position has a significant influence on the chilling tolerance of the plant (Murata, 1983) and so it may now be possible to increase the frost tolerance of plants by the simple addition of one gene. However, it is likely that more than one gene is involved in chilling tolerance.

In addition, a medium-chain specific acyl-ACP thioesterase from California bay laurel has recently been isolated. Introduction of this gene into the *fadD* mutant of *E. coli* has resulted in the production of high quantities of medium-chain fatty acids (T. Volker, personnal communication). It will be interesting to see in the future if the gene encoding this enzyme will be capable of altering the lipid profile of plants.

3.4.3 Potential goals

Some current desirable goals for the fats and oils industry, which may be aided by genetic engineering of plant crops, include:

(1) *An increase in oil yield over other storage products.* It is difficult to envisage the immediate use of molecular biology to this end without further understanding the mechanisms of substrate partitioning between oil, protein and carbohydrate metabolism.

(2) *Production of triacylglycerols containing 100% of one acyl constituent.* This is a possibility if acyltransferases with desired selectivity properties can be cloned and expressed in transgenic plants, e.g. the lysophosphatidate acyltransferase from *Limnanthes*, to put erucate at the *sn*-2 position of triacylglycerol of rape.

(3) *Medium-chain fatty acid synthesis in already established agricultural crops, e.g. rape.* This could be achieved by introducing the Californian bay laurel medium-chain thioesterase into rape. It will, however, require an assessment of whether the glycerol acylating enzymes can utilise the short acyl chain products at a sufficiently rapid rate, as they might simply be extended to long chain fatty acids.

(4) *Production of cocoa butter-type fats in temperate crops*. It has already been shown that given the right proportions of palmitate, stearate and oleate both safflower and rape microsomes can construct cocoa butter-type fats (Bafor *et al*., 1990b). The success of such a project would require the mechanism of stearate biosynthesis to be more fully understood. In recent studies it appears not simply to involve a thioesterase with increased selectivity towards stearoyl-ACP (Griffiths and Harwood, unpublished results). Furthermore, such fats would have to compete with cocoa butter substitutes made by other methods.

(5) *Increased levels of γ-linolenic acid for potential pharmaceutical uses*. The quantity of γ-linolenate is currently limited by the fact that the (Δ6)-desaturase only utilises linoleate at the *sn*-2 position of phosphatidylcholine. A (Δ6)-desaturase with activity to linoleate at the *sn*-1 position would increase the quantity of this product.

The potential targets for genetically manipulating plant lipids are limited because we do not have reliable probes with which to clone the necessary genes. This is because many of the proteins which are involved in plant lipid biosynthesis have not been purified to homogeneity. It is an unfortunate fact that we still rely heavily on traditional methods of protein chemistry, albeit with modern instrumentation, to advance genetic engineering problems. Although there is a large and growing depth of knowledge concerning plant lipid metabolism, the area of membrane-bound enzymes has received little attention. These proteins are all present in low abundance and difficult to purify. In fact, to date no triacylglycerol assembling enzyme has been purified to homogeneity from any developing oil-seed source. The use of alternative procedures for isolating and characterising genes involved in the triglyceride synthesis may prove a more fruitful avenue of approach. Such techniques could include mutant analysis, chromosome walking from restriction fragment length polymorphism information (Somerville and Browse, 1991); mutant generation by T-DNA tagging (Feldmann, 1991) or complementation of equivalent *E. coli* mutants (cf. the complementation of an enolase-deficient mutant by a maize gene (Lal *et al*., 1991)).

The potential contribution of microbial routes to produce activated fatty acids from plant oils may be of paramount importance in the future to provide a method of upgrading plant oils. Due to the versatility of microbes, a whole host of products could be made which, together with plants, could form the basis of a new chemical industry as the industry is already well versed in the use of hydrocarbon sources (petroleums) as its starting material. This could provide the industry with an environmentally friendly way of obtaining renewable raw materials for a variety of products ranging from detergents and cosmetics to plastics, rubber and textiles.

Acknowledgements

The authors would like to thank Dr Sten Stymne for information on petroselinic acid and Dr Kieran Elborough and Mr Bill Simon for providing their artistic and computational skills in the preparation of diagrams used in this chapter.

References

Badami, R.C. and Patil, K.B. (1982) Structure and occurrence of unusual fatty acids in minor seed oils. *Prog. Lipid Res.* **19**: 119.

Bafor, M., Jonsson, L., Stobart, A.K and Stymne, S. (1990a) Regulation of triacylglycerol biosynthesis in embryos and microsomal preparations from the developing seeds of *Cuphea lanceolata*. *Biochem. J.* **272**: 31.

Bafor, M., Stobart, A.K. and Stymne, S. (1990b) Properties of the glycerol acylating enzymes in microsomal preparations from the developing seeds of safflower and turnip rape and their ability to assemble cocoa-butter type fats. *J. Am. Oil Chem. Soc.* **67**: 217.

Bafor, M., Smith, M., Jonsson, L., Stobart, A.K. and Stymne, S. (1991) Ricinoleic acid biosynthesis and triacylglycerol assembly in microsomal preparations from developing castor bean. *Biochem. J.*: in press.

Barron, E.J. and Stumpf, P.K. (1962) The biosynthesis of triglycerides by avocado mesocarp enzymes. *Biochim. Biophys. Acta* **60**: 329.

Bernhard, W.R. and Somerville, C.R. (1989) Co-identity of putative amylase inhibitors from barley and finger millet with phospholipid transfer proteins inferred amino acid sequence homology. *Arch. Biochem. Biophys.* **269**: 695.

Bernhard, W.R., Thoma, S., Botella, J. and Somerville, C.R. (1991) Isolation of a cDNA clone for spinach lipid transfer protein and evidence that the protein is synthesized by the secretory pathway. *Plant Physiol.* **95**: 164.

Bouillon, P., Drischel, C., Vergonolle, C., Duranton, H. and Kader, J.-C. (1987) The primary structure of spinach leaf phospholipid transfer protein. *Eur. J. Biochem.* **166**: 387.

Cahoon, E.B. and Ohlrogge, J.B. (1991) Deposition and synthesis of petroselinic acid in seeds of Umbelliforae. *International news on fats, oils and related materials 2*, p.342.

Campos, F.A.P. and Richardson, M. (1984) The complete amino acid sequence of the α-amylase inhibitor L-2 from seeds of ragi (Indian finger millet, *Eleusine coracana* Gaertn). *FEBS Lett.* **167**: 221.

Cao, Y-Z., Oo, K.-C. and Huang, A.H.C. (1990). Lysophosphatidate acyltransferase in the microsomes from maturing seeds of meadowfoam (*Limnanthes alba*). *Plant Physiol.* **94**: 1199.

Casey, J., Dobb, R. and Mycock, G. (1990) An effective technique for enrichment and isolation of *Candida cloacae* mutants defective in alkane metabolism. *J. Gen. Micriobiol.* **136**: 1197.

Chandler, C.S. and Ballard, F.J. (1986) Multiple biotin-containing proteins in 3T3-LI cells. *Biochem. J.* **237**: 123.

Charles, D.J. and Cherry, J.H. (1986) Purification and characterization of acetyl-CoA carboxylase from developing soybean seeds. *Phytochemistry* **25**: 1067.

Cooper, C.L., Hsu, L., Jackowski, S. and Rock, C.O. (1989). 2-acylglycerol phosphoethanolamine acyltransferase/acyl-acyl carrier protein synthetase is a membrane associated acyl carrier protein binding protein. *J. Biol. Chem.* **264**: 7384.

Coves, J., Joyard, J. and Douce, R. (1988) Lipid requirement and kinetic studies of solubilized UDP-galactose-diacylglycerol galactosyl transferase activity from spinach chloroplast envelope membranes. *Proc. Natl Acad. Sci. USA*. **85**: 4966.

Dai, D.H., Pape, M.E., Lopez-Casillas, F., Luo, X.C., Dixon, J.E. and Kim, K.H. (1986) Molecular cloning of cDNA for acetyl-coenzyme-A carboxylase. *J. Biol. Chem.* **261**: 12395.

Dickman, M.B., Podika, G.K. and Kollatukudy, P.E. (1989) Insertion of a chitinase gene into a wound pathogen enables it to infect intact host. *Nature* **342**: 446.

Douady, D. and Dubacq, J.-P. (1987) Purification of acyl-CoA: glycerol-3-phosphate acyltransferase from peas leaves. *Biochem. Biophys. Acta* **921**: 615.

Douady, D., Guerbette, F., Grosbois, M. and Kader, J.-C. (1985) Purification of phospholipid transfer protein from maize seeds using a 2-step chromatographic procedure. *Physiol. Veg.* **23**: 373.

Dutta, P.C., Appelqvist, L.-A. and Stymne, S. (1991) Utilization of petroselinate by glycerol acylating enzymes in microsomal preparations of developing embryos of carrot (*Daucus carota* L.), safflower (*Carthamus tinctorius* L.) and oil rape (*Brassica napus*). *Plant Sci.*: in press.

Egin-Buhler, B. and Ebel, J. (1983) Improved purification and further characterisation of acetyl CoA carboxylase from cultured cells of parsley (*Petroselinum hortease*). *Eur. J. Biochem* **133**: 335.

Fehling, E., Murphy, D.J. and Mukherjee, K.D. (1990). Biosynthesis of very long chain triacylglycerols in cruciferous oilseeds. In *Plant Lipid Biochemistry, Structure and Utilization* (Quinn, P.J. and Harwood, J.L. eds.) Portland Press Limited, London, p. 201.

Feldmann, K.A. (1991) T-DNA insertion mutagenesis in *Arabidopsis*: mutational spectrum. *The Plant J.* **1**: 71.

Frentzen, M. (1986). Origin of diacylglycerol moities in plants. *J. Plant Physiol.* **124**: 193.

Frentzen, M., Heinz, E., McKeon, A. and Stumpf, P.K. (1983) Specificities and selectivities of glycerol-3-phosphate acyltransferase and monoacylglycerol-3-phosphate acyltransferase from pea and spinach chloroplasts. *Eur. J. Biochem.* **129**: 629.

Genez, A., McCarter, D., Nelson, J., Bleibaum, J.L., Thompson, G.A., Knauf, V.C. and Stalker, D. (1991) Molecular cloning and characterization of β-ketoacyl-ACPsynthase genes from *Ricinus communis*. The International Society for Plant Molecular Biology, 3rd International Congress, Tuscon, Arizona. 6–11th October 1991, Poster 719.

Griffiths, G. and Harwood, J.L. (1991) Regulation of triacylglycerol biosynthesis in cocao. *Planta* **184**: 279.

Griffiths, G., Stobart, A.K. and Stymne, S. (1985) Acylation of glycerol 3-phosphate and the metabolism of phosphatidate in microsomal preparations from the developing cotyledons of safflower seed. *Biochem. J.* **230**: 379.

Griffiths, G., Stobart, A.K. and Stymne, S. (1988a) Delta-6 and delta-12 desaturase activities and phosphatidic acid formation in microsomal preparations from the developing cotyledons of borage. *Biochem. J.* **252**: 641.

Griffiths, G., Stymme, S. and Stobart, A.K. (1988b) Triacylglycerol biosynthesis. In *Proceedings of the World Conference on Biotechnology for the Fats and Oils Industry* (Applewhite, T.H., ed.), American Oil Chemists' Society, p.23.

Griffiths, G., Stobart, A.K. and Stymne, S. (1988c) Phosphatidylcholine and its relationship to triacylglycerol biosynthesis in oil-tissues. *Phytochem.* **27**: 2089.

Griffiths, G., Brechany, E.Y., Christie, W.W., Stymne, S. and Stobart, K. (1989) Synthesis of octadecatetraenoic acid in borage. In *Biological Role of Plant Lipids* (Biacs, P.A. Gruuiz, K. and Kremmer, T., eds.), Akadémiai Kiadó, Budapest and Plenum Press, New York, p. 151.

Gurr, M.I. (1980) The biosynthesis of triacylglycerols. In *The Biochemistry of Plants*, Vol. 4 (Stumpf, P.K. and Conn, E.E.,eds.), Academic Press, New York, p. 205.

Hahlbrock, K. and Grisebach, H. (1979) Enzymic controls in the biosynthesis of lignin and flavanoids. *Ann. Rev. Plant Physiol.* **30**: 105–130.

Hansen, L. (1987) Three cDNA clones for barley leaf acyl carrier proteins I and III. *Carlsberg Res. Commun.* **52**: 381.

Hansen, L. and Kauppinen, S. (1991) Barley acyl carrier protein II:nucleotide sequence of cDNA clones and chromosomal location of the *acl2* gene. *Plant Physiol.* **97**: 472.

Hansen, L. and von Wettstein-Knowles, P. (1991) The barley genes acl1 and acl3 encoding acyl carrier proteins I and III are single copy genes located on different chromosomes. *Mol. Gen. Genet.*: submitted.

Harwood, J.L., Walsh, M.C. and Walker, K.A. (1990) Enzymes of fatty acid synthesis. *Methods Plant Biochem.* **3**:193.

Hellyer, A. (1990) Ph.D. thesis, University of Cambridge.

Hellyer, A. and Slabas, A.R. (1990) Acyl-[acyl-carrier protein] thioesterase from oil seed rape—purification and characterization. In *Plant Lipid Biochemistry, Structure and Utilization* (Quinn, B.J. and Harwood, J.L. eds.), Portland Press p. 157.

Hellyer,A., Bambridge, H.E. and Slabas, A.R. (1986) Plant acetyl-CoA carboxylase. *Biochem. Soc. Trans.* **14**: 565–568.

Hillditch, T.P. and Williams, P.N. (1964) *Chemical Constitution of Natural Lipids*. Chapman & Hall, London.

Hilt, K.L. (1984) Ph.D. thesis, University of California, Davies, CA.

Hoj, P.B. and Svedson, I. (1983) Barley acyl carrier protein:its amino acid sequence and assay using purified malonyl CoA:ACP transacylase. *Carlsberg Res. Commun.* **48**: 285–305.

Hoj, P.B. and Svedson, I. (1984) Barley chloroplasts contain 2-acyl carrier protein coded for by different genes. *Carlsberg Res. Commun.* **49**: 483.

Holloway, P.W. (1983) Fatty acid desaturation. In *The Enzymes: Lipid Enzymology, Vol. 16* (Boyer, P.D., ed.) Academic Press, New York, p. 63.

Horrobin, D.F. and Manku, M.S. (1983) In *Fats for the Future* (Brooker, S.G., Renwick, A., Hannan, S.F. and Eyres, L., eds.), Duromark, New Zealand, p. 99.

Howling, D., Morris, L.J., Gurr, M.I. and James, A.T. (1972) Specificity of fatty acid desaturases and hydroxylases:dehydrogenation and hydroxylation of monoenoic acids. *Biochem. Biophys. Acta* **260**: 10.

Ichihara, K., Asahi, T. and Jufi, S. (1987) 1-acyl-*sn*-glycerol 3 phosphate acyltransferase in maturing safflower seeds and its contribution to the non-random fatty acid distribution of triacylglycerol. *Eur. J. Biochem*. **167**: 339.

Ishizaki, O., Nishida, I., Agata, K., Eguchi, G. and Murata, N. (1988) Cloning and nucleotide sequence of cDNA for the plastid glycerol 3-phosphate acyltransferase from squash. *FEBS Lett.* **283: 424.**

Jackowski, S. and Rock, C.O. (1987) Acetoacetyl-acyl carrier protein synthease, a potential regulator of fatty acid biosynthesis in bacteria. *J. Biol. Chem.* **262** 7927.

Jamieson, G.R. and Reid, E.H. (1969) Leaf lipids of some members of the Boraginaceae family. *Phytochem.* **8**: 1489.

Jaworski, J.G. (1987). Biosynthesis of monoenoic and polyenoic fatty acids. In *The Biochemistry of Plants, Vol. 9* (Stumpf, P.K., ed.), Academic Press, New York, p. 159.

Jaworski, J.G., Clough, R.C. and Barnum, S.R. (1989) A cerulenin insensitive short-chain 3-ketoacyl-acyl carrier protein synthease in *Spinacia oleracea* leaves. *Plant Physiol.* **90**: 41.

Kader, J.-C., Julienne, M. and Vergnolle, C. (1984) Purification and characterisation of spinach-leaf protein capable of transferring phospholids from liposomes to mitochondria or chloroplasts. *Eur. J. Biochem.* **139**: 411.

Kater, M.M., Konningenstein, G.M., Nijkamp, J.J. and Stuitje, A.R. (1991)cDNA cloning and expression of *Brasica napus* enoyl-acyl carrier protein reductase in *Escherichia coli. Plant Mol. Biol.* **17**: 895–909.

Kauppinen, S., Siggaard-Andersen, M. and von Wettstein-Knowles, P. (1988) β-ketoacyl-ACP synthase I of *Escherichia coli*: nucleotide sequence of the *fabB* gene and identification of the cerulenin binding residue. *Carlsberg Res. Commun.* **53**: 357.

Kennedy, E.P. (1961) Biosythesis of complex lipids. *Fed. Proc. Fed. Am. Soc. Exp. Biol.* **20**: 943.

Kleiman, R. and Spencer, G.F. (1982) Search for new industrial oils:XVI. Umbelliflorae—seed oils rich in petroselinic acid. *J. Am. Oil Chem. Soc.* **50**: 29.

Knudsen, J. and Grunnet, I. (1982) Transacylation as a chain-termination mechanism in fatty acid synthesis by mammalian fatty acid synthetase. *Biochem. J.* **202**: 139.

Kornberg, A. (1989) *For the Love of Enzymes*. Harvard University Press.

Kridle, J.C., Knutzon, D.S., Johnson, W.B., Thompson, G.A., Radke, S.E., Turner, J.C. and Knauf, V.C. (1991) Modulation of stearoyl-ACP desaturase levels in transgenic rapeseed. The International Society for Plant Molecular Biology, 3rd International Congress, Tuscon, Arizona. 6–11th October 1991, Poster 723.

Kuijk van, F.J.G.M., Sevanian, A. and Handelman, G.J. (1987) New role for phospholipase A_2:protection of membranes from lipid peroxidation damage, *TIBS* **12**: 31.

Kuo, T.M. and Ohlrogge, J.B. (1984) The primary sequence of spinach acyl carrier protein. *Arch. Biochem. Biophys*. **234**: 290.

Lal, S. K., Johnson, S., Conway, T. and Kelly, P.M. (1991) Characterization of a maize cDNA that complements an enolase deficient mutant of *Escherichia coli. Plant Mol. Biol.* **16**: 787.

Lampa, G. and Jacks, S.C. (1991). Analysis of two linked genes coding for the acyl carrier protein (ACP) from *Arabidopsis thaliania* (columbia). *Plant Mol. Biol.* **16**:469.

Lohden, I., Bernerth, R. and Frentzen, M. (1990) Acyl-CoA:1-acylglycerol-3-phosphate acyltransferase from developing seeds of *Limnanthes douglasii* (R.Br.) and *Brassica napus*(L). In: *Plant Lipid Biochemistry, Structure and Utilization* (Quinn, B.J. and Harwood, J.L., eds.), Portland Press, p. 175.

Lynch, D.V. and Thompson, G.A. (1984) Retailored lipid molecular species: a tactile mechanism for modulating membrane properties. *TIBS* **9**: 442.

Mackintosh, R.W., Hardie, D.G., and Slabas, A.R. (1989) A new assay procedure to study the induction of β-ketoacyl-ACP synthease I and II, and the complete purification of β-ketoacyl-ACP synthease I from developing seeds of oil seed rape (*Brassica napus*). *Biochem. Biophys. Acta* **1002**: 114.

MacRae, A.R. (1989) Tailored triacylglycerols and esters. *Biochem. Soc. Trans.* **17**: 1146.

Masterson, C., Wood, C. and Thomas, D.R. (1990) L-acetylcarnitine, a substrate for chloroplast fatty acid synthesis. *Plant, Cell Environ.* **13**: 755.

Matsumara, S. and Stumpf, P.K. (1968) Fat metabolism in higher plants: XXXV. Partial primary structure of spinach acyl carrier protein. *Arch. Biophys. Biochem.* **125**: 932.

Moreton, R.S. (1988) Technical and economic aspects and feasibility of single cell oil production using yeast technology. In *World Conference on Biotechnology of the Fats and Oils Industry* (Appelwhite, T.H., ed.), American Oil Chemists' Society, Champaign, 111.

Mundy, J. and Rogers, J.C. (1986) Selective expression of a probable amylase/protease inhibitor in barley aleurone cells: Comparison to the barley amylase/subtilisin inhibitor. *Planta* **16**: 51.

Murata, N. (1983) Molecular species composition of phosphatidylglycerols from chilling-sensitive and chilling-resistant plants. *Plant Cell Physiol.* **24**: 81.

Murata, N. (1991) Gene-technological manipulation of fatty-acid unsaturation: desaturase and acyltransferase. The International Society for Plant Molecular Biology, 3rd International Congress, Tuscon, Arizona, 6–11th October 1991.

Murphy, D.J. (1990) Storage lipid bodies in plants and other organisms. *Prog. Lipid Res.* **29**: 299.

Murphy, D.J., Cummins, I. and Kang, S. (1989) Synthesis of the major oil-body membrane protein in developing rapeseed (*Brassica napus*) embryos. *Biochem. J.* **258**: 285.

Nikolau, B.J. and Wurtele, E.S. (1991) Molecular cloning and characterization of acetyl-CoA carboxylase and other biotin enzymes of plants. The International Society for Plant Molecular Biology, 3rd International Congress, Tuscon, Arizona, 6–11th October 1991, Poster 720.

Nishida, I., Frentzen, M., Ishizaki, O. and Murata, N. (1987) Purification of isomeric forms of acyl [acyl-carrier-protein]:glycerol 3-phosphate acyltransferase from greening squash cotyledons. *Plant Cell Physiol.* **18**: 1071.

Nishida, I., Tasaka, Y., Shiraishi, H., Okada, K., Shimura, Y., Beppu, T. and Murata, N. (1990) The genomic gene and cDNA for the plastidial glycerol-3-phosphate acyltransferase of *Arabidopsis*. In *Plant Lipid Biochemistry, Structure and Utilization* (Quinn, P.J. and Harwood, J.L., eds.), Plenum, p 462–464.

Ohlrogge, J.B. (1987) Biochemistry of plant acyl carrier proteins. In *The Biochemistry of Plants, Vol.9.* (Stumpf, P.K. and Conn, E.E., eds.), Academic Press, p.137.

Ohlrogge, J.B., Browse, J. and Somerville, C.R. (1991) The genetics of plant lipids. *Biochim. Biophys. Acta* **1082**: 1.

Ottey, K.A., Takhan, S. and Munday, M.R. (1989) Comparison of two cyclic-nucleotide-independent acetyl-CoA carboxylase kinases from lactating rat mammary glands: identification of the kinase responsible for acetyl-CoA inactivation *in vivo*. *Biochem. Soc. Transac.* **17**: 349.

Parker, W.B., Marshall, L.C., Burton, J.D., Somers, D.A., Wise, D.L., Gronwald, J.D. and Gengenbach, B.G. (1990) Dominant mutations causing alterations in acetyl-coenzyme A carboxylase confer tolerance to cyclohexanedione and aryloxyphenoxyproprionate herbicides in maize. *Proc. Nat. Acad. Sci. USA* **87**: 7175.

Payne, P.I., Holt, L.M., Jackson, E.A. and Law, C.N. (1984) Wheat storage proteins: their genetics and their potential for manipulation by plant breeding. *Phil. Trans. R. Soc. Lond.* **304**: 359.

Pollard, M.R., Anderson, L., Fan, C., Hawkins, D.J. and Evans, H.M. (1991) A specific acyl-ACP-thioesterase implicated in medium chain fatty acid production in immature cotyledons of *Umbellularia california*. *Arch. Biochem. Biophys.* **284**: 306.

Post-Beittenmiller, M.A., Housek-Radojcic, A. and Ohlrogge, J.B. (1989) DNA sequence of a genomic clone encoding an *Arabidopsis* acyl carrier protein. *Nucl. Acids Res.* **17**: 1777.

Poulose, A.J., Rogers, L., Cheesbrough, T.M. and Kolattukudy, P.E. (1985) Cloning and sequencing of the cDNA for *S* acyl fatty acid synthase from the uropygial gland of mallard duck. *J. Biol. Chem.* **260**: 15953.

Qu, R. and Huang, A.H.C. (1990) Oleosin KD 18 on the surface of oil bodies in maize. *J. Biol. Chem.* **265**: 2238.

Qu, R., Wang, S.-M., Liu, Y.-H., Vance, V.B. and Huang, A.H.C. (1986) Characteristics and biosynthesis of membrane proteins of lipid bodies in scutella of maize (*Zea mays*L.) *Biochem. J.* **235**: 57.

Rose, R.E., De Jesus, C.E., Moylan, N.P., Scherer, D.E. and Knauf, V.C. (1987) The nucleotide sequence of a cDNA clone encoding acyl carrier protein (ACP) from *Brassica campestris* seeds. *Nucl. Acids Res.* **15**: 7197.

Roughan, G. (1987) On the control of fatty acid compositions of plant glycerolipid. In *The Metabolism, Structure and Function of Plant Lipids* (Stumpf, P.K., Mudd, J.B., and Nea, W.D., eds), Plenum. p.247

Safford, R., de Silva, J., Lucas, C., Windust, J.H.C., Shedden, J., James, C.M., Sidebottom, C.M., Slabas, A.R., Tombs, M.P. and Hughes, S.G. (1987) Molecular cloning and sequence analysis of complementary DNA encoding rat mammary gland medium-chain *S*-acyl fatty acid synthetase thioester hydrolase. *Biochemistry* **26**: 1358.

Safford, R., Windust, J.H.C., Lucas, C., De Silva, J., James, C.M., Hellyer A., Smith, C.G., Slabas, A.R. and Hughes, S.G. (1988) Plastid-localized seed acyl-carrier protein of *Brassica napus* is encoded by a distinct, nuclear multigene family. *Eur. J. Biochem.* **174**: 287.

Sambanthamurthi, R. and Oo, K.-C. (1990) Thioesterase activity in the oil palm (*Elaeis guineensis*) mesocarp. In *Plant Biochemistry, Structure and Utilization* (Quinn, P.J. and Harwood, J.L., eds.), Portland Press, London, p. 186.

Scherer, D.E. and Knauf, V.C. (1987) Isolation of complementary DNA clone for the acyl carrier protein-I of spinach. *Plant Mol. Biol.* **9**: 127.

Schroder, S., Brown, J.W.S. and Schroder, J. (1988) Molecular analysis of resveratrol synthase cDNA, genomic clones and relationship with chalcone synthase. *Eur. J. Biochem.* **172**: 161.

Shanklin, J. and Somerville, C. (1991) Stearoyl-acyl-carrier protein desaturase from higher plants is structurally unrelated to the animal and fungal homologs. *Proc. Natl. Acad. Sci. USA.* **88**: 2510.

Shanklin, J., Mullins, C. and Somerville, C. (1991) Sequence of a complementary DNA from *Cucumis sativus* (L.) encoding the stearoyl-acyl-carrier protein desaturase. *Plant Physiol.* **97**:467.

Sheldon, P.S., Kekwick, R.G.O., Sidebottom, C., Smith, C.G. and Slabas, A.R. (1990) 3-oxoacyl-(acyl-carrier protein) reductase from avocado (*Persea americana*) fruit mesocarp. *Biochem. J.* **271**: 713.

Sheldon, P.S., Kekwick, R.G.O., Smith, C.G., Sidebottom, C. and Slabas, A.R. 3-Oxoacyl-[ACP] reductase from oilseed rape (*Brassica napus*). *Biochim. Biophys. Acta*: **1130**: 151.

Shimakata, T. and Stumpf, P.K. (1982a). Isolation and function of spinach leaf β-ketoacyl-[acyl-carrier-protein] synthetases. *Proc. Natl. Acad. Sci. USA* **79**: 5808.

Shimakata, T. and Stumpf, P.K. (1982b) Purification and characteristics of β-keto-[acyl-carrier-protein] reductase, β-hydroxyacyl-[acyl-carrier protein] dehydrase and enoyl-[acyl-carrier-protein] reductase from *Spinicia oloracea* leaves. *Arch. Biophys. Biochem.* **218**: 77.

Siggaard-Andersen,M., Kauppinen, S. and von Wettstein-Knowles, P. (1991) Primary structure of a cerulenin binding β-ketoacyl [acyl carrier protein] synthase from barley chloroplasts. *Proc. Natl. Acad. Sci. USA.* **88**: 4114.

de Silva, J., Loader, N.M. Jarman, C., Windust, J.H.C., Hughes, S.G. and Safford, R. (1990) The isolation and sequence analysis of two seed-expressed acyl carrier protein genes from *Brassica napus. Plant Mol. Biol.* **14**: 537.

Simoni, R.D., Criddle, R.S. and Stumpf, P.K., (1967) Fat metabolism in higher plants 31: purification and properties of plant and bacterial acyl carrier protein. *J. Biol. Chem.* **242**: 573.

Slabas, A.R. and Fawcett, T. (1992) The biochemistry and molecular biology of plant lipid biosynthesis. *Plant Mol. Biol.* **19** : 169.

Slabas, A.R. and Hellyer, A. (1985) Rapid purification of a high molecular weight subunit polypeptide form of rape seed acetyl CoA carboxylase. *Plant Sci. Lett.* **39**: 177.

Slabas, A.R., Sidebottom, C.M., Hellyer, A., Kessell, R.M.J. and Tombs, M.P. (1986a) Induction, purification and characterization of NADH-specific enoyl-acyl carrier protein reductase from developing seeds of oil seed rape (*Brassica napus*). *Biochim. Biophys. Acta* **877**: 271.

Slabas, A.R., Hellyer, A. and Bambridge, H.E. (1986b) The basic polypeptide subunit of rape leaf acetyl-CoA carboxylase is a 220 kDa protein. *Biochem. Soc. Trans.* **14**: 714.

Slabas, A.R., Harding, J.,Hellyer, A., Roberts, P. and Bambridge, H.E. (1987a) Induction, purification and characterization of acyl carrier protein from developing seeds of oil seed rape (*Brassica napus*). *Biochim. Biophys. Acta.* **921**: 50.

Slabas, A.R., Hellyer, A., Sidebottom, C., Bambridge, H., Cottingham, I.R., Kessell, R., Smith, C.G., Sheldon, P., Kekwick, R.G.O., de Silva, J., Windust, J., James, C.M., Hughes, S.G. and Safford, R. (1987b) Molecular structure of plant fatty acid synthesis enzymes. In *Plant Molecular Biology* (von Wettstein D. and Chua, N.H., eds.), NATO ASI Series A, Vol. 140, p. 265.

Slabas, A.R., Davies, C., Hellyer, A., MacKintosh, R.W., Sheldon, P., Hardie, D.G., Kekwick, R.G.O. and Safford, F. (1988) Plant lipids: targets for manipulation. In *British Plant Growth Regulators Group*, Monograph 17.

Slabas, A.R., Cottingham, I.R., Austin, A., Hellyer, A., Safford, R. and Smith, C.G. (1990) Immunological detection of NADH-specific enoyl-ACP reductase from rape seed (*Brassica napus*)—induction relationship of α and β polypeptides, mRNA translation and interaction with ACP. *Biochim. Biophys. Acta* **1039**: 181.

Slabas, A.R., Cottingham, I.R., Austin, A., Fawcett, T. and Sidebottom, C.M. (1991) Amino acid sequence analysis of rape seed (*Brassica napus*) NADH-enoyl ACP reductase. *Plant Mol. Biol.* **17**: 915.

Slabas, A.R., Chase, D.A., Nishida, I., Safford, R., Sheldon, P.S., Kekwick, R.G.O., Hardie, D.G. and MacKintosh, R.W. Molecular cloning of higher plant 3-oxyacyl [acyl carrier protein] reductase *Biochem. J.* **283** : 321.

Slack, C.R., Roughan, P.G., and Browse, J. (1979) Evidence for an oleoylphosphatidylcholine desaturase in microsomal preparations from cotyledons of safflower seeds. *Biochem. J.* **179**: 649.

Slack, C.R., Roughan, P.G., Browse, J.A. and Gardiner, S.E. (1985) Some properties of cholinephosphotransferase from developing safflower cotyledons. *Biochim. Biophys. Acta* **883**: 438.

Smith, C.G. (1974) The ultrastructural development of spherosomes and oil bodies in the developing embryo of *Crambe abyssinica*. *Planta* **119**: 125.

Smith, S. and Libertini, C.J. (1979) Specificity and site of action of a mammary gland thioesterase which released acyl moieties from thioester linkages to the fatty acid synthetase. *Arch. Biochem. Biophys.* **196**: 92.

Somerville, C. and Browse, J. (1991) Plant lipids: metabolism, mutants and membrane. *Science* **252**: 80.

Stobart, A.K. and Stymne, S. (1985) Interconversion of diacyglycerol and phosphatidylcholine during triacylglycerol production in microsomal preparations of developing cotyledons of safflower. *Biochem. J.* **232**: 217.

Stobart, A.K., Stymne, S. and Höglund, S. (1986) Safflower microsomes catalyse oil accumulation *in vitro*: a model system. *Planta* **169**: 33.

Stumpf, P.K. (1980) Biosynthesis of saturated and unsaturated fatty acids. In *The Biochemistry of Plants, Vol. 4* (Stumpf, P.K. and Conn, E.E., eds.), Academic Press, p.177.

Stumpf, P.K. (1987) The biosynthesis of saturated fatty acids. In *The Biochemistry of Plants*, Vol. 9. (Stumpf, P.K. and Conn, E.E., eds.), Academic Press, p.121.

Stumpf, P.K. and Shimakata, T. (1983) Molecular structures and functions of plant fatty acid synthetase enzymes. In *Biosynthesis and Function of Plant Lipids* (Mudd, J.B. and Gibbs, M., eds.), Waverly Press, p.1.

Stymne, S. and Appelqvist, L.-Å. (1978) Biosynthesis of linoleate from oleoyl-CoA via oleoyl-phosphatidylcholine in microsomes of developing seeds. *Eur. J. Biochem*. **90**: 223.

Stymne, S. and Appelqvist, L.-Å. (1980) Biosynthesis of linoleate and linolenate in homogenates from developing soya bean cotyledons. *Plant Sci. Lett.* **17**: 287.

Stymne, S. and Stobart, A.K. (1986) Biosynthesis of gamma-linolenic acid in cotyledons and microsomal preparations of the developing seeds of borage. *Biochem. J.* **240**: 385.

Stymne, S. and Stobart, A.K. (1987) Triacylglycerol biosynthesis. In *The Biochemistry of Plants: a Comprehensive Treatise, Vol. 9* (Stumpf, P.K., ed.), Academic Press, New York, p.175.

Stymne, S., Green, A. and Tonnet, M.L. (1990) In *Lipid Synthesis in Developing Cotyledons of Linseed* (Biacs, P.A., Gruiz, K. and Kreemer, T., eds.), Akademia Kiado, Budapest and Plenum Publishing Corporation, New York and London, p.147.

Svennson, B., Asano, K., Jonassen, I., Poulsen, F., Mundy, J. and Svedsen, I. (1986) Isolation and characterization of a barley seed protein with M_W 10 000 showing homology with an α-amylase inhibitor for Indian finger millet. *Carlsberg Res. Commun.* **51**: 493.

Takashima, K., Watanabe, S., Yamada, M. and Maniya, G. (1986) The amino acid sequence of nonspecific lipid transfer protein from germinating castor bean. *Biochem. Biophys. Acta* **870**: 248.

Taylor, D., Weber, N., Barton, D., Underhill, E. and Pomeroy, K. (1990) Biosynthesis of triacylglycerols containing erucic acid in microspore derived embryos of *Brassica napus*. In *Plant Lipid Biochemistry, Structure and Utilization* (Quinn, P.J. and Harwood, J.L., eds.), Portland Press, London, p. 210

Tchang, F., This, P., Stiefel, V., Arondel, V., Morch, M.D., Pages, M., Puigdomech, P., Grellet, F., Delseny, M., Bouillon, P., Huet, J.C., Guerbette, F., Beauvais-Cante, F., Duranton, H., Pernollet, J.C. and Kader, J.C. (1988) Phospholipid transfer protein: full length cDNA and amino acid sequence in maize. *J Biol. Chem.* **263**: 16849.

Thompson, G.A., Scherer, D.E., Foxall-Van Aken, S., Kenny, J.W., Young, H.L. Shintani, D.K., Kridl, J.C. and Knauf, V.C. (1991) Primary structures of the precursors and mature forms of stearoyl-acyl carrier protein desaturase from safflower embryos and requirement of ferredoxin for enzyme activity. *Proc. Natl Acad. Sci. USA* **88**: 2578.

Truchet, G., Roche, P., Leronge, P., Vasse, J., Camet, S., de Billy, F., Prome, J-C. and Denarie, J. (1991) Sulphated lipo-oligosaccharide signals of *Rhizobium meliloti* elicit root nodule organogenesis in alfalfa. *Nature* **351**: 670.

Turnham, E. and Northcote, D.H. (1983) Changes in the activity of acetyl-CoA carboxylase during rape-seed formation. *Biochem. J.* **212**: 223.

Tzen, J.T.C., Lai, Y-K., Chan, K-L. and Huang, A.H.C. (1990) Oleosin isoforms of high and low molecular weights are present in the oil bodies of diverse species. *Plant Physiol.* **94**: 1282.

Vance, V.B. and Huang, A.H.C. (1987) The major protein from lipid bodies of maize. *J. Biol. Chem.* **262**: 11275.

Watanabe, S. and Yamada, M. (1986) Purification and characterisation of a non-specific lipid transfer protein from germinating castor bean endosperms which transfers phospholipids and galactolipids. *Biochem. Biophys. Acta* **876**: 116.

Weber, S., Wolter, F.-P., Buck, F., Frentzen, M. and Heinz, E. (1991) Purification and cDNA sequencing of an oleate-selective acyl-ACP:*sn*-glycerol-3-phosphate acyltransferase from pea chloroplasts. *Plant Mol. Biol.* **17**: 1067.

Wurtele, E.S. and Nikolau, B.J. (1990) Plants contain multiple biotin enzymes: discovery of 3-methylcrotonyl-CoA carboxylase, proprionyl-CoA carboxylase acid pyruvate carboxylase in the plant kingdom. *Arch. Biochem. Biophys.* **278**: 179.

Ykema, A., Verbree, H., Nijkamp, J.J. and Smit, H. (1989) Isolation and characterization of fatty acid auxotrophs from the oleaginous yeast *Apotrichum curvatum*. *Appl. Microbiol. Biotech.* **31**: 76.

4 Carotenoid biosynthesis and manipulation

P. M. BRAMLEY, C. R. BIRD and W. SCHUCH

4.1 Introduction

The carotenoids are the most widespread group of pigments in nature, with an estimated yield of some 100 million tons annually. They are present in all photosynthetic organisms, and are responsible for most of the yellow to red colours of most fruits and flowers. The characteristic colours of many birds, insects and marine invertebrates are also due to the presence of carotenoids which have originated in the diet. Commercially, carotenoids are used as food colorants, nutritional supplements (because of their pro-vitamin A activity) and in the treatment of certain cancers.

The diversity of structures, and hence biosynthetic pathways within the carotenoids, together with their diverse locations in the cell, clearly imply that efficient regulatory mechanisms are present to control the rates of biosynthesis, breakdown and deposition of the pigments at their functional locations. These mechanisms must be understood in order to manipulate carotenogenesis in higher plants or microorganisms. Such an understanding necessitates detailed studies on the carotenogenic enzymes, the genes encoding them and the metabolic fluxes through the pathways.

The aim of this chapter is to review our current understanding of carotenoid biosynthesis in higher plants and microorganisms, with emphasis on the structural diversity of carotenoids, the properties of isolated enzymes, the emerging information on structural features of genes and the experimental techniques available for the manipulation of carotenoid accumulation in tissues.

4.2 Structure and nomenclature of carotenoids

Carotenoids are isoprenoids that generally consist of eight isoprene units joined together so that the linking of units is reversed at the centre of the molecule to give methyl groups 20 and 20′ with a (1,6) positional relationship, whereas the remaining methyl groups are (1,5) (Figure 4.1). The most obvious feature of a carotenoid is the polyene chain, which may extend from 3 to 15 conjugated double bonds. The length of the chromophore determines the absorption spectrum of the molecule, which is often recorded experimentally as an aid to the identification and quantitation of the carotenoid (Davies, 1976; Britton, 1985, 1991).

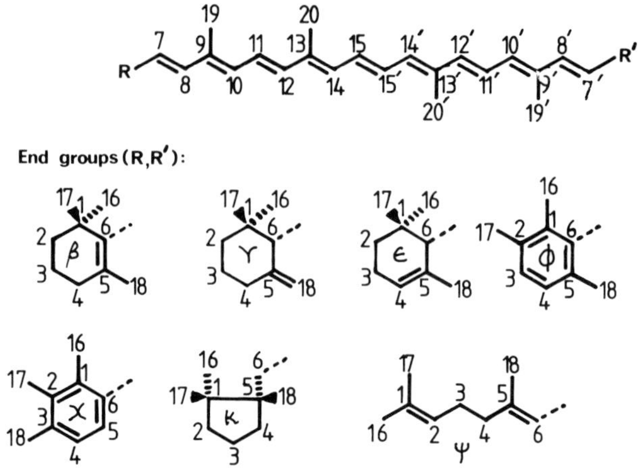

Figure 4.1 Numbering and end groups of carotenoid molecules.

Although over 500 carotenoid structures have been determined, all are based on only 7 different end groups of which only 4 (β, ε, κ and ψ; Figure 4.1) are found in higher plant carotenoids. Cyclisation of the carbon skeleton occurs at one or both ends, while xanthophylls are formed from the hydrocarbon carotenes by the introduction of oxygen functions. In addition, skeletal modifications involving chain elongation or degradation occur.

As many of the trivial names of carotenoids are related more to their original source rather than their chemical structure, a standardised semi-systematic nomenclature was introduced (Commission on Biochemical Nomenclature, 1971, 1975) to rationalise all the structures by using the seven end groups shown in Figure 4.1. The carotene stem names, e.g. β, β-carotene, ψ, ψ-carotene, are then modified by prefixes and suffixes to denote changes in the carbon skeleton. The semi-systematic names of the carotenoids mentioned in this article are listed in Straub (1987).

The structures of most known carotenoids are given in Straub (1987) and Britton (1991), while annual reviews of advances in carotenoid chemistry can be found in publications of the The Chemistry Society of Great Britain (e.g. Britton, 1989).

4.3 Distribution in higher plants

4.3.1 Photosynthetic tissues

Carotenoids are present in the chloroplasts of all green leaves and, irrespective of the type and habitat of the higher plant, all the species so far examined contain the same major carotenoids, β-carotene, lutein, violaxanthin and neoxanthin (Figure 4.2.). Minor components include α-carotene, β-cryptoxanthin, zeaxanthin, antheraxanthin and lutein 5,6-epoxide. It is thought that this constancy of distribution

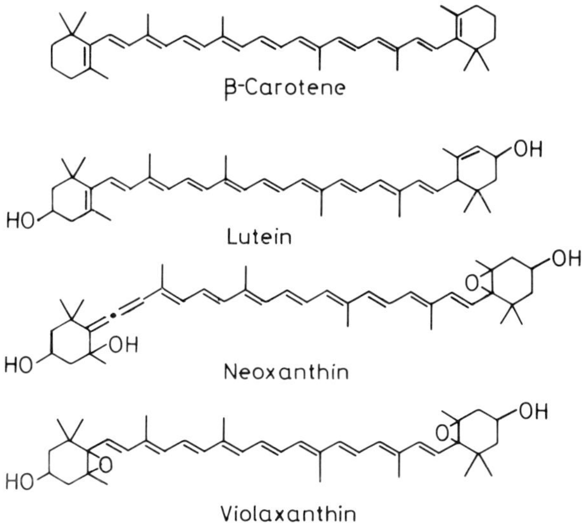

Figure 4.2 Structures of major carotenoids in photosynthetic tissues.

reflects the common ancestry of higher plants (Goodwin, 1973). During leaf senescence chloroplast disintegration is accompanied by esterification of the xanthophylls (Goodwin, 1958).

Quantitative differences between species have been found but, in general, the carotenes are about 25% and lutein about 45% of the total carotenoids present. Apart from the compounds listed above, a number of unusual carotenoids have been identified in various higher plant species (see Goodwin and Britton, 1988; Britton, 1991).

In photosynthetically active chloroplasts, the carotenoids are found as part of photosynthetic pigment–protein complexes (PPCs) in the thylakoid membranes. The core complex, CCI, contains one β-carotene molecule per 40 of chlorophyll a, whereas the light-harvesting complex (LHCI) is associated with lutein, violaxanthin and neoxanthin. β-Carotene is also located in CCII (Tang and Satho, 1985), while LHCII contains xanthophylls (Thornber, 1975; Lichtenthaler *et al.*, 1982). The identification and nomenclature of PPCs has been discussed in detail elsewhere (Siefermann-Harms, 1985; Peter *et al.*, 1988).

There are also small amounts of carotenoids in the chloroplast envelope (Joyard *et al.*, 1991) as well as in the envelope of amyloplasts (Fishwick and Wright, 1980). In all envelope membranes, violaxanthin is the major carotenoid (Jeffrey *et al.*, 1974). Carotenes are also found in plastoglobuli of photosynthetic tissues (Lichtenthaler, 1968), while dark-grown plants contain predominantly xanthophylls in their etioplasts (Goodwin, 1958). Coniferous trees and cycads may accumulate unusual keto-carotenoids or seco-carotenoids in leaves and needles at certain stages of development. These are located in oil droplets (see Britton, 1991).

4.3.2 Flowers

The carotenoids of flower petals can be divided into three main groups (Goodwin, 1980): (a) highly oxygenated carotenoids such as auroxanthin and flavoxanthin; (b) carotenes, sometimes present in high concentrations and (c) species-specific carotenoids such as eschscholtziaxanthin and crocetin (Figure 4.3). They are found in chromoplasts which themselves are of various types (Whately and Whately, 1987). Unlike chloroplast xanthophylls, flower xanthophylls are usually esterified. The recent characterisation of geometric isomers of carotenoids found in petals has been reviewed by Goodwin and Britton (1988).

4.3.3 Fruits

The carotenoid content of fruits has been studied for over 40 years mainly because of the obvious aesthetic appeal of ripe fruit colours but also due to the wide variety

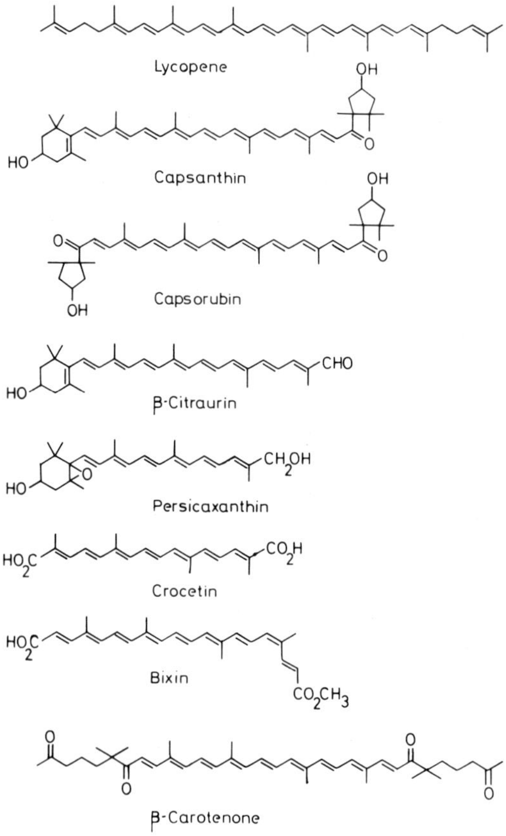

Lycopene

Capsanthin

Capsorubin

β-Citraurin

Persicaxanthin

Crocetin

Bixin

β-Carotenone

Figure 4.3 Carotenoids found in ripe fruit, flowers and seeds.

of structures that are present. The distribution of carotenoids in fruit is extremely complex and subject to considerable variation. Patterns characteristic of each species and variety occur often at each ripening stage. These can vary from relatively few carotenoids up to over 50 carotenoids, as found in citrus fruits. A higher rate of carotenoid biosynthesis often occurs in the peel compared with the pulp, leading to different patterns in the two tissues. In either case biosynthesis is autonomous in most fruit and continues after the fruit has been harvested from the parent plant.

Unripe (green) fruit contains the same pigments as other photosynthetic tissues (see section 4.3.1), but upon ripening the chloroplasts differentiate into chromoplasts and there is often, but not always, *de novo* synthesis of carotenoids. According to Goodwin (1980) eight main groups of fruits can be distinguished according to carotenoid content, although these divisions are not absolute:

1) Insignificant amounts, e.g. strawberry.
2) Large amounts of chloroplast carotenoids, e.g. blueberry.
3) Large amounts of lycopene and its hydroxy derivatives and more saturated acidic carotenoids, e.g. tomato.
4) Large amounts of β-carotene and its hydroxy derivatives, e.g. peach.
5) Large amounts of epoxides, particularly furanoid epoxides, e.g. carambola.
6) Unusual carotenoids, e.g. capsanthin in red pepper.
7) Poly-*Z*-carotenoids, e.g. prolycopene in tangerine tomatoes.
8) Apocarotenoids, e.g. persicaxanthin in *Citrus* species.

Comprehensive tables detailing the carotenoid content of many fruits have been published (Goodwin and Goad, 1971; Goodwin, 1980). In this review only three examples will be discussed: citrus, pepper and tomato.

4.3.3.1 Citrus. Citrus fruits are yellow, orange and red and contain both common carotenoids and genus-specific C_{30}-apocarotenoids such as β-citraurin and also apocarotenoids (Figure 4.3). The biosynthesis of apocarotenoids is discussed in section 4.4.5. Finally it should be noted that the carotenoid content of citrus can be altered by treatment with onium compounds such as CPTA and MPTA (Yokoyama *et al.*, 1982; Benedict *et al.*, 1985).

4.3.3.2 Pepper. The ripe fruit of the sweet pepper, *Capsicum annuum*, and the paprika variety, *C. frutescens*, contain unique keto-carotenoids such as capsanthin and capsorubin (Figure 4.3). Their absolute configuration has been established (Bowden *et al.*, 1983), and *C. annuum* has been used extensively by Camara and co-workers in studies in the enzymology of carotenogenesis (see section 4.4.).

4.3.3.3 Tomato. The carotenoids of the commercially grown tomato, *Lycopersicon esculentum*, were first analysed in 1932, when Kuhn and Grundmann found that the carotenes predominate, with lycopene being the most abundant pigment (up to 90% of total carotenoids). Minor quantities of other acyclic carotenes, β-carotene, xanthophylls, carotene epoxides and apocarotenoids are also present (reviewed by Goodwin and Goad, 1971).

Besides the common red tomatoes there are numerous mutant varieties with a range of colours from white to deep orange, reflecting differences in their carotenoid content. In certain mutants a specific block in the carotenoid pathway is apparent, based upon carotenoid analysis, but other mutations have pleitropic effects on carotenoid levels and also ripening. These mutant varieties are discussed fully in section 4.6.

4.3.4 Other tissues

Anthers and pollen contain carotenoids, especially carotenes and their hydroxy, epoxy and hydroxyepoxy derivatives (Goodwin, 1976, 1980; Goodwin and Britton, 1988), and include novel xanthophylls in *Lilium tigrinum* (Märki-Fischer and Eugster, 1985). Most seeds contain only traces of carotenoids, but maize is an exception in accumulating large amounts of β-carotene, β-cryptoxanthin and zeaxanthin (White *et al.*, 1942). The seeds of *Bixa orellana* contain bixin, the pigment of annatto (Karrer and Jucker, 1950).

The best known example of a root carotenoid is that of β-carotene in the carrot, *Daucus carota*, and in the sweet potato.

4.4 Biosynthetic pathways and enzymology

The biosynthesis of carotenoids has been subjected to numerous investigations over the past 40 years, using a wide range of photosynthetic and non-photosynthetic tissues and organisms. The first unified hypothesis for the biosynthesis of carotenoids in higher plants was proposed by Porter and Lincoln (1950), based upon the inheritance of carotenoids in tomato fruits, and later revised by Porter and Anderson (1962). The subsequent advent of sophisticated analytical techniques, the use of inhibitors of carotenogenesis, microbial mutant strains and, more recently, attempts to isolate the enzymes and their encoding genes, have all advanced our understanding of the biosynthesis of these pigments. In this section, only the salient features of the biosynthetic pathway, relevant to the ensuing discussion on the manipulation of carotenogenesis will be described. Numerous general and comprehensive review articles on carotenoid biosynthesis have been published in recent years (Spurgeon and Porter, 1983; Jones and Porter, 1986; Britton, 1988, 1990; Bramley, 1985, 1991, 1992; Sandmann, 1991).

Since carotenoids are isoprenoids they are related biosynthetically and share a common early pathway with other biologically important isoprenoids such as sterols, gibberellins and terpenoid quinones (Figure 4.4). In the following subsections the pathway is divided into several convenient stages, starting at the conversion of geranylgeranyl diphosphate (GGDP) into phytoene. Reviews on the earlier, common steps of the pathway can be found elsewhere (Qureshi and Porter, 1983; Poulter and Rilling, 1983; Gray, 1987; Kleinig, 1989; Poulter, 1990).

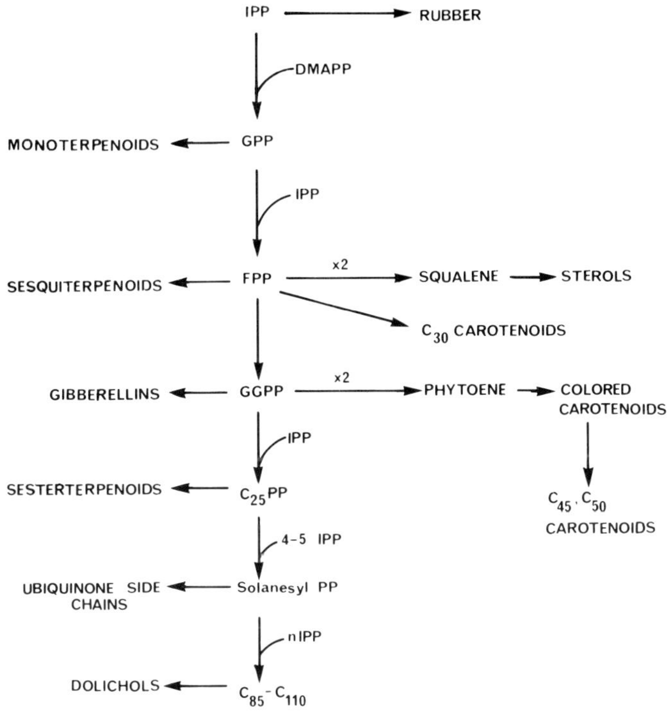

Figure 4.4 Overall pathway of isoprenoid biosynthesis.

4.4.1 Formation of phytoene from geranylgeranyl diphosphate

The first unique step in the formation of carotenoids is the head-to-head condensation of 2 molecules of all-*E* GGDP via the cyclopropylcarbinyl diphosphate, prephytoene diphosphate (PPDP, Figure 4.5). Depending on the stereochemistry of removal of a hydrogen atom from PPDP, either 15-*Z* or all-*E* phytoene can be formed. Phytoene in eukaryotic plants is predominately the 15-*Z* isomer, with only minute quantities of the all-*E* compound (see Jones and Porter, 1986).

Numerous plant cell extracts are able to form phytoene from radioactive precursors such as mevalonic acid (MVA), isopentenyl diphosphate (IDP) and GGDP, and possible cofactor requirements have been reported (reviewed by Bramley, 1992). There is little consensus in the reports, however, as several different pyridine and flavin nucleotides and ATP have been found to be stimulatory *in vitro* (Maudinas *et al.*, 1975, 1977; Clarke *et al.*, 1982). The requirement for a divalent cation has often been reported, especially Mn^{2+} for the plant enzyme.

A partially purified phytoene synthase complex was isolated from tomato plastids with an apparent M_r of 200 kDa (Maudinas *et al.*, 1977). On further purification, the complex was disrupted to yield pure IDP isomerase and prenyl transferase, but only a partially pure preparation which converted GGDP and PPDP into phytoene (Jones and Porter, 1986)

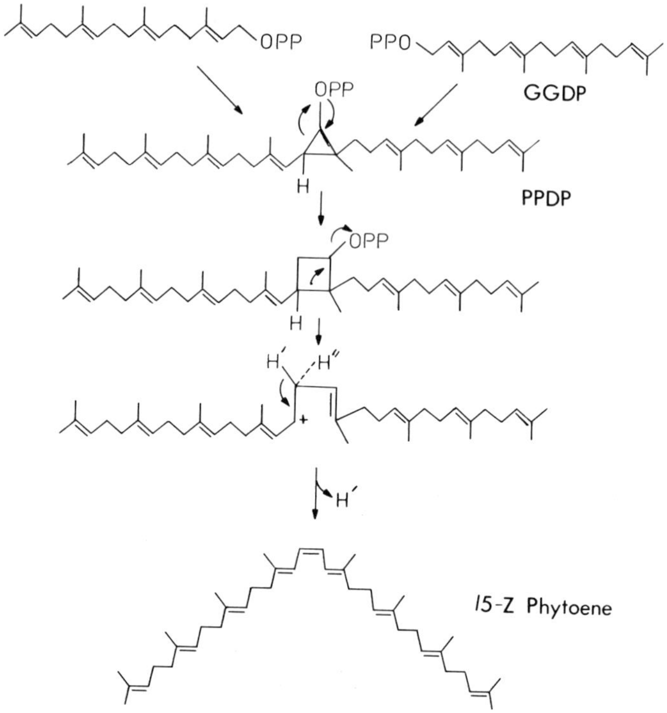

Figure 4.5 The formation of 15-*Z* phytoene from GGDP.

More recently, phytoene synthase has been purified from the chromoplast stroma of *Capsicum annuum*, to yield a single protein, M_r 47.5 kDa, with an absolute requirement for Mn^{2+}, but no other cofactors (Dogbo *et al.*, 1988). The pure enzyme catalyses a two-step, kinetically coupled reaction, with K_m values of 0.3 and 0.27μM for GGDP and PPDP, respectively. The conversion is inhibited by inorganic phosphate.

A single enzyme is thought to catalyse the two-step reaction in the non-photosynthetic bacterium *Erwinia* (Sandmann and Misawa, 1992) but different enzymes for the GGDP→PPDP and PPDP→phytoene conversions are assumed to be present in *Rhodobacter capsulatus* (Armstrong *et al.*, 1990a). Further details of the genes encoding these enzymes are given in section 4.5. A prokaryotic phytoene synthase has yet to be purified.

Phytoene synthase is located in the stroma of chromoplasts, chloroplasts and amyloplasts of *Capsicum* with the highest activity in the non-photosynthetic plastids (Dogbo *et al.*, 1987). It has also been found in the stroma of pea fruits (Graebe, 1968) and *Triticum* leaves (Kleinig, 1989). In contrast, the enzyme is associated with the chloroplast envelope of spinach (Lütke-Brinkhaus *et al.*, 1982) and is a peripheral membrane protein of the inner membrane of *Narcissus* chromoplasts (Kreuz *et al.*, 1982). Whether there is a genuine difference in the intracellular location of phytoene synthase between plant species remains to be confirmed. Arguments that carotenoids associated with the chloroplast envelope are an artifact of the isolation procedure (Grumbach, 1984) have been strongly refuted (see Joyard *et al.*, 1991). It is apparent, however, that the protein(s) is synthesised on 80S ribosomes prior to post-translational processing and entry into the plastid (Camara, 1984).

4.4.2 Desaturation and isomerisation reactions

The desaturation sequence from phytoene to lycopene is by far the most studied of all the carotenogenic reactions. This is partly for historical reasons, as it was the first carotenoid biosynthetic sequence to be deduced (Porter and Lincoln, 1950) but also because of the importance of the desaturation reactions as a prime target site for bleaching herbicides (reviewed by Sandmann and Böger, 1989; Bramley, 1991, 1992).

There is general agreement that the sequence involves four stepwise dehydrogenations, occurring alternatively at either side of the chromophore, to yield in turn phytoene, ζ-carotene, neurosporene and lycopene (Figure 4.6). Desaturation may be continued further to 3, 4-didehydro and 3, 4, 3′, 4′ tetradehydrolycopene, as

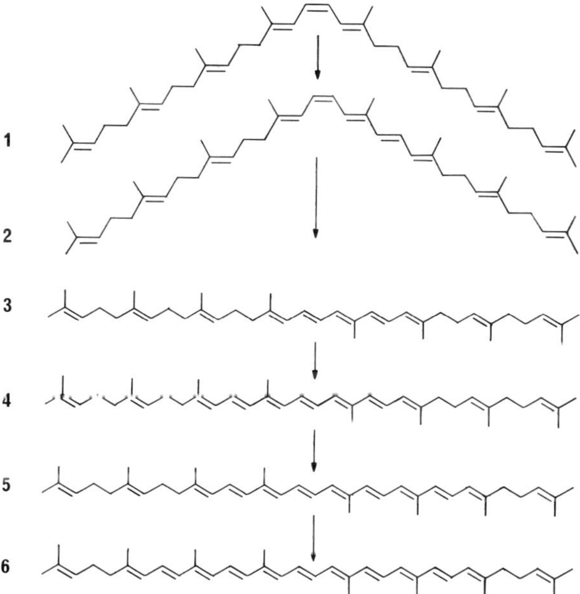

Figure 4.6 The desaturation of 15-*Z* phytoene (1) into all-*E* lycopene (6), via 15-*Z* phytofluene (2), all-*E* phytofluene (3), all *E*ζ-carotene (4), and all-*E* neurosporene (5).

reported to occur in Valencia oranges (Winterstein *et al.*, 1960) and, perhaps, in spinach chloroplast envelopes (Lütke-Brinkhaus *et al.*, 1982). In addition, an isomerization of the 15–15′ double bond from the *Z*- to *E*-configuration must take place, since 15-*Z* phytoene is formed by phytoene synthase yet lycopene is predominately the all-*E* isomer.

As efforts continue to elucidate fully the molecular details of the desaturation sequence, four outstanding questions remain to be answered satisfactorily:

1. How many enzymes are required for the conversion of phytoene into lycopene?
2. At what stage does isomerisation occur?
3. What is the mechanism of desaturation?
4. Where are the enzymes located within the plant cell?

1. The four dehydrogenations and isomerisation raise the possibility of four discrete dehydrogenases, one specific for each step, and an isomerase. From mutants isolated so far it is unlikely that there are specific dehydrogenases for each step. Mutants of algae and higher plants either accumulate phytoene or ζ-carotene, suggesting the presence of two enzymes, a phytoene dehydrogenase and a ζ-carotene dehydrogenase, both of which are encoded by nuclear genes (Kirk and Tilney-Bassett, 1978). Inhibitor studies with herbicides that specifically inhibit phytoene dehydrogenation compared with others that predominately cause the accumulation of ζ-carotene support this view (see Sandmann and Böger, 1989; Bramley, 1991, 1992; Young, 1991). *In vitro* carotenogenic activities of *Narcissus* and *Aphanocapsa* cell extracts also point to two enzymes being involved. The conversions of phytoene into ζ-carotene and ζ-carotene into lycopene by *Narcissus* chromoplasts have different cofactor requirements and involve *E*-removal of hydrogens to form ζ-carotene, but *Z*-elimination during the ζ-carotene to lycopene sequence. The two stages are also sensitive, to different extents, to the herbicidal inhibitor, norflurazon (Mayer *et al.*, 1989). Inhibition constants of herbicides in the *Aphanocapsa* cell-free system also suggest the presence of two such enzymes (see Sandmann and Böger, 1989).

Using a tomato cell extract, different cofactors were required for the phytoene to phytofluene and phytofluene to lycopene steps (Subbarayan *et al.*, 1970), an observation also reported for spinach leaf cell extracts (Kushwaha *et al.*, 1969). A similar difference has been found in the bacterium *Halobacterium cutirubrum* (Kushwaha *et al.*, 1976) and the fungus *Phycomyces blakesleeanus* (Fraser and Bramley, unpublished results). If these reports indicate two enzymes for the complete desaturation sequence, then they do not have the same catalytic activities as the phytoene and ζ-carotene desaturases discussed earlier.

In contrast to the suggestion of two desaturases, genetic complementation studies with *P. blakesleeanus* mutants, defective in phytoene metabolism, have shown that the phytoene to lycopene sequence is catalysed by multiple copies of the same *carB* gene product, in an enzyme aggregate (de la Guardia *et al.*, 1971; Aragon *et al.*, 1976), while recent analyses of the genes from *Erwinia uredovora*

(Misawa *et al.*, 1990) and *E. herbicola* (Schnurr *et al.*, 1991) have demonstrated that *crtI* encodes a single desaturase which catalyses the conversion of phytoene into lycopene.

2. The stage at which isomerisation occurs is also a matter of some debate. It may be that isomerisation is not enzymically catalysed but results from a strained, saturated carotene at the active site of the desaturase (Goodwin, 1983) and is therefore associated with the desaturation reactions. However, isomerisation does appear to be affected in the tangerine tomato mutant (see section 4.3.3.3). The isomerisation of the 15–15′ double bond is thought to take place at the phytofluene stage in tomato plastids (Kushwaha *et al.*, 1970) and *Capsicum* fruits (Camara and Moneger, 1982), although it occurs at the ζ-carotene level in a strain of *Scenedesmus* (Britton and Powls, 1977). The direct conversion of 15-*Z* phytoene into unsaturated carotenoids has been demonstrated with cell extracts of tomato fruit (Kushwaha *et al.*, 1970) *Capsicum* (Camara and Moneger, 1982), daffodil chromoplasts (Beyer *et al.*, 1985) and *Aphanocapsa* (Bramley and Sandmann, 1985). The 3-[(3-cholamidopropyl)-dimethylammonio]-1-propane sulphonate (CHAPS)-solubilised preparation of *Narcissus* catalyses a series of *Z* to *E* isomerisations *in vitro*, although there are no reports of such reactions in the intact flower (Beyer *et al.*, 1989).

A range of poly-*Z*-carotenes are found in the fruit of the tangerine tomatoes (Zechmeister, 1962; Raymundo and Simpson, 1972). Prolycopene and proneurosporene are found in a variety of plant sources (see Goodwin, 1980). The structural analysis of the poly-*Z*-carotenes of the tangerine tomato (Clough and Pattenden, 1979, 1983) suggests that the branch point for synthesis of *E*-carotenes is phytoene, although any enzymes involved have not been characterised apart from a report of *in vitro* incorporation studies (Qureshi *et al.*, 1974).

3. Two mechanisms have been proposed for the removal of two hydrogen atoms and introduction of a new double bond. One possibility is the direct loss of two hydrogens via a dehydrogenase-electron transferase mechanism. Evidence in support of this assumption has come mainly from *in vitro* experiments. In the presence of either NAD or NADP, phytoene desaturation is enhanced in *Anacystis* membranes and is inhibited under anaerobic conditions (Sandmann and Kowalczyk, 1989). These two pyridine nucleotides, as well as FAD, have been implicated in phytoene desaturation in other organisms (reviewed by Bramley, 1992). The alternative mechanism is a two-step process of hydroxylation followed by dehydration. Proponents of this mechanism point to the presence of hydroxylated acyclic carotenes found in cells treated with herbicides (Britton *et al.*, 1987), and to the suggestion that desaturation is an oxidative process requiring a mixed function oxygenase coupled to cytochrome P450 (Britton, 1979). However, no experimental evidence has been reported to substantiate the involvement of a mixed function oxygenase, while the positions of the hydroxy groups in monohydroxy phytoene are not at C11 and C12, where the introduction of a new double bond occurs (Albrecht *et al.*, 1991). On present experimental evidence, therefore,

desaturation of phytoene is likely to be an oxidative dehydrogenation reaction. In daffodil chromoplast membranes desaturation is independent of oxidised pyridine nucleotides and oxygen is the final electron acceptor with an oxidoreductase as a redox mediator (Beyer *et al.*, 1989). Different oxidised quinones can substitute for oxygen *in vitro* (Mayer *et al.*, 1990).

The hydrogens lost are in an *E*-relationship whether in the formation of all-*E* (McDermott *et al.*, 1973) or poly-*Z* lycopene (Williams, Britton and Goodwin, unpublished, quoted by Britton, 1990).

4. Although there is general agreement that phytoene metabolism is catalysed by a membrane-bound enzyme(s), there is no consensus of opinion on the intracellular location of phytoene desaturase. It is reported to be associated with the chloroplast envelope of spinach leaves (Lütke-Brinkhaus *et al.*, 1982) and in chromoplast membranes of daffodil flowers (Kreuz *et al.*, 1982; Kleinig and Beyer, 1985) and *Capsicum* fruits (Camara *et al.*, 1982). In contrast, Grumbach and Britton (1984) have found that it is located in the thylakoid membranes of radish, and not in the chloroplast envelope fraction. Immunogold labelling of thylakoids of *Synechocystis* and *Anabaena* using a polyclonal antibody to phytoene desaturase showed 85% of the enzyme in photosynthetic membranes (Serrano *et al.*, 1990). Technical difficulties in isolating pure envelope and thykaloid fractions may account for these anomalies, although the enzyme may be located in both thykaloid and plastid envelope membranes.

4.4.3 Cyclisation reactions

Alicyclic end rings are a common feature of many carotenoids. They are present not only in higher plants, but also in bacteria, fungi and algae. In higher plants, two ring types are found, the β-ring (e.g. β-carotene, Figure 4.1) and ε-ring (e.g. in lutein, Figure 4.1).

The β- and ε-rings are formed independently, although probably from the same precursor. The mechanism of cyclisation, involving initial proton attack at C2 of the acyclic precursor, has been proved by experiments in which accumulated acyclic precursors were allowed to cyclise *in situ* in the presence of D_2O and the position of the deuterated carbon in the β- or ε-ring determined. The resulting carbonium ion is stabilised by the loss of a proton either from C1 or C4 to yield a β- or ε-ring, respectively. The stereochemistry of cyclisation has been reviewed by Britton (1990).

Historically, there has been a prolonged debate as to whether neurosporene or lycopene is the primary substrate for cyclisation, since cyclised products of both of these acyclic carotenes are found in nature. Mechanistically, cyclisation would be the same in both cases, as the requirement for cyclisation is that one end group has reached the lycopene level of desaturation. The various cyclisation products of both neurosporene and lycopene are shown in Figure 4.7.

The conversion of lycopene into β-carotene has been demonstrated in several

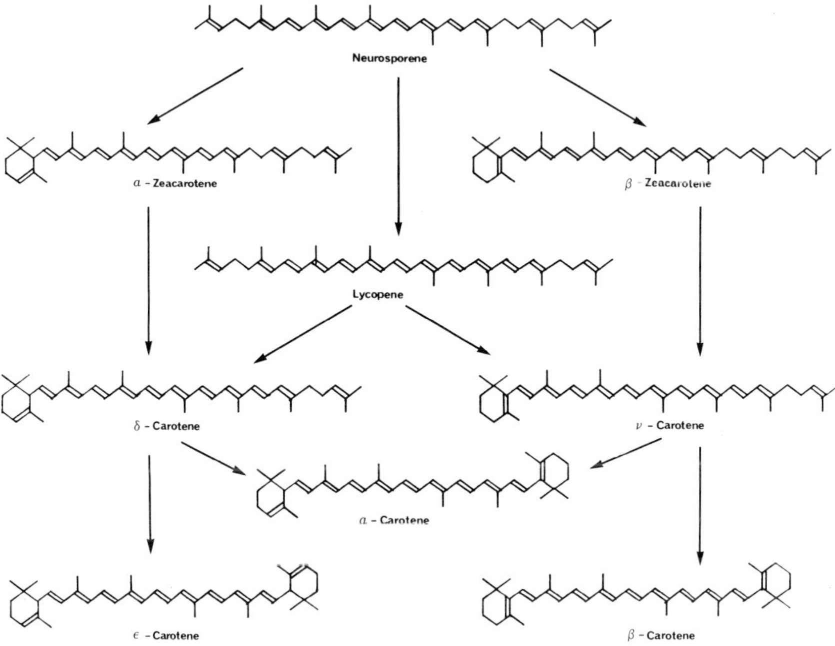

Figure 4.7 Cyclisation reactions in carotenogenesis.

systems, e.g. plastids of French bean and tomato fruit (Kushwaha *et al.*, 1970; Hill *et al.*, 1971), spinach chloroplasts (Kushawha *et al.*, 1969) and cell extracts of *Flavobacterium* (Mohanty, 1988). The formation of α-, β-, γ-, and δ-carotenes requires FAD in tomato plastids (Maudinas *et al.*, 1975) but, in contrast, no nucleotide cofactor is needed for β-carotene synthesis by a membrane-solubilised extract of *Capsicum* (Camara *et al.*, 1985; Camara and Dogbo, 1986).

It is not known with any certainty how many enzymes are required for cyclisation of lycopene into β- or ε-ring-containing carotenes. Gene analysis of *Erwinia* shows only one gene product is required for the two-step reaction via γ-carotene (Sandmann *et al.*, 1990; Schnurr *et al.*, 1991) and a dimer of the *carR* gene product catalyses the same sequence in *Phycomyces* (Candau *et al.*, 1991). In contrast, separate genes encode the enzymes for β-and ε-ring formation in tomato. There is indirect evidence for two enzymes being required for the formation of monocyclic and bicyclic carotenes. The second cyclisation reaction is more sensitive to cyclase inhibitors such as CPTA and nicotine than the first (reviewed by Jones and Porter, 1986), although this could be interpreted as two catalytic sites on the same enzyme.

4.4.4 Xanthophyll formation

A wide range of xanthophylls are present in nature. Alicyclic xanthophylls with oxygen substituents on the end rings are commonly found and in higher plants carotenoids with hydroxy groups at C3 and C3′ of the rings (e.g. lutein and zeaxanthin, Figure 4.1) or epoxy groups at 5, 6, or 5′, 6′ (e.g. violaxanthin and neoxanthin, Figure 4.1) frequently occur. The oxygen moieties originate from molecular oxygen. Hydroxylation at C3 has been studied experimentally in the formation of zeaxanthin and is thought to be catalysed by a mixed-function oxidase, involving a cytochrome P450 (Britton, 1990). Hydroxylation of β-carotene to β-crytoxanthin has been studied using membranes of *Aphanocapsa*. The reaction shows typical characteristics of a mono-oxygenase reaction as it is stimulated by NADPH, dependent upon oxygen and is sensitive to carbon monoxide and plant specific mono-oxygenase inhibitors (Sandmann and Bramley, 1985).

The conversion of zeaxanthin into antheraxanthin and violaxanthin (the 'violaxanthin cycle') requires O_2 and NADPH (Siefermann and Yamamoto, 1975). The epoxidase appears to be an 'external mono-oxygenase' located in a chloroplast component that remains 'neutral' during illumination, i.e. the violaxanthin cycle is a transmembrane system, whereby violaxanthin de-epoxidation occurs on the lumen side of the thylakoid membrane, while epoxidation takes place on the stromal side. The de-epoxidase is a lipoprotein, with monogalactosyl diglyceride as the only lipid, and has been isolated from spinach (Hager and Perz, 1970) and lettuce chloroplasts (Yamamoto *et al.*, 1974; Yamamoto and Higashi, 1978).

During the ripening of *C. annuum* there is a rapid synthesis of keto-carotenoids such as capsanthin and capsorubin (Figure 4.3). The biosynthesis of the five-membered ring is thought to occur by a pinacolic rearrangement of a 3-hydroxy-5, 6-epoxy group (Camara, 1980) and these keto-carotenoids are formed from β-cryptoxanthin, antheraxanthin and violaxanthin (Davies *et al.*, 1970). Isolated chromoplasts of *C. annuum* convert zeaxanthin and antheraxanthin into capsanthin and capsorubin. The cofactors NADP, NADPH and ATP had no effect on the reaction (Camara and Moneger, 1981).

Finally, it should be noted that the purple non-sulphur photosynthetic bacteria, which have been used for carotenoid gene isolation (section 4.5), are characterised by the presence of acyclic carotenoids with tertiary hydroxy and methoxy groups at C1 and C1′ and an unsaturated double bond at C3,4, typified by spheroidene. The biosynthesis of these compounds is discussed in detail elsewhere (Spurgeon and Porter, 1983).

4.4.5 Degraded carotenoids

A range of carotenoids are found in which the carbon skeleton has been modified by removal or rearrangement of carbon atoms. In higher plants, two types of degraded carotenoids occur: the apocarotenoids in which the molecule has been shortened by the loss of carbon atoms, e.g. bixin and crocetin (Figure 4.3), and the secocarotenoids in which the ring has been opened at a position other than C1,6 e.g. β-carotenone (Figure 4.3).

The secocarotenoids are thought to be oxidative products of β-carotene

(Yokoyama and White, 1970) but no enzymic studies have been reported. They are often found in the fruits of citrus.

The apocarotenoids found in many higher plants are likely to be products of the oxidative degradation of C_{40} carotenoids (Weeden, 1971). For example, the C_{20} crocetin is thought to be formed by the degradation of zeaxanthin (Pfander and Schurtenberger, 1982), rather than from GGDP.

4.5 Carotenoid genes in bacteria and lower eukaryotes

The genes involved in the biosynthesis of carotenoids have been most extensively characterised in prokaryotic and lower eukaryotic organisms. The identification of colour mutants has been the major factor facilitating this process. Such mutants may result from alterations of genes encoding either enzymes of the carotenoid biosynthetic pathway or proteins involved in the regulation of carotenogenesis. In the last few years genes involved in these pathways have been isolated and characterised from several bacteria and lower eukaryotes. These genes have proved to be useful for the identification of related genes in higher plants (see section 4.6).

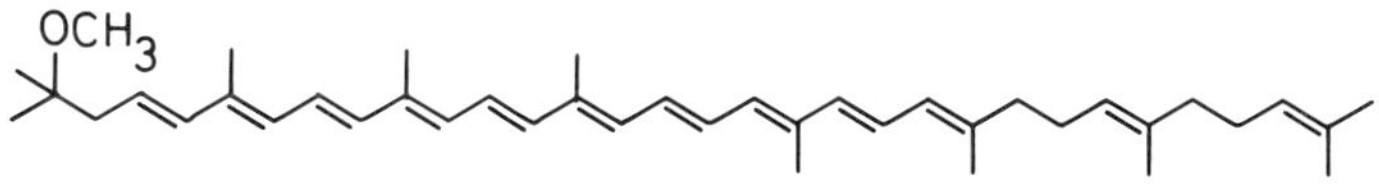

OCH_3

Spheroidene

4.5.1 Carotenoid genes in bacteria

The genes involved in carotenoid biosynthesis in the Gram-negative purple non-sulphur photosynthetic bacterium *Rhodobacter capsulatus* have been extensively analysed. When grown with reduced oxygen, the acyclic carotenoids spheroidene and hydroxyspheroidene are accumulated at the same time as the induction of intracytoplasmic photosynthesis. A 46 kilobase (kb) cluster of photosynthetic genes has been shown to contain at least nine carotenoid biosynthesis (*crt*), eleven bacteriochlorophyll (*bch*) and various structural photosynthetic genes (Yen and

Table 4.1

Gene product	Residues	M_r
crtA	591	64761
crtB	339	37299
crtC	281	31855
crtD	494	52309
crtE	289	30004
crtF	393	43004
crtI	491	54755
crtK	160	17607

Marrs, 1976; Scolnik *et al.*, 1980; Zesbo and Hearst, 1984; Armstrong *et al.*, 1989, 1990a). Eight of the nine *crt* genes (*crtA*, *B*, *C*, *D*, *E*, *F*, *I* and *K*) are located in an 11 kb region. The nucleotide sequence of this region and the organisation of the gene cluster have been determined (Armstrong *et al.*, 1989; Bartley and Scolnik, 1989). From this, the predicted sizes of the gene products have been determined (Table 4.1) (Armstrong *et al.*, 1989). The other known carotenoid biosynthesis gene (*crtJ*) is located about 12 kb from the cluster (Zesbo and Hearst, 1984). The nucleotide sequence of this gene has not yet been determined.

The functions of the *Rhodobacter capsulatus crt* genes have been studied in a variety of point and transposon mutants. Both *in vivo* accumulation of intermediates and *in vitro* biochemical defects in cell-free carotenoid synthesis have been used to assign functions to the genetic loci. The current understanding of the genes involved in the pathway has been summarised by Armstrong and co-workers (1990a). However, further work will be required to confirm the assignments of these genes.

Mutations in the pathway prior to the dimerisation of GGDP would not be specific to carotenoid biosynthesis and may well be lethal since the products of this general isoprenoid biosynthesis are involved in several subsequent pathways (Figure 4.4). The *crtB*, *crtE*, *crtI* and *crtJ* gene products have been associated with the early reactions specific to carotenoid biosynthesis (Zesbo and Hearst, 1984). Mutations in any of these genes result in the inability to accumulate the pigmented carotenoids produced later in the pathway.

Cell-free extracts of *crtB* and *crtJ* mutants accumulate [^{14}C]-GGDP when incubated with [^{14}C]-IDP, while extracts of a *crtE* mutant accumulate [^{14}C]-PPDP (Armstrong *et al.*, 1990a). Thus, it has been proposed that CrtE is phytoene synthase and that CrtB, and possibly CrtJ, are components of the prephytoene diphosphate synthase. However, the *crtE* sequence shows significant homology to the *Neurospora al-3* gene which is thought to encode GGDP synthase (Carattoli *et al.*, 1991) and there is no evidence that CrtB and/or CrtJ do not catalyse both of the reactions involved in the dimerisation of GGDP to form phytoene. Additional evidence for the role of CrtB came from the 23.3% sequence identity with the protein encoded by a tomato cDNA (pTOM5) which has subsequently been shown to be involved in the conversion of GGDP to phytoene (Bramley *et al.*, 1992) (see section 4.6). Confirmation of the functions of these three genes will require further investigation, in particular studies of the gene products.

The colourless phytoene is converted in *Rhodobacter* into the coloured carotenoids by a series of dehydrogenation reactions leading initially to neurosporene. It is not clear how many individual enzymes are required for these reactions in *R. capsulatus*. However, *crtI* mutants accumulate phytoene and mixtures of cell-free extracts from *crtI* and *crtB* mutants are able to convert phytoene to coloured carotenoids (Giuliano *et al.*, 1986). Regions of the *crtI* gene product have strong homology to similar regions of the *R. capsulatus* CrtD protein which is thought to be involved in dehydrogenation reactions after neurosporene (Armstrong *et al.*, 1989). In addition, antibodies raised against an *E. coli lac–crtI* fusion gene product have been shown to inhibit the dehydrogenation of phytoene in cell-free extracts (Schmidt *et al.*, 1989). Thus, *crtI* probably encodes phytoene dehydrogenase which catalyses at least the first dehydrogenation reaction in the pathway. There is as yet no report of additional genes encoding enzymes involved in the further dehydrogenations required for the production of ζ-carotene and neurosporene.

The conversion of neurosporene into the acyclic xanthophylls that accumulate in *R. capsulatus* (see section 4.4) requires a series of hydroxylations, dehydrogenations and methylations. The products of the *crtA*, *crtC*, *crtD* and *crtF* genes are thought to be involved in these reactions. The acyclic carotenoids produced by these reactions are not the major carotenoids found in higher plants which are probably synthesised by different mechanisms.

The genes involved in the early steps of the carotenoid pathway have also been isolated and characterised from other bacterial species. *Erwinia herbicola* and *E. uredovora* are non-photosynthetic epiphytic bacteria which accumulate cyclic carotenoids that are probably derivatives of β-carotene. Thus, different carotenoids are synthesised by the two *Erwinia* species and *R. capsulatus* which implies that the biosynthetic pathways after phytoene dehydrogenation are likely to be divergent. The pathway to cyclic carotenoids in *Erwinia* is closer to that in higher plants.

Clusters of carotenogenic genes have been isolated from both *E. herbicola* and *E. uredovora* (Perry *et al.*, 1986; Misawa *et al.*, 1990). These have included genes that encode products with homology to the *R. capsulatus* CrtB (33.7%:35.1%), CrtE (30.8%:27.8%) and CrtI (41.7%:40.1%) proteins (Armstrong *et al.*, 1990b, Misawa *et al.*, 1990). These proteins have been implicated in the first three reactions specific to carotenoid biosynthesis.

The role of the *Erwinia crtB* and *crtE* genes has been studied by expression in *E. coli* and *Agrobacterium tumefaciens* (Sandmann and Misawa, 1992). Extracts of *E. coli* containing *crtE* showed enhanced conversion of farnesyl diphosphate (FDP) to GGDP, while *A. tumefaciens* containing *crtB* was able to convert GGDP into phytoene. Thus *crtE* and *crtB* are likely to encode GGDP synthase and phytoene synthase respectively.

Functional analysis of the *E. uredovora crtI* gene expressed in *E. coli* indicated that only one gene product was required to catalyse the four dehydrogenation reactions to convert phytoene to lycopene (Misawa *et al.*, 1990). The CrtI (phytoene dehydrogenase) sequences from *Rhodobacter* and *Erwinia* also have significant homology to the *pds* gene which encodes the same enzyme in the unicellular cyanobacteria *Anacystis nidulans* (Chamovitz *et al.*, 1991). The CrtI proteins from the three bacterial genuses all contain a short, highly conserved sequence that has homology to FAD or NAD(P) binding sites.

A phytoene dehydrogenase gene has also been isolated from the cyanobacteria *Aphanocapsa* by using the *R. capsulatus crtI* gene as a hybridisation probe (Schmidt and Sandmann, 1990). The encoded protein is 43 amino acids longer than CrtI from *R. capsulatus*. However, significant similarity is observed in two regions. Unlike the phytoene dehydrogenase from *Erwinia* this enzyme is thought to catalyse the insertion of only two double bonds into phytoene resulting in the

Table 4.2

Reaction	*E. herbicola*	*E. uredovora*
Lycopene cyclisation	*crtZ*	*crtY*
β-Carotene hydroxylation	*crtH*	*crtZ*
Zeaxanthin glycosylation		*crtX*

formation of ζ-carotene. An alternative gene has been postulated to encode ζ-carotene dehydrogenase (Schmidt and Sandmann, 1990).

Anacystis and *Erwinia* both accumulate cyclic carotenoids derived from lycopene. Three genes which have been implicated in these reactions have been isolated from both *Erwinia* species (Misawa *et al.*, 1990; Schnuur *et al.*, 1991). However, the gene nomenclature used for the two species was different (Table 4.2). The functional assignments of these genes is largely based on expression studies in *E. coli*.

4.5.2 Carotenoid genes in fungi

Carotenogenic genes in eukaryotes are less well defined than those in the bacterial systems. Colour mutants of fungi have been the principal source of the characterisation that has been achieved. Four fungal species (*Gibberella fujikuroi*, *Neurospora crassa*, *P. blakesleeanus* and *Ustilago violacea*) have been the major subjects of genetic studies. All of these species accumulate cyclic carotenoids similar to those found in higher plants.

The *albino-3* (*al-3*) gene of *N. crassa*, which encodes GGDP synthetase, has been isolated and characterised (Nelson *et al.*, 1989; Carattoli *et al.*, 1991). It encodes a polypeptide with a molecular mass of 47.8 kDa which has substantial homology to the CrtE proteins from *Rhodobacter* and *Erwinia*. Comparison with other prenyltransferase sequences shows three highly conserved regions that may be involved in the formation of the catalytic site.

Several mutants of *G. fujikuroi* and *N. crassa* that accumulate PPDP have been identified (Avalos *et al.*, 1988; Kushwaha *et al.*, 1978). These are presumed to be mutations in the phytoene synthase genes although the possibility that they are involved in the regulation of the pathway cannot be excluded. The genes involved in the dimerisation of GGDP to form phytoene have not been isolated from these organisms.

Mutations in the *al-1* gene of *Neurospora* resulted in the accumulation of phytoene. This gene has been cloned and sequenced and encodes a 66 kDa polypeptide (Schmidhauser *et al.*, 1990). Alignment of the Al-1 sequence with the sequences of the dehydrogenases CrtI and CrtD from *R. capsulatus* revealed conserved sequences (Bartley *et al.*, 1990). Further evidence that *al-1* encodes phytoene dehydrogenase came from complementation experiments in which the gene was expressed in *R. capsulatus crtI* mutants. The complemented mutant accumulated a carotenoid that co-migrated with lycopene on thin layer chromatography. This carotenoid normally accumulates in *N. crassa* but is not normally found in *R. capsulatus*. This observation supported the previously proposed hypothesis (Bramley and Mackenzie, 1988) that only one enzyme is required to catalyse the four desaturations in the conversion of phytoene to lycopene. However, this may not be the case for all fungi. Studies of *Ustilago* mutants have been interpreted to indicate separate dehydrogenases catalysing the conversion of phytoene to neurosporene and neurosporene to lycopene (Will *et al*., 1984)

Few mutants have been identified which are blocked in cyclisation of lycopene. One example is the *carR* mutant of *P. blakesleeanus* (de la Guardia *et al*., 1971). The *carR* gene product is believed to catalyse the two-step cyclisation from lycopene to β-carotene (Candau *et al*., 1991). It seems likely that a single cyclase is responsible for the conversion of lycopene to β-carotene in fungi (Bramley and Mackenzie, 1992). However, no genes involved in either the cyclisations of further modifications have been isolated.

Mutants of *P. blakesleeanus* defective in some regulatory function with respect to carotenogenesis produce either reduced or increased levels of carotenoids compared to the wildtype. They may also fail to respond to external stimuli. Carotenoid biosynthesis in this fungus is controlled by the end product, β-carotene, light, various chemicals and sexual interactions of the (+) and (-) strains. At least 6 regulatory genes are thought to control carotenoid biosynthesis in *Phycomyces* (see Bramley and Mackenzie, 1992).

4.6 Higher plant genes and their manipulation

In bacteria and lower eukaryotes, mutants have played a significant part in biochemical studies aimed at unravelling the pathway and control of carotenoid biosynthesis. In higher plants a large number of mutants have been described which affect carotenoid accumulation and synthesis. It is perhaps surprising that many of the available mutants have not been extensively used by biochemists and molecular biologists in their studies towards understanding the structure and regulation of the expression of genes involved in carotenoid biosynthesis. For example, a large number of mutants of maize and tomato are available. Recently, genetic and molecular approaches have been used to identify genes involved in carotenoid biosynthesis from higher plants.

4.6.1 Higher plant mutants

For many years, plant breeders have accumulated mutants in carotenoid biosynthesis as the phenotypes are very easily recognisable in most instances, e.g., the lack of pigmentation of the tomato fruit or the maize kernel. Several mutants of higher plants have been referred to by Bramley and Mackenzie (1988). We will concentrate in this review on mutants of maize and tomato. The reason for this is the great variety of mutants which exist in these species. In addition, tomato is a suitable experimental system to explore the manipulation of carotenoid biosynthesis.

Table 4.3 Carotenoid content of class I mutants of maize.

Mutant	Phytoene	Phytofluene	ζ-Carotene	Neurosporene	Lycopene
vp2	+	+			
vp5	+	+	+		
vp9	+	+	+		
ps	+	+			+
w3	+	+	+	+	
	+	+	+		
al	+	+	+		
	+	+			
y9	+	+	+		
z			+		
lc					+
ly					+

4.6.1.1 Maize. Robertson (1975) has listed many of the mutants affecting carotenoid accumulation and biosynthesis in maize. These can be divided into two classes: class I mutants have white or pale yellow endosperm and albino seedlings. In class II mutants the endosperm pigmentation is not affected whereas the plants are albino. Several of the class I mutants are listed in Table 4.3.

4.6.1.2 Tomato. Several of the known carotenoid mutants of tomato fruit are listed in Table 4.4. These have been compiled from Stevens and Rick (1986), Rick (1982) and Tanksley and Mutschler (1990).

A scheme has been proposed for the biochemical pathway and the potential steps which are affected by these mutants (Stevens and Rick, 1986). It is apparent that these mutants represent many different steps in carotenoid synthesis. Further biochemical work is required to establish both the carotenoid complement and the biochemical steps which are inhibited in the different mutants.

Carotenoid accumulation is also affected in ripening mutants of tomato. In addition to changes in carotenoid accumulation, other facets of fruit ripening are

Table 4.4 Carotenoid mutants of the tomato fruit.

Mutant	Phenotype	Chromosome [a]
r (yellow fruit)	No carotenes	3–29
hp (high pigment)	Increased carotenoid levels	12
dg (dark green)		
t (tangerine)	Increased β-carotene, no other carotenoids	10–104
og^c (crimson)	Increased lycopene reduced β-carotene	6–106
B (β-carotene)	Increased β-carotene, reduced lycopene	6–106
del (delta)	Increased δ-carotene, reduced lycopene	
sh (sherry)	Yellow fruit flesh	10L
Mo_b	Modifier of B	
at (apricot)	Apricot fruit colour	5

[a]Chromosome number and map position: S short arm; L long arm.

inhibited, such as ethylene synthesis or firmness. These mutants are listed in Table 4.5.

The availability of those mutants in which carotenoid biosynthesis is affected as part of a general alteration of fruit ripening can also be a useful tool in the identification of genes involved in carotenoid biosynthesis. Molecular techniques such as differential hybridisation of cDNA libraries has identified genes which are

Table 4.5 Fruit mutants in which carotenoid biosynthesis is affected as part of a general delay of ripening.

Mutants	Fruit phenotype	Chromosome [a]
alc (alcobaca)	Dirty orange	10S
nor (non-ripening)	Greenish	10–19
Nr (never ripe)	Dirty orange	9
Nr 2 (never ripe 2)	Yellowish green	1L
rin (ripening inhibitor)	Yellow	5–0

[a]Chromosome number and map position: S short arm; L long arm.

specifically expressed during fruit ripening (Slater *et al.*, 1985). The expression of these cDNA clones has been studied in normal and mutant tomatoes (Knapp *et al.*, 1989). Several clones were identified whose expression was reduced during fruit ripening of the mutants. Among these was pTOM5 (Ray *et al.*, 1987) which was subsequently found to encode phytoene synthase (Bird *et al.*, 1991; see below for details).

Other mutants of tomato exist in which altered leaf or corolla phenotypes are displayed. These may be due to changes in the levels of carotenoids in these organs.

Table 4.6 Mutants affecting carotenoid levels in other organs of the tomato plant.

Mutant	Phenotype	Chromosome[a]
au (aurea)	Foliage bright yellow	1–32
ch (chartreuse)	Corolla greenish yellow	8–28
cn (cana)	Leaves grey-green, tiny plants	3–24
ele (elegans)	Plant tiny, leaves yellow-green	11–30
ful (fulgens)	Leaves bright yellow, turning green	4–24
gh (ghost)	Plant starts green, later white	11–43
hy (homogeneous yellow)	All vegetative parts yellow	10–5
lu (luteola)	Corolla light yellow	1
og (old gold)	Corolla tawny orange	6–106
pdw (pale dwarf)	Leaves pale yellow	7
pli (plicata)	All parts small, leaves dark yellow	3
sy (sunny)	Cotelydons bleached Whitish, true leaves emerge yellow, becoming green	3–46
wf (white flower)	Corolla white to bluff	3–44
Xa (xanthophyllic)	Yellow leaves, growth retarded, homozygous inviable	10–80
Xa-2 (xanthophyllic-2)	Same phenotype as Xa, homozygous viable	10–9
Xa-3 (xanthophyllic-3)	Leaves more greenish than Xa	10–38
yg-6 (yellow–green-6)	Seedling etiolated	11–61
yg-8 (yellow–green-8)	Seedling leaves bright yellow	8L

[a]Chromosome number and map position: S short arm; L long arm.

Further biochemical and molecular work is required to prove this assumption. Some of these mutants are summarised in Table 4.6.

4.6.1.3 Arabidopsis. Many of the mutants referred to above have been available for research for many years. During the past 10 years, *Arabidopsis thaliana* has gained prominence in plant science research due to the ease in which mutant phenotypes can be generated and genetically characterised. In addition, *Arabidopsis thaliana* has a very small genome size, a good genetic map and an ever improving restriction length polymorphism (RFLP) map. The use of molecular techniques to construct libraries of large DNA fragments covering the *Arabidopsis* genome is well advanced (Hauge *et al.*, 1990). Thus, it will be feasible to use RFLP markers to 'walk' to genes which have only been characterised genetically.

Recently Rock and Zeevart (1991) analysed carotenoids in the *aba* mutants of *Arabodipsis*. In the *aba-4* mutant β-carotene accumulated while in the *aba-3* mutant zeaxanthin accumulated. In both mutants the levels of epoxy-carotenoids violaxanthin and neoxanthin were reduced. Thus, these mutants may represent mutations in the later stages of the carotenoid biosynthetic pathway.

4.6.2 Methods used to identify carotenoid genes

In this section progress made on the isolation of genes specific to the carotenoid pathway will be summarised. Genes involved in the early steps of the pathway from melavonic acid to geranylgeranyl diphosphate (GGDP) which are common to the biosynthesis of several isoprenoids have been isolated. HMG-CoA reductase has been isolated from tomato and its role in fruit development has been investigated (Narita and Gruissem, 1989). The isolation of geranylgeranyl diphosphate synthase from pepper has recently been reported (Kuntz *et al.*, 1992). These will not be discussed further.

Several different methods have been used for the identification of carotenoid biosynthesis genes. These are reviewed below.

4.6.2.1 Antisense RNA approach.

Identification of phytoene synthase genes. Antisense RNA techniques permit the inhibition of specific genes, leading to a mutant phenotype in which a targeted gene product is reduced or absent. This has been elegantly documented through the inhibition of chalcone synthase (CHS) in petunia (van der Krol *et al.*, 1988). CHS is a critical enzyme for the biosynthesis of anthocyanins which give petunia flower petals their distinctive coloration. Inhibition of CHS via antisense RNA led to mutant flower petal phenotypes ranging from anthocyanin deficient petals to petals with novel coloration patterns.

Bird *et al.* (1991) argued that it should be possible to use this technique to identify a gene involved in carotenoid biosynthesis. These authors suggested that among genes isolated from a ripe tomato cDNA library there should be genes involved in lycopene synthesis. For this purpose they constructed antisense genes to ripening enhanced cDNA clones. One of these clones was pTOM5 which is expressed highly during tomato fruit ripening (Ray *et al.*, 1987). Sequence determination of this clone and database searches had not revealed the biochemical function of protein encoded by the clone. The pTOM5 antisense gene was inserted into wildtype tomato plants using an *Agrobacterium* mediated transformation method. The analysis of the modified primary transformants revealed many plants with ripe yellow fruit (29 out of 33 analysed). In addition to the altered fruit colour phenotype the plants also exhibited reduced corolla pigmentation. This novel phenotype was inherited and segregated with the presence of the antisense gene. In this study it was also shown that pTOM5 mRNA was absent in plants with the yellow ripe fruit phenotype. Modified plants which had red ripe fruit showed normal levels of pTOM5 mRNA in the fruit. Fruits with the yellow fruit phenotype had greatly reduced overall levels of carotenoids and lycopene in particular. In one plant, for example, no lycopene was found. Thus these data indicated that pTOM5 encodes a protein involved in lycopene biosynthesis.

In order to identify the step in the carotenoid pathway that was inhibited, biochemical feeding experiments were carried out using radiolabelled MVA as a precursor of carotenoids (Bramley *et al.*, 1992). Extracts from ripe control fruit were able to efficiently convert the radiolabelled precursor to phytoene. When radiolabelled phytoene was used as precursor, the extract was able to generate unsaturated carotenes. The extract of the mutant yellow fruit was equally well able to metabolise radiolabelled phytoene indicating that the desaturation reactions leading to lycopene synthesis were not inhibited. However, radiolabelled MVA was not converted into phytoene. Instead, large amounts of GGDP accumulated as an intermediate indicating that the repression of pTOM5 gene expression led to the inhibition of the conversion of GGDP to phytoene, the step encoded by phytoene synthase.

This enzyme has been purified by Dogbo *et al.* (1988) from pepper. It is interesting as it carries out two reactions: the single polypeptide of 47.5 kDa molecular mass catalyses the conversion of two molecules of GGDP to form prephytoene diphosphate (PPDP) and the synthesis of phytoene from PPDP. There have been no reports of separate enzymes catalysing these reactions in higher plants. The enzyme is localised in the stroma of pepper chromoplasts. The pTOM5 encoded protein has a molecular weight of 46.7 kDa (Ray *et al.*, 1987) and can be imported into isolated pea chloroplasts (Moore, personal communication). Thus it was concluded (Bramley *et al.*, 1992) that pTOM5 may encode the tomato homologue of the pepper phytoene synthase gene.

An analysis of carotenoid content in leaves, green fruit, ripe fruit and flower petals of the yellow fruited and control tomato lines has been carried out (Bramley *et al.*, 1992). The expression of the pTOM5 antisense gene did not affect the composition and levels of carotenoids in the leaves or green tissues whereas carotenoid levels in the flower petals were reduced. This result may be interpreted to mean that pTOM5-related gene expression occurs in fruit and flower petals only. The expression of the pTOM5 encoding gene has been studied (Maunders *et al.*, 1987; Knapp *et al.*, 1989). It was demonstrated that pTOM5 related sequences were found during fruit ripening but not in green fruit, stems, roots or leaves. Expression in flower petals was not investigated. Thus the lack of effect of the pTOM5 antisense construct on the levels of carotenoids in the leaves can be explained by the apparent absence of the target RNA in this tissue. These experiments raise an intriguing question: what is the nature of the phytoene synthase in leaves? The data can be interpreted in two ways: (i) the biosynthetic pathway in leaves involves a different phytoene synthase to the one found in fruit or flower petal tissues, or (ii) in leaves and green fruit tissues a gene distantly related to pTOM5 but with the same structure is expressed which is not targeted by the antisense gene. Further molecular and biochemical work is needed to clarify these issues. In addition some of the mutants referred to above may prove useful in these studies.

Analysis of the structure of genes encoding pTOM5 has demonstrated the presence of two gene loci with sequences complementary to pTOM5 which are found on chromosome 2 and chromosome 3 (Kinzer *et al.*, 1990). Sequences complementary to pTOM5 have been isolated from a tomato genomic library (Ray *et al.*, 1992). Two genes were identified: one gene, represented by the clone gTOM5, is identical to pTOM5; a second gene is closely related but not identical to pTOM5. No expression of this gene has been detected leading to the suggestion that it may represent a pseudogene. Further work is now in progress to analyse the tissue specificity of expression of these two genes.

It is also interesting to note that sequences found on chromosomes 2 and 3 of tomato surrounding the pTOM5 loci appear to be duplicated. A genetic locus (*r*) has been identified on chromosome three (Table 4.4) which gives rise to yellow fruit similar in appearance to the pTOM5 antisense fruit. It is intriguing to speculate that the pTOM5 gene represents the wildtype allele of the gene whose mutations led to the *r* mutation.

It is clear that the antisense approach has been very successful in identifying the phytoene synthase gene. One of the major advantages of this technique is that once the mutant has been generated, the gene is immediately available. Thus it should prove a powerful tool in the identification of other genes in the carotenoid biosynthesis pathway.

4.6.2.2 The use of heterologous probes and complementation analysis

Identification of the phytoene desaturase gene. Another approach has been used by Chamovitz *et al.* (1991) and Bartley *et al.* (1991). Chamovitz *et al.* isolated and characterised a DNA clone encoding phytoene desaturase from *Synechococcus* PCC7942. A series of mutants were generated which showed resistance to the bleaching herbicide norflurazon. A DNA fragment conferring resistance to this herbicide was transferred from the mutant cells to wildtype cells. The mutation was mapped allowing the identification of a 2.8 kb BamH1-Sal1 restriction fragment. Sequencing this fragment from the mutant and control strains led to the identification of the phytoene desaturase gene. This clone has been used to identify a clone from the algae *Dunaliella* and from tomato. The tomato phytoene desaturase clone is approximately 80% homologous to the *Synechococcus* clone. However, the details of this sequence similarity have not yet been published.

Bartley *et al.* (1991) used a cDNA clone encoding phytoene desaturase from *Synechococcus* PCC6301 to screen a soybean immature cotyledon cDNA library. A clone was identified which encodes a phytoene desaturase polypeptide of approximately 63.8 kDa. Chloroplast uptake experiments have shown that this protein is efficiently imported into chloroplasts. The putative leader sequence of phytoene desaturase is 94 amino acids giving the mature protein a molecular weight of approximately 54 kDa which is in good agreement with the size of carotenoid desaturases from *N. crassa* and *R. capsulatus*. Comparison of the nucleotide sequence of the soybean desaturase with other carotenoid desaturases (*crtI* and *crtD*

R. capsulatus, *N. crassa al-1*) revealed sequence conservation in several regions of the peptide with particularly high conservation in the putative FAD binding region found near the N-terminus.

The soybean clone was then used to functionally complement an *R. capsulatus* mutant strain deficient in *crtI*. This strain normally accumulates phytoene. However, when a *pds1* expression construct was introduced, a yellow carotenoid was synthesised which was identified as 15-*cis*-ζ-carotene. Thus, the *pds1* clone encodes a desaturase that is able to carry out two successive desaturation reactions without any isomerisation. This does not account for geometric isomers of carotenoids found in leaves and may imply either the presence of other genes involved in the isomerisation reactions or a difference in the folding of the plant enzyme expressed in *R. capsulatus* which does not allow the isomerisation reaction to be carried out.

Using genomic Southern hybridisations the presence of a small multigene family encoding phytoene desaturase in soybean has been demonstrated. However, the exact number of members of this family has not yet been revealed. Using the soya bean *pds1* clone, Bartley *et al.* have also isolated a phytoene desaturase from tomato (G.E. Bartley, personal communication). Thus, it is clear that significant sequence conservation exists between the phytoene desaturase genes from bacteria and lower eukaryotes and those of higher plants. This will permit the isolation of these genes enabling their structure and expression to be studied during plant development. This will also facilitate the detailed biochemical analysis of the enzymic reactions involved as well as the manipulation of the genes and the pathway.

4.6.2.3 Transposon tagging

Identification of the Y1 gene from maize. Transposon tagging has been used in maize to isolate a number of genes (Federoff *et al.*, 1984; O'Reilly *et al.*, 1985; McLaughlin and Walbot, 1987; Theres *et al.*, 1987). Considerable efforts are, being employed to transfer suitably modified transposable elements to other plant species and adapt them for efficient gene tagging (Dean *et al.*, 1991). The hope is that such a system will permit the isolation of any gene which has been mutated and tagged by a transposon.

In maize several naturally occurring transposable elements are available for tagging. Buckner *et al.*, (1990) have used a Robertson's mutator element to tag the *y1* locus of maize. The dominant Y1 allele determines the presence of β-carotene in the endosperm leading to yellow kernels. The mutant *y1* allele leads to a lack of β-carotene accumulation and white kernels. There are several *y1* alleles which determine both white kernels, and a reduction in the amount of β-carotene in the leaves when the plants are grown at elevated temperatures (e.g. 35 °C) (Robertson and Anderson, 1961). Studies with these mutant alleles which demonstrated tissue specificity of the expression of the *y1* gene led to the suggestion that *Y1* is a potential regulator of the carotenoid biosynthetic pathway in maize (Robertson and Anderson, 1961).

The *y1* gene was tagged with the Robertson's mutator Mu3 element. For this purpose the Mu3 element was introduced into a line of maize homozygous for the dominant wildtype allele *Y1*. These were crossed with plants containing the homozygous recessive *y1* gene. The progeny of this cross is expected to be heterozygous *Y1/y1* (yellow kernels) unless the Mu3 element moves into the *Y1* gene which would be expected to be white. Plants which had white kernels were identified in the progeny of this cross and the Mu3 induced mutant allele was called *y1-mum*. Plants homozygous for the *y1-mum* allele were separated into two classes: kernels which germinated as green seedlings and kernels which germinated as green seedlings when grown at 25°C and light-green seedlings when grown at 35°C. Kernels giving rise to to these 'pastel' seedlings were chosen as the starting material for cloning experiments. Plants carrying the 'pastel' allele, *y1-mum2053 (Y1/y1-mum 2053)*, were crossed to homozygous recessive *y1* plants. This resulted in plants carrying yellow *(Y1/y1)* and white *(y1-mum-2053/y1)* kernels.

DNA was isolated from these plants and analysed by Southern blotting using a HindIII-XbaI fragment specific to the Mu3 element. The restriction fragments containing the Mu3 gene sequences inserted into the putative *y1* gene sequences were identified. In BamH1 restriction digests two fragments of 4.4 kb and 5.1 kb were identified which were present both in the parent mutant plant *(Y1/y1-mum 2053)* and in all plants derived from white endosperm kernels with the genotype *(y1-mum 2053/y1)*. One of these two restriction fragments (4.4 kb) was cloned using an internal fragment of the Mu3 element as hybridisation probe. The location of the Mu3 sequences within this fragment was determined by restriction mapping and a fragment flanking the Mu3 sequences (150 bp BamH1-PvuII) was isolated and used as a probe in further Southern hybridisation experiments. This demonstrated the co-segregation of this DNA with Mu3 hybridising fragments and putative *Y1* DNA fragments which gave rise to fragments of 2.4 kb and 3.1 kb as well as a 2.9 kb fragment associated with the *y1* allele.

Revertants of the *y1-mum 2035* allele were identified in further crosses and Southern blot analysis was performed. The 150 bp BamH1-PvuII fragment identified DNA fragments of the same size as seen in the *Y1* or *y1* genes, thus supporting the conclusion that the fragment did indeed represent part of the maize *y1* gene.

This probe (150 bp fragment) was then used to isolate the wild-type *Y1* allele. A 11.4 kb EcoR1 fragment was identified which cross-hybridised to the probe. Further analysis demonstrated that a transcript of 2 kb hybridised to the probe indicating that an active gene is encoded in the genomic fragment. So far no further work has been published on the characterisation of the *Y1* gene.

These data clearly demonstrate the power of the genetic and transposon tagging approach. As many carotenoid mutants are available in maize (see above) it is likely that this approach will be used for the isolation of other genes involved in carotenoid biosynthesis.

4.6.3 Manipulation of carotenoid genes

The identification of phytoene synthase (Bird *et al.*, 1991; Bramley *et al.*, 1992) represents an excellent example of the genetic methods now available for the manipulation of the carotenoid biosynthesis. Not only can gene function be determined but novel mutant phenotypes can be generated. This may be particularly useful in plants in which naturally occurring mutants are not available.

One of the challenges for the future now lies in the use of the carotenogenic genes for the overproduction of carotenoids, such as an increase in lycopene in tomatoes. The genetic elements for these experiments are now available and these experiments are in progress (Schuch *et al.*, unpublished).

When other genes of carotenoid biosynthesis become available, through the application of the techniques discussed above, other manipulations will become possible. Of particular interest will be the insertion of genes into plants which will permit the production of carotenoids not normally found in target plants or the production of completely novel carotenoid structures.

4.7 Industrial production of carotenoids

Carotenoids are produced commercially in three ways: chemical synthesis, fermentation and extraction from natural sources, especially higher plants.

4.7.1 Chemical synthesis

The wide variety of natural carotenoids presents chemists with a fascinating range of challenges to their ingenuity in synthesis, and a continual search is maintained for new and better methods of polyene synthesis. The first total synthesis of β-carotene was reported in 1950 (Karrer and Eugster, 1950) and in the proceeding decades a range of synthetic methods has been developed for the large-scale production of vitamin A, β-carotene, apocarotenoids, apocarotenoid esters, canthaxanthin, citranaxanthin and astaxanthin (Mayer and Isler, 1971; Kienzle, 1976; Klaui, 1982; Bernhard, 1990). The commercial production of these compounds employs relatively few key reactions, most of which proceed via β-ionone and often vitamin A alcohol (Kienzle, 1976; Bernhard, 1990). The major companies involved in commercial synthesis are F. Hoffman-LaRoche, BASF and Rhone-Poulenc.

4.7.2 Fermentation

Fungi (both filamentous and yeast) and algae have been used as commercial sources of carotenoids. During the 1960s interest in the biosynthesis of carotenoids by fungi resulted in several patents but commercialisation has proved difficult due to the relatively low carotenoid content. Strain improvement of wildtype

P. blakesleeanus has resulted in yields of some 25 mg β-carotene per gram of mycelia (Cerdá-Olmedo, 1987; Bramley and Mackenzie, 1992). Carotenoid-containing yeasts such as *Rhodotorula* and *Phaffia rhodozyma* have also been used, the latter producing astaxanthin. In general, however, the high costs of production have been commercially prohibitive (Nonomura, 1990). It may be that the use of genetic transformation techniques with fungi will lead to increased yields of carotenoids.

A more recent development, largely over the past seven years, has been the use of mass algal cultures open to sunlight (Klausner, 1986). In particular, the unicellular algae *D. salina* and *D. bardawil* have been cultured outside for the production of β-carotene. High accumulations of carotene are achieved when the algae is grown in high salinity, high light intensity and at a low nitrogen content (Ben-Amotz and Avron, 1983). Under these conditions, the algae grows slowly and turns orange-red (Ben-Amotz and Avron, 1980), yielding some 13% of dry weight as carotene, and 2–9% of the algal dry weight under commercial growing conditions (Ben-Amotz *et al.*, 1982). Controlled high density cultures have been developed in Israel and the USA, with capacities of 10 million litres, cell densities of 1 g/litre and yields of 50 mg carotene per litre (Nonomura, 1990).

After harvest, the algae are either washed and dried as a powder or the carotenoids may be extracted with petroleum solvents or vegetable oil. The extracts are not extensively purified and contain several natural products besides carotenoids, e.g. sterols, fatty acids and phospholipids. The final extract must be free of solvent residues in order to meet regulations and good manufacturing practices. Further details of the extraction process are given by Nonomura (1990), and the algal carotene products produced commercially are listed in Table 4.7.

4.7.3 Extraction from higher plants

There are a number of extracts of higher plants which yield carotenoid mixtures that are used commercially (Table 4.8). Generally, the low concentrations of

Table 4.7 Algal carotene products.

Company	Location	Product
Microbio Resources Inc.	USA	1–3% *D. salina* carotene in oil (Provatene) and algal astaxanthin
Cyanotech Inc.	USA	Dry algal powder (Konatene)
Betalene Ltd	Australia	4% *D. salina* carotene in soybean oil
Western Biotechnology Ltd	Australia	1–3% *D. salina* carotene in oil (Bionova)
Salt Research Institute	China	Crystalline and oil suspensions of *D. salina* carotene
Nature Beta Technology	Israel	Dry algal powder
Phycotene	USA	Mixture of carotenes and xanthophylss (Phycotene)
Earthrise Farms	USA	Dry mixture of *Spirulina* (Lina)

Table 4.8 Extracts of higher plants used commercially.

Plant	Carotenoids	Colour of extract
Bixa orellana seeds	Bixin, norbixin ('annatto')	Red/orange
Carrot root	Carotenes, mainly β-carotene	Yellow
C. annuum fruit	Capsanthin, capsorubin (paprika)	Red
Crocus sativus petals	Crocin, crocetin (saffron)	Yellow
Marigold petals	Lutein, zeaxanthin	Yellow
Tomato fruit	Lycopene	Red
Palm oil	Carotenes	Yellow
Citrus peel and DCPTA	Lycopene	Red

carotenoids in vascular plants result in high production costs. Extractions are generally undertaken with petroleum solvents. A novel method of production of mixed carotenoids from waste citrus fruit has been developed by Yokoyama and co-workers (1975, 1982). Onium compounds such as CPTA and DCPTA, sprayed onto orange peel, increased the carotenoid content of the peel turning it red as a result of the accumulation of lycopene (see also section 4.4).

Increased concentrations of carotenoids in higher plants, especially in fruits, flowers and roots, may be achieved using transformation protocols (see section 4.6).

4.8 Commercial utilisation of carotenoids

The commercial value of carotenoids, estimated at some $105 million worldwide in 1989 (Taylor, 1990), is related to the aesthetic appeal of their characteristic colours, their biological role as precursors of vitamin A and their potential value as dietary chemo-preventative agents of some types of cancer.

4.8.1 Food, pharmaceutical and cosmetic colorants

The colour of food is a significant factor in determining its acceptability. Foods are expected to be a 'natural' colour, and consumers become cautious when a product has an unexpected colour since they interpret this as a sign of spoilage, poor processing or adulteration. As food colorants, carotenoids possess several advantages including high colour quality (thus needing only small amounts to achieve the desired effect), lack of toxicity, and ease of formulation in aqueous and lipid-soluble systems. Such formulations are also relatively stable. In addition, carotenoids are natural pigments, a fact that increasingly appeals to consumers and is reflected in the number of 'natural extracts' of plant and algal tissues available commercially (Tables 4.7 and 4.8) and the use of the term 'nature-identical' for synthetically produced carotenoids used to pigment food.

Fat-based foods are usually coloured with β-carotene using an oily suspension which is added to the product. The stability of the β-carotene under these conditions is very good (Bauernfeind *et al.*, 1971). When required for colouring water-based foods, such as juices and beverages, the carotenoids are formulated into water-dispersable products, such as colloidal suspensions, fine dispersions in surface-active agents, such as sucrose fatty-acid esters, or into a gelatin-sugar matrix.

Water-dispersable carotenoids are used for colouring sugar-coated tablets and also cosmetic products such as lipsticks, lotions and powder bases. Further details of the formulation processes can be found in reviews by Bauernfeind and co-workers (1971), Klaui (1982) and Bernhard (1990).

4.8.2 Nutritional supplements

The primary nutritional value of carotenoids is from their pro-vitamin A activity. All-*E*-β-carotene has a theoretical vitamin A activity of 1.7 million international units per gram, although limited absorption and poor transformation of β-carotene into vitamin A reduces the actual yield of the vitamin from foods. Carotenoids with only one β-ring, e.g. apocarotenoids, have only half the vitamin A activity, while acyclic carotenes, e.g. lycopene, have no such nutritional value.

4.8.3 Pigmentation of animal and fish feeds

Although carotenoids are abundant in the animal kingdom (see Goodwin, 1986), they are not synthesised *de novo* by the animal and must be ingested in the diet. Many carotenoids are responsible for colour in fish, animals and animal products used as human food, and such pigmentation is viewed as a sign of quality. Therefore many feeds contain additional amounts of carotenoids in order to colour the skin and fat of animals and fish, or products such as eggs. Consequently, such dietary supplements are an indirect method of colouring food. For example, the addition of carotenoid mixtures, either from natural sources (marigold leaves, lucerne) or synthetic compounds such as β-apo-carotenoic acid and canthaxanthin, to poultry feed results in enhanced skin and egg yolk pigmentation while a formulated astaxanthin ('Carophyll pink') is added to feed for trout and salmon raised in fish farms (Bernhard, 1990).

Whatever additive is used in feeds, extensive toxicological tests and registration procedures are required. A detailed account of the legislation as applied to the addition of synthetic astaxanthin to fish feed is given by Bernhard (1990).

4.8.4 Carotenoids in medicine

Carotenoids are currently recommended as a treatment for light-sensitive diseases such as erythropoietic protoporphyria and congenital porphyria (Mathews-Roth, 1982, 1989), and also for vitamin A deficiency (Krinsky, 1989). There is, however, an increasing body of evidence that shows retinoids and carotenoids can prevent some types of epithelial and colonic cancer (see Matthews-Roth, 1989; Krinsky, 1990). If epidemiological studies in human populations confirm a role for carotenoids in preventing certain cancers, then the market demand is estimated to increase to some $600 million by 1995 (Taylor, 1990). This demand would largely be due to the extra consumption of β-carotene and other carotenoids as supplements to the normal diet, as the projected daily requirement of 60–300 mg per day could not be obtained from the diet alone (Taylor, 1990).

4.9 Conclusions and future studies

It is clear from the preceding sections that although we have a solid understanding of the structural diversity of carotenoids our understanding of carotenoid biosynthesis, in terms of the properties of the enzymes and their associated genes, is just emerging. There is only one report of a purified carotenogenic enzyme, and the majority of studies of carotenoid genes are with microorganisms in which little enzymology has been attempted.

Although the genetic approaches reviewed here have allowed the isolation of gene fragments encoding several of the carotenogenic genes, the biochemical identification of proteins encoded is by no means complete. It is therefore important to combine the powerful molecular approaches with biochemical techniques for the purification and characterisation of carotenogenic enzymes. There is a need to isolate the enzymes of the pathway, especially from higher plants. This will permit the accumulation of data on the biosynthesis of carotenoids such as the estimation of the levels of enzymic activity or gene expression in higher plants, especially during plastid development. Suitable molecular probes are now becoming available and these should be used with suitable plant tissues in order to quantify the rates of enzyme synthesis and flux of metabolites through the carotenoid pathway.

Once the regulatory steps within carotenogenesis are known and the interaction with isoprenoid pathways is clear, it will be feasible to manipulate the pathway both in microorganisms and in higher plants to enhance carotenoid production. This could simply lead to increased yield of carotenoids in commercial production in microorganisms, but may also facilitate the production of carotenoids if expression of rate limiting enzymes in the pathway can be enhanced. Alternatively, it could lead to the production of carotenoids not found in higher plants, especially those formed from β-carotene or lycopene. These could be synthesised through the introduction of genes isolated from microorganisms. Enhanced carotenoid accumulation in plants would potentially allow tissues to be more nutritious, e.g. with increased β-carotene content, or else make the extraction of carotenoids from higher plants a commercially viable proposition. The combination of techniques described here will make a major contribution to our understanding of carotenoid biosynthesis and should encourage scientists to take up the challenge to work in this area.

References

Albrecht, M., Sandmann, G., Musker, D. and Britton, G. (1991) Identification of epoxy-and hydroxy-phytoene from norflurazon-treated *Scenedesmus*. *J. Agric. Food Chem.* **39**: 566.

Aragon, C.M.G., Murillo, F.J., De La Guardia, M.D. and Cerdá-Olmedo, E. (1976) An enzyme complex for the dehydrogenation of phytoene in *Phycomyces*. *Eur. J. Biochem.* **63**: 71.

Armstrong, G.A., Alberti, M., Leach, F. and Hearst, J.E. (1989) Nucleotide sequence, organisation and nature of the protein products of the carotenoid biosynthesis gene cluster of *Rhodobacter capsulatus*. *Mol. Gen. Genet.* **216**: 254.

Armstrong, G.A., Schmidt, A., Sandmann, G. and Hearst, J.E. (1990a) Genetic and biochemical characterisation of carotenoid biosynthesis mutants of *Rhodobacter capsulatus*. *J. Biol. Chem.* **265**: 8329.

Armstrong, G.A., Alberti, M. and Hearst, J.E. (1990b) Conserved enzymes mediate the early reactions of carotenoid biosynthesis in nonphotosynthetic and photosynthetic prokaryotes. *Proc. Natl. Acad. Sci. USA*. **87**: 9975.

Avalos, J., MacKenzie, A., Nelkie, D.S. and Bramley, P.M. (1988) Terpenoid biosynthesis in cell extracts of wild type and mutant strains of *Gibberella fujikuroi*. *Biochem. Biophys. Acta* **966**: 257.

Bartley, G.E. and Scolnik, P.A. (1989) Carotenoid biosynthesis in photosynthetic bacteria. *J. Biol. Chem.* **264**: 13109.

Bartley, G.E., Schmidhauser, T.J., Yanofsky, C. and Scolnik, P.A. (1990) Carotenoid desaturases from *Rhodobacter capsulatus* and *Neurospora crassa* are structurally and functionally conserved and contain domains homologous to flavoprotein disulfide oxidoreductases. *J. Biol. Chem.* **265**: 16020.

Bartley, G.E., Viitanen, P.V., Pecker, I., Chamovitz, D., Hirschberg, J. and Scolnik, P.A. (1991) Molecular cloning and expression in photosynthetic bacteria of a soybean cDNA coding for phytoene desaturase, an enzyme of the carotenoid biosynthesis pathway. *Proc. Natl. Acad. Sci. USA*. **88**: 6532.

Bauernfeind, J.C., Brubacher, G.B., Klaui, H.M. and Marusich, W.L. (1971) Use of carotenoids. In *Carotenoids* (Isler, O., ed.), Birkhauser Verlag, Basel, pp. 743–770.

Ben-Amotz, A. and Avron, M. (1980) Glycerol, β-carotene and dry algal meal production by *Dunaliella*. In *The Production and Use of Micro-algal Biomass* (Shelef, G., and Soeder, C.J., eds), Elsevier, Amsterdam, pp. 603–610.

Ben-Amotz, A. and Avron, M. (1983) On the factors which determine massive β-carotene accumulation in the halotolerant algae. *Dunaliella bardawil*. *Plant Physiol*. **72**: 593.

Ben-Amotz, A., Katz, A. and Avron, M. (1982) Accumulation of β carotene in halotolerant algae. purification and characterisation of β-carotene rich globules from *Dunaliella bardawil*. (*Chlorophyceae*). *J. Phycol.* **18**: 529.

Benedict, C.R, Rosenfield, C.L, Mahan, J.R, Madha van, S. and Yoloyama, H. (1985) The chemical regulation of carotenoid biosynthesis. *Citrus*. *Plant Science* **41**: 169.

Bernhard, K. (1990) Synthetic astaxanthin. The route of a carotenoid from research to commercialisation. In *Carotenoids: Chemistry and Biology* (Krinsky, N.I., Mathews-Roth, M.M. and Taylor, R.F., eds.), Plenum Press, New York, pp. 337–363.

Beyer, P., Weiss, G. and Kleinig, H. (1985) Solubilization and reconstitution of the membrane bound carotenogenic enzymes from daffodil chromoplasts. *Eur. J. Biochem*. **153**: 341.

Beyer, P., Weiss, G. and Kleinig, H. (1989) Molecular oxygen and the state of geometric isomerisation of intermediates are essential in the carotene desaturation and cyclisation reactions in daffodil chromoplasts. *Eur. J. Biochem*. **184**: 141.

Bird, C.R., Ray, J.A., Fletcher, J.D., Boniwell, J.M., Bird, A.S., Teulieres, C., Blain, I., Bramley, P.M. and Schuch, W. (1991) Using antisense RNA to study gene function: inhibition of carotenoid biosynthesis in transgenic tomatoes. *BioTechnology* **9**: 635.

Bowden, R.D., Cooper, R.D.G., Harris, C.J., Moss, G.P. Weedon, B.C.L and Jackman, L.M. (1983). *J. Chem. Soc. Perkin Trans* **1**: 1465.

Bramley, P.M. (1985) The *in vitro* biosynthesis of carotenoids. *Adv. Lipid Res.* **21**: 243.

Bramley, P.M. (1991) Carotenoid biosynthesis. In *Target Sites for Herbicide Action* (Kirkwood, R.C., ed.), Plenum, New York, pp.95–122.

Bramley, P.M. (1992) Carotenoid biosynthesis. In *Methods in Plant Biochemistry* (Lea, P.J., ed.), Academic Press, London (in press).

Bramley, P.M. and Mackenzie A. (1988) Regulation of carotenoid biosynthesis. *Curr. Top. Cellul. Regul.* **29**: 291.

Bramley, P.M. and Mackenzie, A. (1992). Carotenoid biosynthesis and its regulation in fungi. In *Handbook of Applied Mycology, Vol 4, Fungal Biotechnology*. (Arora, D.K., Elander, R.P. and Mukerjii, K.G., eds.), Marcel Dekker, New York, pp. 407–444.

Bramley, P.M. and Sandmann, G. (1985). *In vitro* and *in vivo* biosynthesis of xanthophylls by the cyanobacterium *Aphanocapsa Phytochemistry* **24**: 2919.

Bramley, P., Teulieres, C., Blain, I., Bird, C.R. and Schuch, W. (1992) Biochemical characterisation of transgenic tomato plants in which carotenoid synthesis has been inhibited through the expression of antisense RNA to pTOM5. *Plant J.* **2** : 343.

Britton, G. (1979) Carotenoid biosynthesis—a target for herbicidal activity. *Z. Naturforsch.* **34C**: 979.

Britton, G. (1985) General carotenoid methods. *Methods Enzymol.* **111**: 113.

Britton, G. (1988) Biosynthesis of carotenoids. In *Plant Pigments* (Goodwin, T.W., ed.), Academic Press, London, pp. 133–182.

Britton, G. (1989) Carotenoids and polyterpenoids. *Nat. Prod. Rep.* **6**: 359.

Britton, G. (1990) Carotenoid biosynthesis—an overview. In *Carotenoids: Chemistry and Biology* (Krinsky, N.I., Mathews-Roth, M.M., and Taylor, R.F., eds.), Plenum Press, New York, pp. 167–184.

Britton, G. (1991) Carotenoids. In *Methods in Plant Biochemistry, Vol. 7* (Charlwood, B.V. and Banthorpe, D.V., eds.), Academic Press, London, pp. 473–518.

Britton, G. and Powls, R. (1977) Phytoene, phytofluene and ζ-carotene isomers from a *Scenedesmus obliquus* mutant. *Phytochemistry.* **16**: 1253.

Britton, G., Barry, P., and Young, A.J. (1987) The mode of action of diflufenican: its evaluation by HPLC. *Proc. Crop Prot. Conf. Weeds*. BLPC Publications, Thornton Heath, UK, p. 1015.

Buckner, B., Kelson, T.L. and Robertson, D.S. (1990) Cloning of the *y1* locus of maize, a gene involved in the biosynthesis of carotenoids. *The Plant Cell.* **2**: 867.

Camara, B. (1980). Biosynthesis of ketocarotenoids in *Capsicum annuum* fruits. *FEBS Lett.* **118**: 315.

Camara, B. (1984) Terpenoid metabolism in plastids: sites of phytoene synthase activity and synthesis in plant cells. *Plant Physiol.* **74**: 112.

Camara, B. and Dogbo, O. (1986) Demonstration and solubilisation of lycopene cyclase from *Capsicum* chromoplast membranes. *Plant Physiol.* **80**: 172.

Camara, B. and Moneger, R. (1981) Carotenoid biosynthesis, *in vitro* conversion of antheraxanthin to capsanthin by a chromoplast enriched fraction of *Capsicum* fruits. *Biochem. Biophys. Res. Commun.* **99**: 1117.

Camara, B. and Moneger, R. (1982) Biosynthetic capabilities and localization of enzymatic activities in carotenoid metabolism of *Capiscum annuum* isolated chromoplasts. *Physiol. Veg.* **20**: 757.

Camara, B., Bardat, F. and Moneger, R. (1982) Sites of biosynthesis of carotenoids in *Capsicum* chromoplasts. *Eur. J. Biochem.* **127**: 255.

Camara, B., Dogbo, O., d'Harlingue, A., Kleinig, H. and Moneger, R. (1985) Metabolism of plastid terpenoids: lycopene cyclisation by *Capsicum* chromoplast membranes. *Biochem. Biophys. Acta.* **836**: 262.

Candau, R., Bejarano, E.R. and Cerdá-Olmedo, E. (1991) In *vivo* channelling of substrates in an enzyme aggregate for β-carotene biosynthesis. *Proc. Natl. Acad. Sci. USA* **88**: 4936.

Carattoli, A., Romano, N., Ballario, P., Morelli, G. and Macino, G. (1991) The *Neurospora crassa* carotenoid biosynthetic gene (Albino 3) reveals highly conserved regions among prenyltransferases. *J. Biol. Chem.* **266**: 5854.

Cerdá-Olemdo, E. (1987). Production of carotenoids with fungi. In *Biotechnology of Vitamin, Growth Factor and Pigment Production* (Vandamme, E., ed.), Elsevier, Amsterdam, pp. 27–42.

Chamovitz, D., Pecker, I. and Hirschberg, J. (1991) The molecular basis of resistance to the herbicide norflurazon. *Plant Mol. Biol.* **16**:967.

Clarke, I.E., Sandmann, G., Bramley, P.M. and Böger, P. (1982) Carotene biosynthesis with isolated photosynthetic membranes. *FEBS Lett.* **140**: 203.

Clough, J.M. and Pattenden, G. (1979) Naturally occurring poly-*cis* carotenoids. Stereochemistry of poly-*cis* lycopene and its cogeners in 'Tangerine' tomato fruits. *J. Chem. Soc. Chem. Commun.* 616.

Clough, J.M. and Pattenden, G. (1983) Stereochemical assignment of prolycopene and other poly-*Z*-isomeric carotenoids in fruits of the tangerine tomato *Lycopersicon esculentum* var. Tangella. *J. Chem. Soc, Perkin Trans.* **1**: 3011.

Commission on Biochemical Nomenclature (1971) *Biochemistry.* **10**: 4827.

Commission on Biochemical Nomenclature (1975) *Biochemistry.* **14**: 1803.

Davies, B.H. (1976) Carotenoids. In *Chemistry and Biochemistry of Plant Pigments, Vol. 2*, 2nd ed (Goodwin, T.W. ed.), Academic Press, London, pp. 38–165.

Davies, B.H., Matthews, S. and Kirk, J.T.O. (1970) The nature and biosynthesis of the carotenoids of different colour varieties of *Capsicum annuum. Phytochemistry.* **9**: 797.

de la Guardia, M.D., Aragon, C.M.G., Murillo, F.J. and Cerdá-Olmedo, E. (1971) A carotenogenic enzyme aggregate in *Phycomyces*: evidence from quantitative complementation. *Proc. Natl. Acad. Sci. USA* **68**: 2012.

Dean, C., Sjodin, C., Bancroft, I., Lawson, E., Lister, C., Scofield, S. and Jones, J. (1991) Development of an efficient transposon tagging system in *Arabidopsis thaliana.* In *Molecular Biology of Plant Development* (Jenkins, G.I. and Schuch, W., eds.), The Company of Biologists, Cambridge, UK, pp. 63–75.

Dogbo, O., Bardat, J., Laferriere, A., Quennemet, J., Brangeon, J. and Camara, B. (1987) Metabolism of terpenoids I. Biosynthesis of phytoene in plastid stroma isolated from higher plants. *Plant Sci.* **49**: 89.

Dogbo, O., Laferriere, A., d'Harlingue, A. and Camara, B. (1988) Carotenoid biosynthesis: isolation and characterisation of a bifunctional enzyme catalysing the synthesis of phytoene. *Proc. Natl. Acad. Sci. USA* **85**: 7054.

Federoff, N.V., Furtek, D.B. and Nelson, O.E. Jr. (1984) Cloning of the *bronze* locus in maize by a simple and generalizable procedure using the transposable controlling element *Activator* (Ac). *Proc. Natl Acad. Sci. USA* **81**: 3825.

Fishwick, M.J. and Wright, A.J. (1980) Isolation and characterisation of amyloplast envelope membranes from *Solanum tuberosum. Phytochemistry* **19**: 55.

Giuliano, G., Pollock, D. and Scolnik, P.A. (1986) The gene *crtI* mediates the conversion of phytoene into colored carotenoids in *Rhodopseudomonas capsulata. J. Biol. Chem.* **261**: 12925.

Goodwin, T.W. (1958) Studies in carotenogenesis 24. The changes in carotenoid and chlorophyll pigments in the leaves of deciduous trees during autumn necrosis. *Biochem. J.* **68**: 503.

Goodwin, T.W. (1973) Carotenoids. In *Phytochemistry, Vol. 1* (Miller, L. P., ed.), Van Nostrand Reinhold, New York. pp. 112–142.

Goodwin, T.W. (1976) *Chemistry and Biochemistry of Plant Pigments*, Academic Press, London.

Goodwin, T.W. (1980) *Biochemistry of Carotenoids, Vol. 1*, Chapman and Hall, London.

Goodwin, T.W. (1983) Developments in carotenoid biochemistry over 40 years. *Biochem. Soc. Trans.* **11**: 473.

Goodwin, T.W. (1986). Metabolism, nutrition and function of carotenoids. *Annu-Rev. Nutr.* **6**: 273.

Goodwin, T.W. and Britton, G. (1988) Distribution and analysis of carotenoids. In *Plant Pigments* (Goodwin, T.W., ed.), Academic Press, London, pp. 61–131.

Goodwin, T.W. and Goad, L.J. (1971) Carotenoids and triterpenoids. In *The Biochemistry of Fruits and Their Products, Vol. 1,* (Hulme, A. C., ed.), Academic Press, London, pp. 305–368.

Graebe, J.E. (1968) Biosynthesis of kaurene, squalene and phytoene from MVA 2-^{14}C in a cell free system from pea fruits. *Phytochemistry* 7. 2003.

Gray, J.C. (1987) Control of isoprenoid biosynthesis in higher plants. *Adv. Bot. Res.* **14**: 25.

Grumbach, K.H. (1984) Does the chloroplast envelope contain carotenoids and quinones *in vivo*? *Physiol. Plant* **60**: 180.

Grumbach, K.H. and Britton, G. (1984) Carotenoid localization and biosynthesis in radish seedlings (*Raphanus sativus*) grown in the presence of absence of bleaching herbicides. In *Advances in Photosynthesis Research Vol. IV* (Sybesma, C., ed.) Maritinus Nijoff/Dr W. Junk Publishers, The Hague, Netherlands, pp.69–75.

Hager, A. and Perz, H. (1970) Veranderung der Lichtabsorption eines Carotinoids in Enzym (De-Epoxidue) Substrat (Violaxanthin) Komplex. *Planta* **93**: 314.

Hauge, B.M., Giraudat, J., Hanley, S., Hwang, I., Kohchi, T. and Goodman, H.M. (1990) Physical mapping of the *Arabidopsis* genome and its applications. *J. Cell. Biol.* **14E**: 259.

Hill, H.M., Calderwood, S.K. and Rogers, L.J. (1971) Conversion of lycopene to β-carotene by plastids isolated from higher plants. *Phytochemistry* **10**: 2051.

Jeffrey, S.W., Douce, R. and Benson, A.A. (1974) Carotenoid transformations in the chloroplast envelope. *Proc. Natl Acad. Sci. USA* **71**: 807.

Jones, B.L. and Porter, J.W. (1986) Biosynthesis of carotenoids in higher plants. *CRC Critical Reviews in Plant Science* **3**: 295.

Joyard, J., Block, M.A. and Douce, R. (1991) Molecular aspects of plastid envelope biochemistry. *Eur. J. Biochem.* **199**: 489.

Karrer, P. and Eugster, C.H. (1950) The occurrence of carotenoids in pollen and anthers of various fruits. *Helv. Chem. Acta.* **33**: 300.

Karrer, P. and Jucker, E. (1950) *Carotenoids*, Elsevier, New York.

Kienzle, F. (1976) The technical synthesis of carotenoids. *Pure Appl. Chem.* **47**: 183.

Kinzer, S.M., Schwager, S.J. and Mutschler, M.A. (1990) Mapping of ripening-related or specific cDNA clones of tomato. *Theor. Appl. Genet.* **79**: 489.

Kirk, J.T.O. and Tilney-Bassett, R. (1978) *The Plastids* 2nd ed, Freeman, San Francisco.

Klaui, H. (1982) Industrial and commercial uses of carotenoids. In *Carotenoid Chemistry and Biochemistry.* (Britton, G. and Goodwin, T.W., eds.), Pergamon Press, Oxford, pp. 309–317.

Klausner, A. (1986) Algacultures: food for thought. *BioTechnology.* **4**: 947.

Kleinig, H. (1989) The role of plastids in isoprenoid biosynthesis. *Annu. Rev. Plant Physiol. Plant Mol. Biol.* **40**: 39.

Kleinig, H. and Beyer, P. (1985) Carotene synthesis in spinach (*Spinacia oleracae* L.) chloroplasts and daffodil (*Narcissuspseudonarcissuss* L.) chromoplasts. *Methods Enzymol.* **110**: 267.

Knapp, J., Moureau, P., Schuch, W. and Grierson, D. (1989) Organisation and expression of polygalacturonase and other ripening related genes in Ailsa Craig 'Never ripe' and 'Ripening inhibitor' tomato mutants. *Plant Mol. Biol.* **12**: 105.

Kreuz, K., Beyer, P. and Kleinig, P. (1982) The site of carotenogenic enzymes in chromoplasts from *Narcissus pseudonarcissus. Planta.* **154**: 66.

Krinsky, N.I. (1989). β-Carotene: functions. In *New Protective Roles for Selected Nutrients.* (Spiller, G.A., ed.), AR Liss, New York, pp. 1–16.

Krinsky, N.I. (1990) Carotenoids in medicine. In *Carotenoids : Chemistry and Biology.* (Krinsky, N.I., Mathews-Roth, M.M. and Taylor, R.F., eds.), Plenum Press, New York, pp. 279–291.

Kuhn, R. and Grundmann, C. (1932) Die Konstitution des lycopins. *Ber. Deutsch. Chem. Ges.* **65**: 1880.

Kuntz, M., Römer, S., Suire, C., Hugueney, P., Weil, J.H., Schantz, R. and Camara, B. (1992) Identification of a cDNA for the plastid-located geranyleranyl pyrophosphate synthase from *Capsicum annuum*: correlative increase in enzyme activity and transcript level during fruit ripening. *Plant J.* **2** : 25.

Kushwaha, S.C., Subbarayan, C., Beler, D.A. and Porter, J.W. (1969) The conversion of lycopene-15,15′ ^{3}H to cyclic carotenes by soluble extracts of higher plant plastids. *J. Biol. Chem.* **244**: 3635.

Kushwaha, S.C., Suzue, G., Subbarayan, C. and Porter, J.W. (1970) The conversion of phytoene-^{14}C to acyclic, monocyclic, and dicyclic carotenes by soluble enzyme systems obtained from the plastids of tomato fruit. *J. Biol. Chem.* **245**: 4708.

Kushwaha, S.C., Kates, M. and Porter, J.W. (1976) Enzymatic synthesis of C_{40} carotenes by cell-free preparation from *Halobacterium cutirubrum. Can. J. Biochem.* **54**: 816.

Kushwaha, S.C., Kates, M., Renaud, R.L. and Subden, R.E. (1978) The terpenyl pyrophosphates of wild type and tetraterpene mutants of *N. crassa. Lipids* **13**: 332.

Lichtenthaler, H.K., Prenzel, U. and Kuhn, G. (1982) Carotenoid composition of chlorophyll-carotenoids proteins from radish cotyledons. *Z. Naturforsch.* **37C**: 10.

Lütke-Brinkhaus, F., Liedvogel, B., Kreuz, K. and Kleinig, H. (1982) Phytoene synthase and phytoene dehydrogenase associated with envelope membranes from spinach chloroplasts. *Planta* **156**: 176.

Marki-Fischer, E. and Eugster, C.H. (1985) Carotenoids from anthers and petals of *Lilium tigrium* cv 'Red Night'. *Helv. Chem. Acta.* **68**: 1704.

Mathews-Roth, M.M. (1982) Medical applications and uses of carotenoids. In *Carotenoid Chemistry and Biochemistry.* (Britton, G., and Goodwin, T.W., eds.) Pergamon Press, Oxford, pp.297–*307.*

Mathews-Roth, M.M. (1989) β-Carotene: clinical aspects. In *New Protective Roles for Selected Nutrients.* (Spiller, G.A., ed.), Liss Inc., New York, pp. 17–38.

Maudinas, B., Bucholtz, M.L., Papastephanou, C., Katiyar, S.S., Briedis, A.V. and Porter, J.W. (1975) ATP stimulation of the activity of partially purified phytoene synthase complex. *Biochem. Biophys. Res. Commun.* **66**: 430.

Maudinas, B., Bucholtz, M.L., Papastephanou, C., Katiyar, S.S., Briedis, A.V. and Porter, J.W. (1977) The partial purification and properties of a phytoene synthesising enzyme system. *Archiv. Biochem. Biophys.* **180**: 354.

Maunders, M.J., Holdsworth, M.J., Slater, A., Knapp, J., Bird, C.R., Schuch, W. and Grierson, D. (1987) Ethylene stimulates the accumulation of ripening-related mRNAs in tomatoes. *Plant Cell Environ.* **10**: 177.

Mayer, H., and Isler, O. (1971). Total synthesis. In *Carotenoids.* (Isler, O., ed.), Birkhauser Verlag, Basel, pp. 325–575.

Mayer, M.P., Bartlett, D.L., Beyer, P. and Kleinig, H. (1989) The *in vitro* mode of action of bleaching herbicides on the desaturation of 15-*cis*-phytoene and *cis*-ζ-carotene in isolated daffodil chromoplasts. *Pestic. Biochem. Physiol.* **34**: 111.

Mayer, M.P., Beyer, P. and Kleinig, H. (1990) Quinone compounds are able to replace molecular oxygen as terminal electron acceptors in phytoene desaturation in chromoplasts of *Narcissus pseudonarcissus* L. *Eur. J. Biochem.* **191**: 359.

McDermott, J.C.B., Britton, G. and Goodwin, T.W. (1973) Carotenoid biosynthesis in a *Flavobacterium* sp. Stereochemistry of H elimination in the desaturation of phytoene to lycopene, rubixanthin and zeaxanthin. *Biochem. J.* **134**: 1115.

McLaughlin, M. and Walbot, V. (1987) Cloning of a mutable *bz*2 allele of maize by transposon tagging and differential hybridisation. *Genetics* **117**: 771.

Misawa, N., Nakagawa, M., Kobayashi, K., Yamano, S., Izawa, Y., Nakamura, K. and Harashima, K. (1990) Elucidation of the *Erwinia uredovora* carotenoid biosynthetic pathway by functional analysis of gene products expressed in *Escherichia coli. J. Bacteriol.* **172**: 6704.

Mohanty, S.S. (1988) Stereochemie des biologischen Ringschlussess bei Carotinen: synthese von spezifisch markierten deuterierten Vorlaufern und Einbauexperimente. Ph.D. thesis, University of Zurich.

Narita, O.J. and Gruissem, W. (1989) Tomato hydroxymethylglutaryl-CoA reductase is required early in fruit development but not during ripening. *The Plant Cell* **1**: 181.

Nelson, M.A., Morelli, G., Carattolli, A., Romano, N. and Macino, G. (1989) Molecular cloning of a *Neurospora crassa* carotenoid biosynthetic gene (albino-3) regulated by blue light and the products of the white collar genes. *Mol. Cell. Biol.* **9**: 1271.

Nonomura, A.M. (1990). Industrial biosynthesis of carotenoids. In *Carotenoids: Chemistry and Biology* (Krinsky, N.I., Mathews-Roth, M.M., and Taylor, R.F., eds.), Plenum Press, New York, pp. 365–375.

O'Reilly, C., Shepherd, N.S., Pereira, A., Schwarz-Sommer, Z., Bertram, I., Robertson, D.S., Peterson, P.A., and Saedler, H. (1985) Molecular cloning of the *a*1 locus of *Zea mays* using the transposable elements. *EMBO J.* **41**: 887.

Perry K.L., Simonitch, T.A., Harrison-Lavoie, K.J. and Liu, S.-H. (1986) Cloning and regulation of *Erwinia herbicola* pigment genes. *J. Bacteriol.* **168**: 607.

Peter, G.F., Machold, D. and Thornber, J.P. (1988) In *Plant Membranes, Structure, Assembly and Function* (Harwood, J. L. and Walton, T. J., eds.), Biochemical Society, London, p. 17.

Pfander, H. and Schurtenberger, H. (1982) Biosynthesis of C_{20} carotenoids in *Crocus sativus. Phytochemistry* **21**: 1039.

Porter, J.W. and Anderson, D.G. (1962) The biosynthesis of carotenes. *Arch. Biochem. Biophys.* **97**: 520.

Porter, J.W. and Lincoln, R.E. (1950) *Lycopersicon* selections containing a high content of carotenes and colourless polyenes. II. The mechanism of carotene biosynthesis. *Arch. Biochem. Biophys.* **27**: 390.

Poulter, C.D. (1990) Isopentenyl diphosphate to squalene—enzymology and inhibition. In *Biochemistry of Cell Walls and Membranes in Fungi* (Kuhn, P.J., Trinci, A.P.J., Jung, M.J., Goosey, M.W. and Copping, L.G., eds.), Springer, New York, pp. 169–188.

Poulter, C.D. and Rilling, H.C. (1983) Prenyl transferases and isomerase. In *Biosynthesis of Isoprenoid Compounds, Vol. 1* (Porter, J.W. and Spurgeon, S.L., eds.), John Wiley and Sons, New York, pp. 161–224.

Qureshi, N. and Porter, J.W. (1983) Conversion of acetyl CoA to isopentenyl pyrophosphate. In *Biosynthesis of Isoprenoid Compounds, Vol.1* (Porter, J.W., and Spurgeon, S.L., eds.), Wiley, New York, pp. 47–94.

Qureshi, A.A., Andrews, A.G., Qureshi, N. and Porter, J.W. (1974) The enzymatic conversion of *cis*-[^{14}C] phytofluene, *trans*-[^{14}C] phytofluene and -ζ-carotene to more unsaturated acyclic, monocyclic and dicyclic carotenes by a cell-free preparation of red tomato fruits. *Arch. Biochem. Biophys.* **162**: 93.

Ray, J., Bird, C.R., Maunders, M., Grierson, D. and Schuch, W. (1987) Sequence of pTOM5, a ripening related cDNA from tomato. *Nucl. Acids Res.* **24**: 10587.

Ray, J., Moureau, P., Bird, C.R., Bird, A., Grierson, D., Maunders, M., Truesdale, M., Bramley, P., and Schuch, W. (1992) Cloning and characterisation of a gene involved in phytoene synthesis from tomato. *Plant Mol. Biol.* **19** : 401.

Raymundo, L.C. and Simpson, K.L. (1972) The isolation of poly-*cis*-ζ-carotene from the tangerine tomato. *Phytochemistry* **11**: 397.

Rick, C.M. (1982) Linkage map of the tomato (*Lycopersicon esculentum*). *Genetic Maps* **2**: 360.

Robertson, D.S. (1975) Survey of the albino and white-endosperm mutants of maize. Their phenotype and gene symbols. *J. Hered.* **66**: 67.

Robertson, D.S., and Anderson, I.C. (1961) Carotenoid protection of porphyrins from photo destruction. *Progress in Photobiology. Third International Congress on Photobiology, Copenhagen (Christensen, B.C. and Buchanan, B. eds.), Elsevier Publishing, Amsterdam p. 477.*

Rock, C.D. and Zeevart, J.A. (1991) The *aba* mutant of *Arabidopsis thaliana* is impaired in epoxy-carotenoid biosynthesis. *Proc. Natl Acad. Sci.* USA **88**: 7496.

Sandmann, G. (1991) Biosynthesis of cyclic carotenoids: biochemistry and molecular genetics of the reaction sequence. *Physiologia Plantarum.* **83**: 186.

Sandmann, G. and Böger, P. (1989) Inhibition of carotenoid biosynthesis by herbicides. In *Target Sites of Herbicide Action* (Böger, P. and Sandmann, G., eds.), CRC Press, Boca Raton, Florida, pp. 25–44.

Sandmann, G. and Bramley, P.M. (1985) The *in vitro* biosynthesis of β-cryptoxanthin and related xanthophylls with *Aphanocapsa* membranes. *Biochem. Biophys. Acta* **843**: 73.

Sandmann, G. and Kowalczyk, S. (1989) *In vitro* carotenogenesis and characterization of the phytoene desaturase reaction in *Anacystis. Biochem. Biophys. Res. Commun.* **163**: 916.

Sandmann, G. and Misawa, N. (1992) New functional assignment of the carotenogenic genes *crtB* and *crtE* with constructs of these genes from *Erwinia* species. *FEMS Microbiol. Lett.* (in press).

Sandmann, G., Woods, W.S. and Tuveson, R.W. (1990) Identification of carotenoids in *Erwinia herbicola* and in a transformed *E. coli* strain. *FEMS Microbiol. Lett.* **71**: 77.

Schmidhauser, T.J., Lauter, F.R., Russo, V.E.A. and Yanofsky, C. (1990) Cloning, sequence and photoregulation of *al-1,* a carotenoid biosynthetic gene of *Neurospora crassa. Mol. Cell. Biol.* **10**: 5064.

Schmidt, A. and Sandmann, G. (1990) Cloning and nucleotide sequence of the *crtI* gene encoding phytoene dehydrogenase from the cyanobacterium *Aphanocapsa* PCC6714. *Gene* **91**: 113.

Schmidt, A., Sandmann, G., Armstrong, G.A., Hearst, J.E. and Böger, P. (1989) Immunological detection of phytoene desaturase in algae and higher plants using an antiserum raised against a bacterial fusion-gene construct. *Eur. J. Biochem.* **184**: 375.

Schnurr, G., Schmidt, A. and Sandmann, G. (1991) Mapping of a carotenogenic gene cluster from *Erwinia herbicola* and functional identification of six genes. *FEMS Microbiol. Lett.* **78**: 157.

Scolnik, P.A., Walker, M.A. and Marrs, B.L. (1980) Biosynthesis of carotenoids derived from neurosporene in *Rhodopseudomonas capsulata. J. Biol. Chem.* **255**: 2427.

Serrano, A., Gimenez, P., Schmidt, A. and Sandmann, G. (1990) Immunocytochemical localisation and functional determination of phytoene desaturase in photoautotrophic prokaryotes. *J. Gen. Microbiol.* **136**: 2465.

Siefermann, D. and Yamamoto, H.Y. (1975) Light induced de-expoxidation of violaxanthin in lettuce chloroplasts IV. The effect of electron transport conditions on violaxanthin availability. *Biochim. Biophys. Acta* **387**: 149.

Siefermann-Harms, D. (1985) Carotenoids in photosynthesis. I Location in photosynthetic membranes and light-harvesting function. *Biochim. Biophys. Acta* **811**: 325.

Slater, A., Maunders, M.J., Edwards, K., Schuch, W. and Grierson, D. (1985) Isolation and characterisation of cDNA clones for tomato polygalacturonase and other ripening-related proteins. *Plant Mol. Biol.* **5**: 137.

Spurgoen, S.L. and Porter, J.W. (1983) Biosynthesis of carotenoids. In *Biosynthesis of Isoprenoids, Vol. 2* (Porter, J.W. and Spurgeon, S.L., eds.), Wiley, New York, pp. 1–122.

Stevens, M.A. and Rick, C.M. (1986) Genetics and breeding. In *The Tomato Crop* (Atherton, J.G., and Rudich, J., eds.), Chapman and Hall, London and New York, pp. 35–109.

Straub, O. (1976) *Key to Carotenoids*. Birkhauser Verlag, Basel.

Straub, O. (1987) *Key to Carotenoids*. 2nd edn, Birkhauser Verlag, Basel.

Subbarayan, C., Kushwaha, S.C., Suzue, G. and Porter, J.W. (1970) Enzymatic conversion of 1PP-4-^{14}C and phytoene-^{14}C to acyclic carotenes by an ammonium sulphate-precipitated spinach enzyme *system. Archiv. Biochem. Biophys.* **137**:547.

Tang, X-S. and Sahto, K. (1985) The oxygen-evolving photosystem II core complex. *FEBS Lett.* **179**:6.

Tanksley, S.D and Mutschler, M.A. (1990) Linkage map of tomato. In *Genetics Maps*, 5th edn (J.O'Brien, ed.) Cold Spring Harbour Press, New York.
Taylor, R.F. (1990) Carotenoids: products, applications and markets. In *Spectrum: Food Industry,* Decision Resources Inc., Burlington MA, pp. 12-1–12-11.
Theres, N., Scheele, T. and Starlinger, P. (1987) Cloning of the *Bz2* locus of *Zea mays* using the transposable element *Ds* as a gene tag. *Mol. Gen. Genet.* **209**:193.
Thornber, J.P. (1975) Chlorophyll proteins: light harvesting and reaction center components of plants. *Annu. Rev. Plant Physiol.* **26**:127.
van der Krol, A.R., Lentung, P.E., Veenstra, J., van der Meer, I.M., Koes, R.E., Gerats, A.G.M., Mol, J.N.M. and Stuitje, A.R. (1988) An antisense chalcone synthase gene in transgenic plants inhibits flower pigmentation. *Nature* **333**:866.
Weedon, B.C.L. (1971) Occurrence. In *Carotenoids* (Isler, O., ed.), Birkhauser Verlag, Basel, pp. 29–59.
Whately, J.H. and Whately, F.R. (1987) What is a chromoplast? *New Phytol.* **106**:667.
White, J.W., Zscheile, F.P. and Brunson, A.M. (1942) The carotenoids of yellow corn grain. *J. Amer. Chem. Soc.* **64**:2603.
Will, O.H., Ruddat, M., Garber, E.D. and Kezdy, F.J. (1984) Characterisation of carotene accumulation in *Ustilago violacea* using high performance liquid chromatography. *Curr. Microbiol.* **10**:57.
Winterstein, A., Studer, A. and Rüegg, R. (1960) Neuere Ergerbruisse der Carotinoidforschung. *Ber. Deutch Chem. Ges.* **93**:2951.
Yamamoto, H.Y. and Higashi, R.M. (1978) Violaxanthin de-epoxidase, lipid composition and substrate specifity. *Archiv. Biochem. Biophys.* **190**:514.
Yamamoto, H.Y., Chenchin, E.E. and Yamada, D.K. (1974) Effect of chloroplast lipids on violaxanthin de-epoxidase activity. *Proc. Third Int. Congr. Photosyn* (Avron, M., ed.), Elsevier, Amsterdam, p. 1999.
Yen, H-C. and Marrs, B. (1976) Map of genes for carotenoid and bacteriochlorophyll biosynthesis in *Rhodopseudomonas capsulata. J. Bacteriol.* **126**: 619.
Yokoyama, H. and White, M.J. (1970) Carotenone formation in *Triphasia trifolia. Phytochemistry* **9**:1795.
Yokoyama, H., Hsu, W.J. and Poling, S.M. (1975). Method for colouring fruits and vegetables. US Patent 3 91 148, Washington DC.
Yokoyama, H., Hsu, W.J., Poling, S.M. and Hayman, E. (1982) Chemical regulation of carotenoid biosynthesis. In *Carotenoid Chemistry and Biochemistry* (Britton, G. and Goodwin, T.W, eds.), Pergamon Press, Oxford, p. 371.
Young, A.J. (1991) Inhibition of carotenoid biosynthesis. In *Herbicides* (Baker, N.R. and Percival, M.P., eds.), Elsevier, Amsterdam, pp. 131–171.
Zechmeister, L. (1962) *Cis-Trans Isomeric Compounds, Vitamins A and Arylpolyenes*. Academic Press, New York.
Zsebo, K.M. and Hearst, J.E. (1984) Genetic-physical mapping of a photosynthetic gene cluster from *R. capsulata. Cell* **37**:937.

5 Manipulating secondary metabolism in culture

J.D. HAMILL and M.J.C RHODES

5.1 Introduction

Secondary metabolites represent large and diverse groups of chemicals produced by plants. Many thousands of chemical structures are known (Luckner, 1990) and dozens of novel structures are reported each month in phytochemical journals. Often products are synthesised at a particular developmental stage in the life cycle of the plant, usually within specialised cell types or in a specific organ, suggesting that sophisticated control mechanisms exist to regulate their synthesis. Their role in the ecological adaption of plants to their environment is becoming clearer (Wink, 1988; Harborne, 1988). In some cases, they appear to have a protective role against predation and the ingress of pathogens, or function as attractants to assist pollination. In other cases, they are used by insects as precursors of sex pheromones or as part of the insect's own defence mechanisms (Nahrstedt, 1989). In addition to their intrinsic value to the plant, many phytochemicals are commercially valuable and find uses as pharmaceuticals in medicine or are used in agriculture. A number of them are used also in foods as colourings or flavourings and as fragrances in perfumes (Table 5.1).

5.1.1 Tissue culture as an aid to studying secondary metabolism

Phytochemicals are often synthesised as the result of the concerted activity of many enzymic steps, e.g. at least 11 enzymic steps in the formation of the antileukaemic compound vincristine, from the amino acid tryptophan, in *Catharanthus roseus*, and approximately 10 steps from ornithine to the muscle relaxant, scopolamine (I), in *Datura* species (Luckner, 1990). Intermediates from two or more pathways may be combined during biosynthesis. For example, in the formation of scopolamine, the tropane ring is formed from ornithine or arginine yet the esterifying acid (tropic acid) is derived from phenylalanine (Figure 5.1). The prospect of transferring the genes coding for numerous enzymic steps and coordinately expressing them in a microorganism is daunting and at present cannot be contemplated as many of the genes have yet to be isolated and characterised.

A comprehensive account of secondary metabolites produced by plant tissues grown *in vitro* was reported recently (Parr, 1989). Plant cell culture has been pursued for more than twenty years as a tool for the elucidation of secondary metabolite biosynthesis and as a potential means for their biotechnological production to rival extraction processes based on whole plant material. There have been

Table 5.1 Examples of important secondary metabolites.

Major chemical	Genus	Use
Medical		
Hyoscyamine[a], scopolamine	*Datura, Hyoscyamus, Duboisia, Atropa, Scopolia*	Anaesthetics, travel sickness
Digitoxin	*Digitalis*	Cardiac arrhythmia
Solasodine, Diosgenin	*Solanum, Costus, Dioscorea*	Steroid drugs
Quinine, artemesinin	*Cinchona, Artemesia*	Antimalarial
Castenospermine	*Castenospermum*	Antiviral, anti-HIV[b]
Taxol	*Taxus*	Antitumour[b]
Vinblastinine, vincristine	*Catharanthus*	Antileukaemic
Shikonin, lignans	*Lithospermum, Piper, Ginko*	Anti-inflammatory
Ajmalicine/serpentine	*Catharanthus*	Circulatory disorders
Podophyllotoxins	*Podophyllum*	Antitumour[b]
Forskolin	*Coleus*	Cardiotonic, antihypertensive
Morphine, codeine	*Papaver*	Pain relief, local anaesthetic
Saponin, polyacetylenes	*Panax*	Circulatory stimulants
Lobeline, glycyrrhizin	*Lobelia, Glycyrrhiza*	Respiratory stimulants
Agrochemicals		
Thiophenes	*Tagetes*	Biocides, antinematode
Pyrethrin	*Pyrethrum*	Insecticides
Foods and fragrances		
Betacyanin	*Beta*	Food colorant
Quinine, quassin	*Cinchona, Quassia*	Food bittering agents
Essential oils (mint, lavender, rose)	*Mentha, Lavendula, Rosa*	Perfumes and food flavouring
Vanillin, capsaicin	*Vanilla, Capsicum*	Food flavouring
Hallucinogens and stimulants		
Caffeine, nicotine theophylline	*Coffea, Nicotiana, Thea*	Stimulants
Cocaine, cannibinol	*Erythroxylum, Cannabis*	Hallucinogens

[a]The natural product is L-hyoscyamine. Atropine is a racemic mixture of D- and L-hyoscyamine.
[b]Trials being undertaken.

some notable successes in achieving the former objective; for example the enzymic steps required to synthesise berberine were elucidated using extracts of cells grown in culture (Zenk *et al.*, 1985; Zenk, 1985, 1988). Progress towards the latter goal, however, has been somewhat disappointing with only a limited number of successes where it matters most—in the market place. These include the often quoted example of the red dye, shikonin (II), produced by cell suspension cultures of *Lithospermum erythrorhizon* (Fujita and Tabata, 1984; Fujita *et al.*, 1987) and the yellow antimicrobial dye, berberine (III), produced by cell cultures of *Coptis japonica* (Fujita, 1990). The success of these operations owes much to the fact that these compounds are coloured, making selection of stable productive clones relatively straightforward, and also to the tenacity of workers in the laboratories involved (Fujita, 1990). Stable cell lines of *Coleus blumeii*, with high levels of rosmarinic acid (IV) biosynthesis, have also been reported, though it is unclear whether the process is commercially viable (Ulbricht *et al.*, 1985).

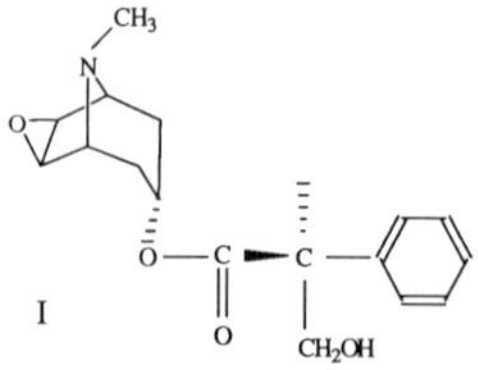

I

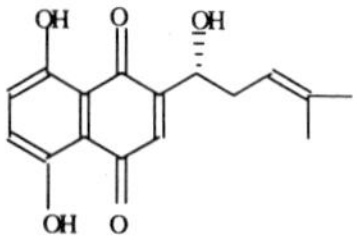

II

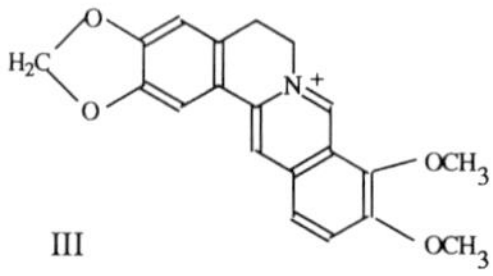

III

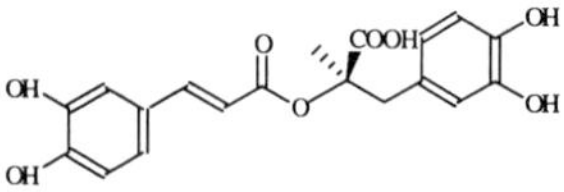

IV

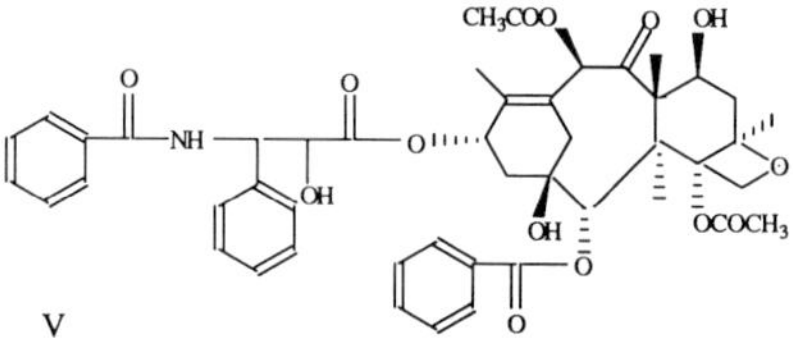

V

VI

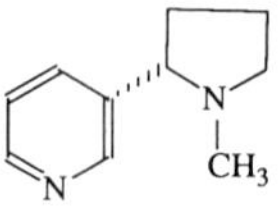

VII

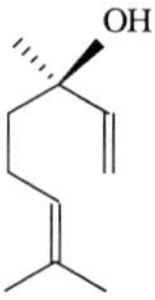

VIII

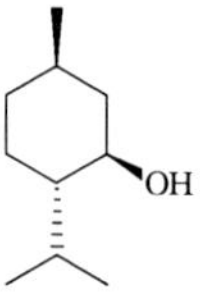

IX

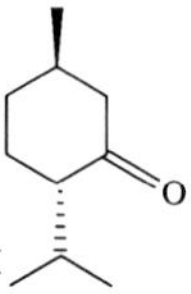

X

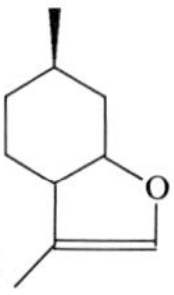

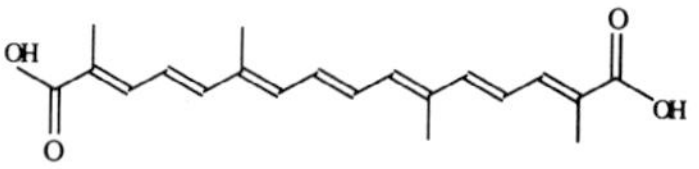

XI

XII

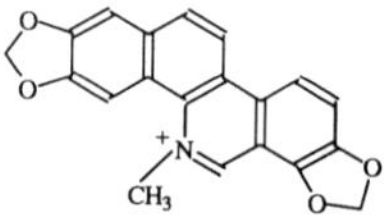

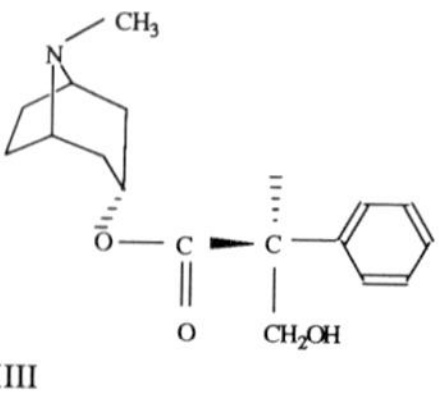

XIII

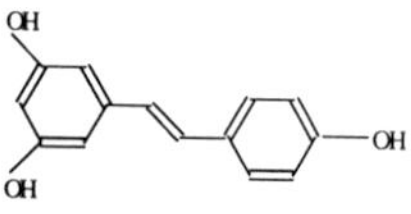

XIV

Disorganised cell cultures, however, generally do not reliably produce high levels of secondary metabolites. This is evident from the large number of disappointingly low levels of biosynthesis by disorganised cells grown in culture reported in the general literature over the past few years. In addition, biochemical instability is a feature of many cell lines even when significant production of secondary metabolites is achieved (Deus-Neumann and Zenk, 1984). Genetic instability, manifested by chromosomal drift and restructuring, which is common in long term cell suspensions (D'Amato, 1985), probably underlies this biochemical variation and must raise doubts regarding the development of industrial processes for secondary product biosynthesis based upon the use of disorganised cell suspensions for many plant species. Indeed, perhaps the most interesting question to be addressed is the basis of the stability of secondary metabolite production in lines of *Coptis, Coleus* and *Lithospermum* noted above.

However, there is little doubt that efficient and productive *in vitro* cultures are highly desirable and will have an important role in the commercial production of important phytochemicals. For example taxol (V), extracted from the bark of the Pacific yew (*Taxus brevifolia*), is now in phase III clinical trials in the USA as an antiovarian cancer drug. Taxol is extracted from the bark of 50–60-year-old trees and as many as 12 trees are required to produce enough taxol to treat one patient (Hermann, 1991). With only about 10^6 trees left in nature and current requirements being about 10^5 trees per year, it is clear that a plant tissue culture approach for the production of taxol is an extremely attractive proposition from both a health and ecological perspective.

5.2 Transformed cultures

5.2.1 Transformed root cultures

In the mid 1980s workers in a number of laboratories independently realised the potential of using transformed root cultures as *in vitro* systems for secondary metabolite production (Flores and Filner, 1985; Hamill *et al.*, 1986; Kamada *et al.*, 1986). The concept itself was not new with several reports already in the literature on the use of non-transformed root cultures as useful organs for studying secondary product biosynthesis. Indeed, the largely forgotten but pioneering work of Dawson in 1942 demonstrated that nicotine (VI) was made in root cultures of *Nicotiana tabacum* (Dawson, 1942). However, the great advantage of using roots transformed by *Agrobacterium rhizogenes* lies in the fact that these root cultures are usually fast growing and are easy to maintain in a hormone-free medium. Furthermore they

---

Figure 5.1 Chemical structures of some of the secondary metabolites discussed in the text. I, scopolamine; II, shikonin; III berberine; IV, rosmarinic acid; V, taxol; VI, nicotine; VII, linalool; VIII, menthol; IX, menthone; X, menthofuran; XI, crocetin, the aglycone of crocin; XII, sanguinarine; XIII, hyoscyamine; XIV, resveratrol.

offer direct methods for the introduction of foreign genes to influence the biosynthetic capacity of the cultures.

The value of transformed root cultures for the study of secondary product biosynthesis is evident from the rapidly increasing number of reports of products formed at high levels by these cultures (Table 5.2). Some cultures release their secondary metabolites into the surrounding medium and the use of adsorbents has been demonstrated to enable recovery of products from this medium, thereby enabling a prolonged period of production by the cultures. Such experiments originally were carried out by Rhodes *et al.* (1986) who used XAD-4 to adsorb nicotine from the medium of a transformed root culture of *Nicotiana rustica*. A recent report by Shimomura *et al.* (1991) has extended the use of this methodology

Table 5.2 Examples of secondary metabolites produced by transformed root cultures and reported in the recent literature.

Metabolite	Genus	Reference
Alkaloids		
Pyridine	*Nicotiana*	Hamill *et al.*, 1986; Parr and Hamill, 1987
Indole	*Catharanthus*	Parr *et al.*, 1988; Toivonen *et al.*, 1989
	Cinchona	Hamill *et al.*, 1989
Tropane	*Duboisia*	Deno *et al.*, 1987; Mano *et al.*, 1989
	Atropa	Kamada *et al.*, 1986; Jung and Tepfer, 1987; Sharp and Doran, 1990
	Datura	Payne *et al.*, 1987; Christen *et al.*, 1989; Robins *et al.*, 1990
	Hyoscyamus	Flores and Filner, 1985
	Scopolia	Mano *et al.*, 1986; Nabeshima *et al.*, 1986
Piperdine	*Lobelia*	Yonemitsu *et al.*, 1990
Alkamides	*Echinacea*	Trypsteen *et al.*, 1991
Pyrrolizidine[a]	*Senecio*	Toppel *et al.*, 1987; Hartmann and Toppel, 1987
Others		
Betalain pigments	*Beta*	Hamill *et al.*, 1986
Sesquiterpenes	*Lippia*	Sauerwein *et al.*, 1991
	Datura	Furze *et al.*, 1991
Polyacetylenes	*Coreopsis, Bidens*	Marchant, 1988
Thiophenes	*Tagetes*	Westcott, 1988; Croes *et al.*, 1989
	Ambrosia, Carthamus, Rudbeckia	Flores *et al.*, 1988
Cardioactive glycosides	*Digitalis*	Saito *et al.*, 1990
Anthraquinones	*Cassia*	Asamizu *et al.*, 1988
Saponins	*Panax*	Yoshikawa and Furuya, 1987
Polyines	*Chaenactis*	Constabel and Towers, 1988
Lignans	*Podophyllum*	Berlin *et al.*, 1988
	Linum	Berlin *et al.*, 1988
Napthoquinones	*Lithospermum*	Shimomura *et al.*, 1991
Steroids	*Solanum*	R.J. Robins, unpublished

[a]In this case the roots were not transformed by *A. rhizogenes* but growth rates comparable to those of transformed root cultures were obtained.

Figure 5.2 (a) Transformed root cultures of *Datura stramonium* growing in a 15 l air-sparyed bioreactor. This culture produces high levels of the muscle relaxant drug, hyoscyamine. (b) Transformed shoot cultures of *Mentha piperita* growing in a 1 l bioreactor. The main product of this culture is menthol.

to enable continuous production of shikonin from a transformed root culture of *L. erythrorhizon*. The use of an XAD-2 column stimulated production by three-fold and enabled shikonin to be produced at about 5 mg/day for a period of more than 220 days from a root culture grown in a 2 l air lift fermenter (Shimomura *et al.*, 1991). A recent search of the world patents index revealed more than 20 applications with the specific aim of using root cultures, transformed with *Agrobacterium rhizogenes*, for the production of phytochemicals. This is an indication of the commercial potential of processes based on the use of differentiated roots grown *in vitro* for production of important secondary metabolites.

Culture of differentiated roots has two major advantages compared to non-differentiated cell or callus cultures. The first relates to the fact that these cultures retain the organisation of the root meristem and this promotes chromosomal stability over prolonged growth *in vitro*. Aird *et al.* (1988a) examined meristem cells of transformed root tissue of eight species, representing five families, and found no evidence of aneuploidy or polyploidy in these cultures. Normal chromosome numbers were also found in transformed root lines of *Crepis capillaris* (Ambros *et al.*, 1986) and plants of *Brassica napus* regenerated from transformed roots (Ooms *et al.*, 1985a; Guerche *et al.*, 1987). Evidence of chromosomal change has been reported to occur in some species following transformation with *A. rhizogenes*; altered chromosome structure but not aneuploidy was reported in karyotypes of transformed root cultures of *Lycopersicon esculentum* (Banerjee-Chattopadhyay *et al.*, 1985) and rearrangements and aneuploidy were found among 6% of karyotypes examined from transformed roots of *Vicia faba* (Ramsay and Kumar, 1990). A significant number of polyploid cells was also observed among transformed roots of *Vicia faba*. However, as pointed out by Ramsay and Kumar (1990), these may have resulted from the transformation of cells which had undergone endoreduplication *in vivo*, before contact with *A. rhizogenes*. Some aneuploidy and polyploidy was also reported in potato plants regenerated from transformed roots (Ooms *et al.*, 1985b; Hanisch ten Cate *et al.*, 1987, 1988; De Vries-Uijtewaal *et al.*, 1988). However it is unclear, in these latter cases, whether these alterations in chromosome number were induced by tissue culture procedures. Plant regeneration, especially in solanaceous species, often proceeds via an intermediate callus stage and Aird *et al.* (1988b) have clearly shown that aneuploidy can occur in transformed root cultures which are induced to form callus after addition of phytohormones and then allowed to redifferentiate root tissue on subsequent removal of the phytohormones. Thus, although some examples of aneuploidy, polyploidy and chromosome rearrangements have been reported to occur among transformed root cultures, in general these cultures exhibit much less chromosomal variation than disorganised cell cultures grown *in vitro*.

Differentiation also brings about the co-ordination of expression of the enzymic steps necessary to ensure synthesis of phytochemicals. Roots are, of course, heterogenous organs with zones of cell division and cell expansion, with differentiated vascular systems, a pericycle, etc., and we are still some way from understanding the factors linking cell differentiation and secondary product biosynthesis in roots. An elegant experiment carried out by Flores (1987) showed that addition of phytohormones caused non-differentiated callus to grow at the expense of differentiated transformed roots of *Hyoscyamus muticus*. This callus produced as much biomass per growth cycle as did the growing transformed root control culture but had less than 5% of the hyoscyamine level of intact roots. Removal of the phytohormones enabled root redifferentiation and this resulted in the restoration of biosynthesis of hyoscyamine to high levels. The experiment could be repeated numerous times demonstrating the link between root differentiation and the expression of the tropane alkaloid biosynthetic pathway. A similar set of experiments, reported by Rhodes *et al.* (1989), demonstrated that biosynthesis of nicotine in *Nicotiana* transformed roots also required the intact differentiated root structure. Other workers have noted the link with differentiation, non-differentiated cultures generally showing low levels of production (Wink, 1987; Rhodes *et al.*, 1991, 1992). Why this phenomenon is so widely observed is unclear but may be partially explained by interesting experiments reported by Hashimoto *et al.* (1991) which indicated that the enzyme hyoscyamine β-6-hydroxylase (Hβ6-H), the enzyme responsible for converting hyoscyamine to scopolamine (Figure 5.3), is localised in the pericycle of roots of *Hyoscyamus muticus*. The gene for Hβ6-H has been cloned (Matsuda *et al.*, 1991) and transcripts have been shown to be present at much higher levels in cultured roots of *Hyoscyamus niger* than in cultured cells, older roots, stems and leaves. It will bc interesting to study the expression profiles of genes coding for secondary metabolite biosynthetic enzymes of other pathways to ascertain whether they too show enhanced or root-specific expression. Root-specific expression of one or more genes coding for enzymes in the pathway would explain why disorganised callus or cell suspensions often have such low levels of productivity compared to differentiated root cultures.

5.2.1.1 Limitations to the use of transformed root cultures. One limitation of transformed roots in the study of secondary metabolism is that they usually produce only compounds characteristic of root tissue and not those synthesised in the aerial parts of the plant. For example, it was found that catharanthine, a precursor of the antitumour drug vinblastine, was produced in transformed roots of *Catharanthus roseus* but the other precursor, vindoline, was not produced in these cultures. Consequently the yield of vinblastine was very low in transformed roots of *C. roseus* (Parr *et al.*, 1988). Several strategies exist to compensate for these limitations. In some cases it is possible to induce the formation of green roots by exposure of cultures to light at high levels (Flores *et al.*, 1988). This phenomenon is unusual but has been reported in transformed roots of *Digitalis purpurea*, producing cardenolides, (Saito *et al.*, 1990), *Lippia dulcis* , producing hernandulcin, (Marchant, 1988) and species of *Coreopis* and *Bidens* genera, producing polyacetylenes,(Sauerwein *et al.*, 1991). An alternative and more generally applicable approach is to attempt the establishment of a culture system analogous to transformed root culture, i.e. transformed shoot cultures, for production of shoot-based compounds. Once again, the concept behind this approach is not entirely novel—differentiation of shoots has been shown to allow enhanced levels of shoot-derived phytochemicals to be produced in culture, for example, ginger oil in *Zingiber* (Charlwood *et al.*, 1988), terpenes in *Pelargonium* (Charlwood and Moustou, 1988) and vinblastine in multiple shoot cultures of *Catharanthus roseus* (Endo *et al.*, 1987).

5.2.2 Transformed shoot cultures

Introduction of the isopentenyl transferase (*ipt*) gene from *A. tumefaciens* has been shown to promote multiple shoot proliferation in species such as tobacco (Medford *et al.*, 1989). However, attempts to replicate these results using several terpene-producing species were unsuccessful (Spencer, 1991). When the gene was placed under the powerful CaMV35S promoter with duplicated upstream enhancers (Kay *et al.*, 1987) it induced some shoot proliferation on *Mentha citrata* and *M. piperita* following transformation (Spencer, 1991). Interestingly, however, the use of nopaline strains of *A. tumefaciens* T37, C58 and N273 was found to be an effective way of inducing multiple shoot proliferation on these species of mint (Spencer *et al.*, 1990a,b; Spencer, 1991). These cultures grew rapidly *in vitro* in medium without exogenous phytohormones. Gas chromatographic analysis of extracts from these shooty teratoma cultures indicated that they produced terpenes characteristic of their species, e.g. linalool (VII) and linalyl acetate in *Mentha citrata* and menthol (VIII) and menthone (IX) in *Mentha piperita*. However, in the case of *M. piperita*, significant amounts of menthofuran (X) were also made in culture whereas this component is only a minor component in the intact plant (Spencer, 1991; Spencer *et al.*, 1992). Non-differentiated callus cultures of mint species did not synthesise appreciable quantities of terpene products. This is not surprising as they are produced in oil glands. These were formed on the epidermal layers of the leaf-like structures in shooty teratoma cultures but were not present in undifferentiated callus (Spencer *et al.*, 1990a; Spencer, 1991). Shooty teratoma production was also induced in tobacco cultures by using strains of *A. tumefaciens* mutated in auxin biosynthetic genes and thus leading to an elevated level of cytokinin in transformed tissues (Saito *et al.*, 1989). Analysis of these cultures showed they were unable to synthesise nicotine but could, if fed with nicotine, transform it to nornicotine (Saito *et al.*, 1989).

While the methodology for shooty teratoma production for many species is not as straightforward as that for producing transformed root cultures, it has demonstrated the principle of using these types of cultures to study shoot-derived secondary product formation by cultures grown *in vitro*. The ability of strain T37 to induce shoot teratoma formation in mint does not depend solely on the presence of cytokinin (*ipt*) and auxin biosynthesis genes (*onc* genes). These genes, inserted in combination into pBin19 (Bevan, 1984) and transferred into mint, caused non-differentiated callus to form (Spencer, 1991; Spencer *et al.*, 1992). It is likely that other genes, outside the *onc* region, are partially responsible for shoot teratoma formation and one likely candidate is the 6b gene which is known to modify the effects of both auxin and cytokinin genes (Tinland *et al.*, 1989; Spanier *et al.*, 1989). Manipulation of the expression of the *ipt* gene and/or other genes from *A. tumefaciens* such as the 6b gene may enable this methodology to become applicable to a wider range of species than is the case at present.

5.2.3 Flower-specific metabolites

Plant flowers produce both pigments and many volatile compounds which are highly prized as perfumes and fragrances, e.g. jasmine and rose oils. Attempts have been made to induce proliferation of stigmata in cell cultures of saffron, *Crocus sativa*, growing in a medium containing benzyladenine and kinetin. Although growth was limited, the stigmata developed and produced the characteristic pigments of the anthers of saffron, crocin (XI) and picrocrocin (Himeno and Sano, 1987; Sano and Himeno, 1987). Floral differentiation can be induced *in vitro* using thin cell layer epidermal peels from floral stems in *Nicotiana* species by careful adjustment of the hormone levels (Tran Thanh Van *et al.*, 1974). Though floral differentiation is obviously a complicated process requiring co-ordinated expression of numerous regulatory genes, this methodology could, in theory, be applicable to other species so that floral organ cultures could be induced by manipulation of expression of appropriate genes. Recent advances in elucidating the genetic control of flower differentiation (Coen and Meyerowitz, 1991) may ultimately enable such experiments to be contemplated. Unfortunately no *Agrobacterium* species has yet been reported which results in the proliferation of floral meristem tissue as a result of transformation.

5.3 Manipulation of secondary metabolism

Cultures of plant tissues grown *in vitro* are well suited to experiments involving the addition of metabolic precursors, biochemical inhibitors and elicitor treatments, all of which may be used to identify enzyme(s) which govern the flux through the metabolic pathways involved in secondary product biosynthesis. The advent of molecular biology and gene transfer techniques have made possible experiments aimed directly at manipulating secondary metabolism by altering the expression of important genes in metabolic pathways.

5.3.1 Exogenous feeding of precursors

Modelling the effects of de-regulated gene expression has been attempted using precursor feeding strategies. The aim of these experiments is to study the effect on the subsequent steps of the metabolic pathway with particular emphasis on the degree to which flux can be stimulated. Using this approach, Zenk *et al.* (1977) achieved a three-fold increase in alkaloid yield of a cell line of *Catharanthus roseus* by feeding tryptophan at 1 mM. Feeding tryptophan to cell suspensions of *Cinchona* species was reported to lead to enhanced levels of the quinoline alkaloids, quinidine and quinine (Hunter *et al.*, 1982; Koblitz *et al.*, 1983). In the experiments reported by Koblitz *et al.* (1983), a 90-fold increase in alkaloid was reported for cell suspensions of *C. pubescens* although Hunter *et al.* (1982), working with *C. ledgeriana* cell suspensions, reported a much lower degree of stimulation. Hay *et al.* (1986) developed a cell suspension of *C. ledgeriana* which showed partial root differentiation. Previous work had shown that the presence of these root-like structures conferred a much greater capacity to synthesise quinoline alkaloids than in comparable non-organised cell suspensions (Anderson *et al.*, 1982). Tryptophan added at 250 mg l^{1} (~1.2 mM) was not toxic to these cultures and led to a 90% increase in synthesis of alkaloids compared with controls. At 500 mg^{1}(~ 2.5 mM), tissues showed a five-fold increase in alkaloid content (Hay *et al.*, 1986). Furthermore, these authors showed that tracer levels of ^{14}C-labelled tryptophan were efficiently taken up and incorporated into quinine and quinidine, the main quinoline alkaloids. In other experiments, from the same laboratory, Anderson *et al.* (1986)

showed that feeding tryptophan at 500 mg^{1} (~ 2.5 mM) resulted in a 76% increase in tryptophan-derived alkaloids in a cell suspension of *Ailanthus altissima*. Experiments with *Nicotiana* species, in which putrescine was fed to cell suspensions of *N. tabacum*, at a concentration of about 1 mM, led to a 50% decrease in nicotine content in these cell suspensions (Lockwood and Essa, 1984). Experiments using transformed roots instead of cell suspensions, showed that feeding putrecine at 1–5 mM caused a 100% increase in the nicotine content of transformed roots of *N. rustica* compared to controls (Walton *et al.*, 1988).

5.3.2 The use of elicitors to increase secondary product formation

Plant tissues respond to the presence of pathogens by an increase in the synthesis of secondary metabolites involved in the defence of the plant against infection. This biosynthesis involves the formation of enhanced levels of constitutive compounds and the *de novo* synthesis of specific defence compounds, the phytoalexins, which are essentially absent in unchallenged tissues (Dixon, 1986). The phytoalexins represent a chemically diverse group of compounds each with a broad spectrum of activity against a range of pathogens and these defence responses can be elicited by the intact pathogen, by extracts of the pathogen cell wall and even by abiotic factors, such as UV light and heavy metals. The response is often observed in cultures of plant cells or plant organs as well as with intact plant tissue.

5.3.2.1 Stimulation in cell cultures. The best studied systems are the elicitation of the formation of isoflavonoid compounds in cell cultures of various leguminaceous species by cell wall fragments of fungi. In *Phaseolus vulgaris*, the induction of the biosynthesis of the isoflavonoids, phaseollin and kievitone following elicitation with extracts of the fungus *Colletotrichum lindemuthianum* (Dixon *et al.*, 1989) involved transitory increases in the rates of transcription of genes coding for key enzymes in isoflavonoid biosynthesis, phenylalanine ammonia-lyase (PAL), chalcone synthase (CHS) and chalcone isomerase (CHI). This led, within 2–4 hours of the addition of elicitor, to a peak in the steady state level of mRNA specific for these enzymes. This peak preceded a peak in the rate of synthesis of the individual enzymes, leading to a maximal measurable enzyme activity 6–16 hours after the start of elicitation and to the accumulation of the isoflavonoid products. Similarly the formation of the pterocarpan phytoalexins, glyceollins I, II and III in *Glycine max* cell cultures (Hahlbrock *et al.*, 1984), induced by treatment with an elicitor prepared from *Phytophthora megasperma*, was preceded by increases in the extractable activity of enzymes of general phenylpropanoid metabolism (PAL), hydroxycinnamate CoA ligase (4-CL), and cinnamate 4-hydroxylase (CA4H), enzymes in flavonoid anabolism (CHS, CHI) as well as enzymes committed to isoflavonoid biosynthesis, isoflavone synthase (Hagemann and Grisebach, 1984), pterocarpan 6α-hydroxylase (Hagemann *et al.* 1984) and an isoflavonoid-specific prenyltransferase (Leube and Grisebach, 1983). In the case of PAL, 4CL and CHS, the transient increase in enzyme activity was preceded by the induction of transcription and increased steady state levels of the relevant mRNAs. Elicitors can also induce divergence in a metabolic pathway involving inactivation of enzymes operating in unchallenged tissue as well as induction of *de novo* enzyme synthesis. In tobacco cell suspensions, induction of sesquiterpene production following treatments with a biotic elicitor, cellulase, involves the diversion of farnesyl pyrophosphate (FPP) from membrane sterol biosynthesis into the formation of the sesquiterpenoid products, capsidiol and debreyol. This involves a transitory stimulation of the activity of the enzymes utilising FPP for sesquiterpene formation and a transitory inhibition of the activity of the enzyme, squalene synthase, using FPP for sterol biosynthesis (Threlfall and Whitehead, 1988).

5.3.2.2 Secretion of products. A feature of phytoalexin formation in response to elicitors *in vitro* is that the major part of the phytoalexin formed is often secreted from the plant cells into the medium. This appears to be a widespread phenomenon (Barz *et al.*, 1988) and where it occurs it has important consequences for the design of biotechnological processes to produce these compounds. The ideal would be to induce production of phytoalexins by repeated cycles of elicitation and recovery of the cells, with the products of elicitation being recovered from the medium without loss of biomass. Phytoalexin products which have been studied in some detail are the benzophenathridine alkaloids produced by *Eschscholtzia californica* (Schumacher *et al.*, 1987) and *Papaver bracteatum* (Cline and Coscia, 1988). The main alkaloid product in *Papaver* culture, sanguinarine (XII), has commercial potential as an oral hygiene product and Tyler *et al.* (1989) have achieved high levels of production of benzophenathridine compounds in cell cultures of *Papaver somniferum* elicited with a sterilised *Botrytis* culture homogenate. Under optimal conditions, the level of sanguinarine and the related compound dihydrosanguinarine may reach 3% of the dry cell weight (up to 200 mg/l). A semi-continuous protocol was designed with three cycles of elicitation and recovery and the sanguinarine (up to 50% of the total formed at each cycle) collected from the growth medium. The elicitation treatment was judged sufficiently mild to maintain cells viable during these treatments.

Brodelius *et al.* (1989) induced sanguinarine formation in cell lines of *Eschscholtzia californica* with a purified glucan fraction isolated from yeast by the method of Hahn and Albersheim (1978). Sanguinarine levels rose to 50 $\mu g\ g^{-1}$ dry weight in elicitated cultures compared to non-detectable levels in the non-elicitated culture. Elicitation can also be a valuable treatment to increase the formation of constitutive secondary compounds which may or may not have a role in the plant's defences. Brodelius *et al.* (1989) showed that treatment of cell cultures of *Thalictrium rugosum* with the yeast glucan fraction increased the level of berberine substantially (four-fold) above the level present in unelicited cells (50 $\mu g\ g^{-1}$ dry weight). In other cases the production of individual constitutive compounds such as furanocoumarins in *Ruta graveolens* cell cultures (Eilert, 1989) is greatly stimulated by elicitor treatments. However, not all secondary metabolites of interest are stimulated by such treatments. Indeed van der Heijden *et al.*

(1988a,b) showed that treatment of cell cultures of *Tabernaemontana divaricata* with a *Candida albicans* elicitor preparation inhibited the formation of the indole alkaloids, vallesamine and *O*-acetylvallesamine even though it promoted the formation of triterpenoid phytoalexins characteristic of this species. These changes in constitutive compounds are only part of major changes in the metabolism of elicited tissue which may occur in primary as well as secondary metabolism. The changes in primary metabolism, such as increased respiration, and rises in ATP levels by two- to three-fold within 1–2 hours of elicitor treatment, occur even with purified elicitor preparations (Brodelius *et al.*, 1989). Analysis of the proteins of elicited and control cells on two-dimensional gels show major changes in protein patterns associated with the major metabolic changes induced by the elicitor (Hahlbrock *et al.*, 1984; Gugler *et al.*, 1988).

5.3.2.3 Elicitation in organ culture. Similar responses to that observed in plant cell cultures are observed in plant organ cultures. Induction of the formation of polyacetylenic phytoalexins has been achieved in transformed root cultures of *Bidens sulphureus* using a fungal culture filtrate as an elicitor (Flores *et al.*, 1988) while transformed root cultures of *Hyoscyamus muticus* produce sesquiterpenoid phytoalexins following elicitation (Signs and Flores, 1989). Furze *et al.* (1991) showed that elicitations of transformed roots of *Datura stramonium* with Cu^{2+} at 1 mM induced high levels of production of the sesquiterpenoid compounds, lubimin and 3-hydroxy lubimin which were secreted into the culture medium. In contrast, the levels of tropane alkaloids such as hyoscyamine were unaffected by the elicitor treatment. However, up to 75% of these alkaloids, which are normally tightly held within the roots, was released into the medium following elicitor treatment. Attempts to achieve repeated cycles of elicitation were unsuccessful, however, and it appears that the elicitor treatment induced irreversible permeabilisation of some of the cells in the root cultures.

5.3.2.4 Value of elicitation. Elicitation may have practical value in processes to produce phytoalexin products and in some cases may be useful in increasing the levels of constitutive compounds. The value of elicitation in biotechnological processes would be enhanced if repeated cycling between elicitation and recovery phases without loss of cell viability could be achieved and in which the phytoalexin product was recovered from this medium without the need to harvest the biomass. However, the conditions required for reversible elicitation without loss of intactness and viability have rarely, if ever, been achieved. In more fundamental studies, elicitation has been a valuable tool in studying the enzymology of phytoalexin production and of changes at the gene and enzyme level upon which the *de novo* synthesis of the phytoalexin products depend. Further, the purified elicitors are recognised presumably at the plant cell surface and institute a series of transductive changes which ultimately induce the co-ordinated expression of the genes involved in phytoalexin biosynthesis. Their study thus enables us to probe the fundamental signal transduction process involved (Dixon and Lamb, 1990; Ryan and Farmer, 1991).

5.3.3 Biochemical studies to identify rate limiting steps in metabolic pathways

The metabolic control theories of Kacser (1987) provide a theoretical background to attempts to manipulate the flux through metabolic pathways. Perturbations of individual enzyme activities *in vivo* and measurement of the effect on flux enable the contributions of individual enzymes to flux to be evaluated and the identification of steps in a pathway at which flux is sensitive to enzyme concentration (activity). Clearly steps at which flux is sensitive to enzyme levels are good targets for programmes of metabolic manipulation. In the past, specific alteration in enzyme activities in plants has been difficult since mutants are rarely available and metabolic inhibitors usually lack absolute specificity. However, the advent of molecular biological approaches to specifically over- or underexpress individual enzymes by single gene alterations enable these control theories to be tested in plant systems. In the absence of the relevant genes, however, the approach to identifying suitable enzymic targets for manipulation must continue to rely on classical biochemical approaches.

5.3.3.1 Putrescine-derived alkaloids. The biosynthetic pathway leading to hyoscyamine (XIII) and scopolamine (I) biosynthesis is known in outline (see Figure 5.3). Transformed root cultures of *Datura stramonium*, and several other members of the Solanaceae, have the capacity to synthesise these alkaloids at similar levels

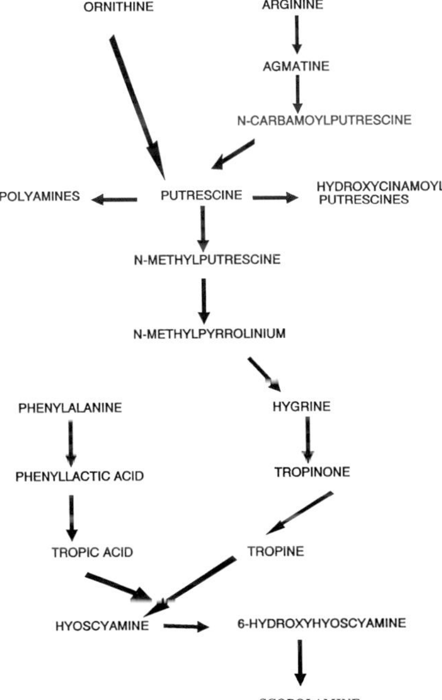

Figure 5.3 Pathway for tropane alkaloid biosynthesis in solanaceous plants.

to those found in the plant and have proved valuable systems in which to investigate biosynthetic problems (Payne *et al.*, 1987; Robins *et al.*, 1990; Parr *et al.*, 1990) and to gain a deeper insight into the biochemical and molecular regulation of tropane alkaloid biosynthesis *in planta*. The pathway leading to scopolamine is shown in Figure 5.3 and a detailed analysis of the kinetics of alkaloid production, particularly in the early part of the pathway, has been reported (Robins *et al.*, 1990, 1991 a,b; Walton *et al.*, 1990). These authors made use of a cloned transformed root line of *Datura stramonium* (designated D15/5) and a root clone generated by transformation of a hybrid between *D. candida* × *D. aurea* (designated DB5). The DB5 hybrid was bred to contain significant levels of scopolamine (El-Dabbas and Evans, 1982) which is reflected in the production of alkaloid by the transformed root lines developed from it (Rhodes *et al.*, 1989).

Putrescine can be derived either from ornithine or arginine (Figure 5.3). Biochemical inhibitors of ornithine decarboxylase (ODC) and arginine decarboxylase (ADC) have been developed by the Merrell Dow Company (Bey *et al.*, 1987). Difluoromethylornithine (DFMO) and difluoromethylarginine (DFMA) act as suicide inhibitors of ODC and ADC respectively. Feeding these inhibitors to a transformed root culture of *D. stramonium* (D15/5) enabled the relative importance of ornithine and arginine in contributing to the formation of the tropane ring to be deduced. It was apparent that the route from arginine via ADC is more important in providing putrescine for alkaloid production in *Datura stramonium* than the ornithine–putrescine route which involves ODC (Walton *et al.*, 1990; Robins *et al.*, 1991a). Similar conclusions were derived for the production of putrescine used as a precursor of nicotine in *Nicotiana* species (Tiburcio and Galston, 1986) and pyrrolizidine alkaloids in *Senecio* species (Hartmann *et al.*, 1987, 1988).

5.3.3.2 Feeding intermediates in tropane alkaloid biosynthesis. Robins *et al.* (1991b) have also reported the effects of feeding key intermediates in tropane alkaloid biosynthesis to transformed roots of *D. stramonium* line D15/5. Putrescine at 1–10 mM did not affect growth but caused a significant increase in the concentration of several intermediates. For example, feeding putrescine at 5 mM caused a 13-fold increase in hygrine, a 30-fold increase in cusgohygrine, a five-fold increase in tropine and a four-fold increase in acetyltropine levels. Interestingly, the concentrations of apohyoscyamine and hyoscyamine remained unchanged in these cultures relative to the control tissues. Feeding either tropinone or tropine at 1–5 mM to the cultures failed to induce a significant increase in the concentration of hyoscyamine or scopolamine. However, feeding tropinone at 10 mM stimulated tropine and acetyltropine levels by about 45-fold (Robins *et al.*, 1991b). The stimulation of acetyltropine formation, which is normally found at low levels in these root cultures, is interesting and may suggest that limitations in the tropyl esterification allows diversion of tropine into the acetyl derivative. An enzyme preparation from *Datura* root cultures had been shown to catalyse the acetylation of tropine using acetyl-CoA as the donor (Robins *et al.*, 1991d). Feeding tropic acid or phenyllactic acid also failed to stimulate apohyoscyamine or hyoscyamine formation and proved to be toxic to growth when added at levels above 2.5 mM. These results suggest that there are at least two important control points which control flux through the tropane alkaloid pathway. The first is in the supply of putrescine and early intermediates, the second is the esterification of tropine with tropic acid to form hyoscyamine.

5.3.3.3 Disruption of the differentiated state. Biochemical inhibitor and precursor feeding experiments can provide valuable information regarding the likely target enzymes for genetic manipulation experiments aimed at increasing flux through pathways. An additional, and complementary, approach is to perturb the pathway and ascertain which enzymes are most sensitive to the perturbation effects. As was noted previously, root cultures of several species can be induced to form callus by supplying phytohormones to the cultures. This often leads to a concomitant reduction in alkaloid content of tissues. Using transformed roots of *Nicotiana rustica*, Rhodes *et al.* (1989) investigated the activity of enzymes involved in nicotine biosynthesis when roots were placed in medium containing callus-inducing phytohormones. One enzyme in particular, putrescine methyl transferase (PMT), showed a fall in activity to less than 10% of the initial value within 24 hours of the onset of phytohormone treatment and before any obvious signs of callus formation were apparent. Similar experiments have been carried out using *D. stramonium* (Robins *et al.*, 1991c) and in this case also hormone treatments led to rapid loss of PMT activity in tissues as they began to form callus. Transformed cultures, maintained as in the dispersed condition for several cycles, lacked measurable PMT activity. However, when roots were allowed to redifferentiate by removal of the phytohormones, PMT activity was fully restored. Thus PMT, which is the first committed step to nicotine and tropane alkaloid biosynthesis in *Nicotiana* and *Datura* species respectively, may represent one important point of regulatory control for alkaloid production in species of both genera.

5.3.4 Selection procedures to increase secondary product formation

Selection procedures, using biochemical inhibitors, have been attempted with the aim of recovering mutants altered in secondary metabolism. Widholm (1972a) isolated cell lines of *N. tabacum* resistant to the tryptophan analogue, 5-methyl tryptophan (5MT), which normally is toxic to plant cells at levels around 1 mM. He found these lines accumulated levels of tryptophan which were about 10-fold higher than control tissues due to the presence of an altered anthranilate synthetase enzyme which was much less sensitive to feedback inhibition by tryptophan than the wildtype enzyme. Similar results were obtained with *Daucus carota* cell cultures where the free L-tryptophan level in 5MT resistant lines was found to be 27 times greater than in the control (Widholm, 1972b). Schallenberg and Berlin (1979) used the same approach with *Catharanthus roseus* cell suspensions which have the potential to synthesise a number of tryptophan-derived indole alkaloids such as ajmalicine, catharanthine, vinblastine and vincristine. Although the experiments were successful, in that mutants were recovered which contained high levels of tryptophan (30-fold increase), no increase in tryptophan-derived indole alkaloids was observed. Sasse *et al.* (1983) used 4-methyl tryptophan (4-MT), which is equally toxic to plant cells but is a much better substrate for tryptophan decarboxylase than is 5-methyl tryptophan. This enzyme catalyses the first committed step to indole alkaloid synthesis and it was predicted that among the 4-MT-resistant mutants recovered there would be lines possessing high levels of tryptophan decarboxylase, due to enhanced detoxification of 4-MT. It was thought that these lines would show a greater flux down the pathway leading to alkaloid formation. In fact Sasse *et al.* (1983) did produce *Catharanthus* cell lines showing levels of tryptophan decarboxylase activity which were 3–10 times higher than that found in control tissues. Tryptamine levels in these lines were three- to five-fold greater than in controls; however, the majority of resistant lines did not exhibit a stimulation of indole alkaloid biosynthesis. It is likely that this was due to a biochemical limitation further down the pathway to alkaloid biosynthesis and it would be worthwhile repeating these experiments using callus or cell suspensions derived from transformed roots of *C. roseus* which make ajmalacine (Parr *et al.*, 1988), then allowing roots to re-differentiate from mutant cells with high levels of resistance to 4-MT.

Other biochemical selection experiments have, however, resulted in lines with an increased concentration of secondary product. For example cell lines *N. tabacum* were selected for resistance to *p*-fluorophenylalanine, a toxic analogue of phenylalanine (Berlin and Witte, 1982), and were shown to have a 10-fold increase in PAL and concentrations of cinnamyl putrescines which were six-fold higher than control tissues.

5.4 Genetic manipulation of secondary metabolism

5.4.1 Overexpression of key genes to increase flux through metabolic pathways

Biochemical inhibitor, precursor feeding and elicitation experiments provide useful background information on which to base genetic manipulation experiments. They can only act as a guide, however, and it is important to attempt the overexpression of genes encoding key enzymic steps to determine the degree to which secondary metabolite levels can be stimulated by directly manipulating expression. The concept of metabolic engineering is one which is receiving increasing attention (Bailey, 1991) and secondary metabolism lends itself to this approach since major alterations are unlikely to be lethal in the majority of cases. However, since our knowledge of regulation of the pathways we wish to manipulate is often limited, an iterative approach using cycles of genetic manipulation and biochemical evaluation is necessary to achieve the desired outcome. This approach will allow the plasticity of metabolism to be explored. Also, the possibility that the metabolic perturbations established may induce quite unanticipated secondary effects may lead to increased knowledge of the inter-relations within metabolism (Bailey, 1991). Significant advances in this subject area have been made recently involving the flavonoid biosynthesis pathway in plants. This subject has been extensively reviewed recently (Van Tunen and Mol, 1990) and will not be discussed in this chapter.

5.4.1.1 Ornithine decarboxylase. Using alkaloid-producing cultures, biochemical experiments suggested that if it were possible to increase endogenous putrescine this might cause an increase in nicotine in *Nicotiana* species and perhaps hygrine and tropine, but probably not hyoscyamine, in *Datura* species (Walton *et al.*, 1988; Robins *et al.*, 1990). The gene coding for ornithine decarboxylase, which converts ornithine to putrescine, has been characterised from mammals, yeast and other eukaryotes (McConlogue *et al.*, 1984; Phillips *et al.*, 1987; Fonzi and Sypherd, 1987; Bassez *et al.*, 1990). So far, however, this gene has not been isolated from plants, although several groups are working toward this end, and the yeast gene has been used for genetic manipulation of plants, described below. Recently, however, the related gene for arginine decarboxylase has been characterised from a plant source, *Avena sativa* (Bell and Malmberg, 1990), as well as from *E. coli* (Moore and Boyle, 1990).

The coding sequence of the yeast ODC gene was placed under the control of the CaMV35S promoter with or without the duplicated upstream enhancer sequences (Kay *et al.*, 1987) and was co-transferred with the Ri T-DNA of *A. rhizogenes* into *N. rustica* (Hamill *et al.*, 1990a). Analysis of a number of transformed root lines containing the ODC gene driven by the CaMV35 promoter with the duplicated enhancer sequence, showed some contained significantly elevated levels of ornithine decarboxylase activity relative to controls. Several such lines also had increased concentrations of nicotine which, in the best lines, were about two-fold higher than the average of the controls (Hamill *et al.*, 1990a). Although the magnitude of the changes in nicotine content was not great, being two-fold at maximum, the results are in agreement with those predicted by precursor feeding experiments (Walton *et al.*, 1988) and suggest that elevating putrescine levels *in vivo* can increase the flux into nicotine. The fact that ODC was used to oversupply putrescine while evidence, noted previously, suggests that the preferred endogenous supply of putrescine in transformed roots for alkaloid biosynthesis *in vivo* is via ADC, may appear contradictory. However, both arginine and ornithine have been shown to be effective precursors of putrescine in *Datura* and *Nicotiana* (Walton *et al.*, 1990; Tiburcio and Galston, 1986) and even when ADC activity is blocked by DFMA, hyoscyamine accumulation is not entirely inhibited. If a mechanism does exist for the compartmentalisation of the two putrescine-synthesising enzymes *in vivo*, it is unlikely that the engineered ODC gene under the control of the CaMV35S promoter would have been subject to this control. Analysis of putrescine conjugates and spermine/spermidine concentrations showed little difference in lines containing the yeast ODC gene relative to those which did not (Hamill *et al.*, 1990a). Putrescine is, of course, an important intermediate in polyamine biosynthesis and its levels are regulated by post-transcriptional (Kameji and Pegg, 1987) and post-translational mechanisms (Koromilas and Kryiakidis, 1988). It is possible that the expression of the yeast ODC may also be subject to these post-transcriptional control mechanisms. It is also possible that other enzymic steps in nicotine biosynthesis, such as PMT (noted above) or possibly the enzyme complex responsible for the condensation of nicotinic acid with *N*-methylpyrolline, to produce nicotine, represent points at which control of flux is exerted. It would be of interest to stimulate the levels of these enzymes, in conjunction with an increased supply of putrescine *in vivo*, to ascertain whether further increases in nicotine accumulation can be induced.

5.4.1.2 ODC enzyme turnover in vivo. The yeast ODC gene also encodes an amino acid sequence which can be considered to be similar to a PEST-rich region. PEST regions are rich in the amino acids proline (P), glutamic acid (E), serine (S) and threonine (T) and are known to cause rapid intracellular degradation of proteins containing these regions (Rogers *et al.*, 1986). The ODC gene from *Trypanosoma brucei* does not encode PEST-rich amino acid sequences and it was suggested that overexpression of this gene in transformed roots of tobacco may lead to yet higher levels of ODC activity in transformed roots of tobacco than was achieved using the yeast ODC gene (Hamill *et al.*, 1990b). Currently such studies are in progress, but recent results of Minocha and De Scenzo (1991) suggest this hypothesis may be correct. These authors reported preliminary results of experiments to alter expression of ODC in tobacco using an ODC gene from mouse. Using a full length ODC gene (which encodes a PEST-rich region), they were unable to achieve an increase in putrescine content of transformed tobacco whereas the use of a truncated ODC gene (lacking PEST sequences) enabled a two-fold increase in putrescine content of plants. In both cases, the CaMV35S promoter was used to express the murine ODC in tobacco. It is also of interest that these authors found that callus from transformed plants contained much more putrescine compared to controls (up to a ten-fold increase in the case of the PEST minus transformant) suggesting that additional control mechanism(s) exist in intact, organised plant tissue which limit the extent to which putrescine levels can be stimulated.

5.4.1.3 Lysine decarboxylase. Other groups have attempted to increase flux through secondary metabolic pathways by overexpression of important genes. The gene for lysine decarboxylase (LDC), from the bacterium *Hafnia alvei*, was introduced into *N. tabacum*, using a disarmed binary vector/leaf disc transformation procedure (Herminghaus *et al.*, 1991). The gene was engineered such that in one group of transformants protein would accumulate in the cytoplasm while a second group of transformants were generated with the gene fused to a DNA sequence encoding a chloroplast-targeting amino acid sequence (see Robinson, 1991). The protein was thus directed to the chloroplast where the decarboxylation of lysine normally occurs to produce cadaverine, a precursor of the alkaloid anabasine. Northern blotting showed both classes of transformant expressed the bacterial gene but the LDC enzyme was only detected in plants containing the gene targeted to the chloroplast. Untransformed plants contained barely detectable cadaverine levels whereas this group of transformants contained cadaverine at 0.3–1% dry weight in leaf tissue. As might be expected, root tissues did not synthesise cadaverine presumably due to the lack of a differentiated chloroplast in these tissues (Herminghaus *et al.*, 1991).

5.4.1.4 Tryptophan decarboxylase. The gene for tryptophan decarboxylase (TDC) from *Catharanthus roseus* was placed under the control of the CaMV35S promoter and transferred into *Peganum harmala* transformed roots using *A. rhizogenes* (Berlin *et al.*, 1991) and also to *N. tabacum* using disarmed *A. tumefaciens* (Songstad *et al.*, 1990). In the *Peganum* cultures, the TDC activity was stimulated by a factor of 10 or more and seratonin levels were increased from barely detectable levels in controls to 15–20 mg g^{-1} dry weight in transformants (Berlin *et al.*, 1991). In the case of tobacco, the TDC activity could be increased more than 40-fold and tryptamine levels were stimulated by up to 260-fold compared to controls so that the concentration of tryptamine exceeded 1 mg g^{-1} fresh weight in these TDC transformants. Despite high tryptamine levels, the transformed tobacco plants were morphologically normal and were fertile. Indole acetic acid levels were very similar in controls and in TDC transformants demonstrating that an increased tryptamine pool did not affect auxin biosynthesis. Songstad *et al.* (1990) suggest that stimulation of tryptamine levels in *Catharanthus roseus*, by overexpression of TDC, might enable a greater flux into indole alkaloid biosynthesis and experiments to test this are underway. Chavadez *et al.* (1991) also transferred the same CaMV35S-TDC construct to *Brassica napus* and *Solanum tuberosum* and were able to stimulate tryptamine levels in these species, though to a lesser extent than in the case of tobacco.

5.4.1.5 Tropane alkaloid biosynthesis. In tropane alkaloid biosynthesis, a c-DNA clone for hyoscyamine 6β-hydroxylase (H6βH) from *H. muticus* (Matsuda *et al.*, 1991) has been expressed in other Solanaceous species. Transfer of this c-DNA into *Atropa belladonna*, under control of the CaMV35S promoter, led to the production of a 38 kDa protein in various tissues throughout the plant, which was recognised by antibody raised against the H6βH protein from *Hyoscyamus.* This manipulation resulted in plants which converted essentially all their hyoscyamine to scopolamine contrasting with non-transformed plants which accumulate high levels of hyoscyamine (Yamada *et al.*, 1991).

5.4.1.6 Resveratrol in tobacco. A gene from *Arachis*, coding for stilbene synthase, was expressed in *Nicotiana* under control of its endogenous promoter. Stilbene synthase and the product of its activity, resveratrol (XIV), was detected using antibodies and HPLC methods (Hain *et al.*, 1990). Resveratrol acts as a phytoalexin in *Arachis* and in the transgenic tobacco its expression was induced by UV and by fungal elicitors as in peanut. This work raises the possibility of transferring chemical defence mechanisms between plants using genetic transformation.

5.4.2. Manipulation of secondary metabolism by antisense technology

A number of experiments involving the use of antisense methodology have indicated the validity of this approach for selectively inhibiting expression of particular genes. Introduction of the chalcone synthase (CHS) gene into *Petunia* or tobacco in an antisense orientation caused a reduction in CHS mRNA and resulted in plants with reduced flower pigmentation (Van der Krol *et al.*, 1988). Several experiments have confirmed antisense RNA methodology to be of general value in reducing or eliminating gene expression, in secondary metabolic pathways and in other areas of cellular metabolism (Van de Krol *et al.*, 1990a; Mol *et al.*, 1990; Smith *et al.*, 1988, 1990b). Cannon *et al.* (1990) showed that small antisense fragments as short as 41 base pairs were sufficient to produce up to 95% inhibition of expression of β-glucuronidase (GUS) activity when transformed in an antisense orientation into GUS positive plants. As secondary metabolic pathways involve numerous enzymic steps and have many branch points leading to the synthesis of a range of metabolites, antisense technology may be used to inhibit either entire pathways or to remove undesirable secondary products from plants. Alternatively it could be used to block a particular branch of a pathway and thus divert precursors along desirable branches of pathways.

Clearly workers in numerous laboratories have recognised the potential of transformation technology to manipulate secondary metabolism in plants and in tissues grown *in vitro*. As additional genes encoding enzymes involved in secondary product biosynthesis are being isolated and characterised, we can expect to see an increased number of experiments aimed at exploring the extent to which these metabolites can be increased or altered *in vivo* by overexpression or down regulation of specific enzyme activities.

5.5 Possible limitations to long-term stability in genetically manipulated culture lines

5.5.1 Co-suppression

The phenomenon of co-suppression, in which integration of a transgene leads to suppressed expression of both the transgene and the endogenous copy, may have important consequences for the design of experiments aimed at increasing secondary metabolite levels in plants by overexpression of key genes in pathways. Genes for either chalcone synthase or dihydroflavonol reductase isolated from *Petunia* introduced into transformed *Petunia* caused a reduction in flavonoid pigment accumulation in the floral tissues of some transformants (Napoli *et al.*, 1990; Van der Krol *et al.*, 1990b). In tomato, transformation with part of the polygalacturonase gene, in the sense orientation with expression driven by the CaMV35S promoter, caused inhibition of the endogenous polygalacturonase gene activity and led to reduced levels of enzyme in the fruit of transformants (Smith *et al.*, 1990a). In each of these cases, reduced levels of expression of transgenes and endogenous genes were observed. Introduction of a homologous gene, in the sense orientation, caused a reduction in PAL activity, though not of mRNA levels, in transgenic tobacco with concomitant reduction in the phenylpropanoid products of PAL activity (Elkind *et al.*, 1990). In tobacco an *nptIII*-positive plant was re-transformed with an additional *nptIII* gene leading to suppression of neomycin phosphotransferase activity and kanamycin resistance (Goring *et al.*, 1991).

A number of hypotheses have been suggested to account for co-suppression. Grierson *et al.* (1991) suggest that co-suppression may be due to the formation of antisense RNA (discussed below), formed by the inserted gene being transcribed from another promoter (i.e. in the reverse orientation), near the point where the gene is inserted. Many commonly used plant vectors are arranged with the *nptll* gene controlled by the nopaline synthase (*nos*) promoter being transcribed in the opposite orientation to that of the transgene and read-through of the message from the *nos* promoter may allow small fragments of antisense RNA to be formed from the gene of interest. Grierson *et al.* (1991) point to the observation that in several co-suppression experiments, and the equivalent antisense experiments, reduced or zero levels of mRNA were observed in the transformed tissues, e.g. *Petunia* (van der Krol *et al.*, 1988, 1990a; Napoli *et al.*, 1990; Smith *et al.*, 1988, 1990a). Mol *et al.* (1991) offer some support to the view of Grierson *et al.* (1991) but raise a number of points to be addressed before the antisense model can be accepted. Jorgensen (1991) also suggests that some examples of co-suppression cannot be easily explained by the hypothesis of Grierson *et al.* (1991) as suppression of unlinked genes has been observed in transgenic plants. Jorgensen (1991) suggested a model involving ectopic pairing between a gene and a non-allelic homologous copy to explain at least some examples of co-suppression. Clearly many critical experiments must be carried out to enable this issue to be resolved.

Although co-suppression does have important implications for the design of experiments aimed at increasing flux in metabolic pathways by attempting to overexpress genes coding for key enzymes, it is clear that high levels of mRNA and gene product can occur as a result of genetic transformation. In addition to experiments noted above (section 5.4.1), other examples include glyphosate resistant cell lines and plants produced which had a 40-fold increase in levels of EPSP synthase as a result of overexpression of the EPSP synthase gene driven by a CaMV35S promoter (Shah *et al.*, 1986). Similarly, expression of the genes for glutamine synthase (GS) from alfalfa in tobacco led to a 35-fold increase in GS activity in the transgenic plants and GS protein represented 5% of the soluble protein in the transformed tobacco (Eckes *et al.*, 1989). Again in tobacco, overexpression of a nuclear gene for superoxide dismutase (SOD) from *Petunia* led to a 50-fold increase in SOD activity in the transformants (Tepperman and Dunsmuir, 1990).

5.5.2 Methylation

A number of studies on gene expression in transformed plants have indicated that the foreign gene can become methylated and thus its expression can be reduced or totally silenced. Methylation, particularly of cytosine residues, plays an important role in regulation of gene expression in plants and it has even been suggested that the process could act as a defence mechanism against DNA from microorganisms, with the plant effectively neutralising the effects of the foreign gene (Linn *et al.*, 1990). In this respect, it is of interest that a number of *Nicotiana* species have homologous sequences to *Agrobacterium rhizogenes* T-DNA within their genomes, presumably as a result of a transformation of an ancestral species (Furner *et al.*, 1986). Methylation of genes is also an important consideration in the manipulation of secondary metabolism and recent experiments have studied this phenomena with respect to the stability of expression of the maize A1 gene in transgenic *Petunia* plants (Linn *et al.*, 1990). This gene encodes the enzyme dihydroflavonol reductase (DFR) and catalyses the conversion of dihydrokaempferol to leuco-pelargonidin, the precursor of pelargonidin pigments (see Van Tunen and Mol, 1991). When transferred into a white flowered mutant of *Petunia* lacking flavonoid 3′- and 3′,5′-hydroxylase activity, the transformants had flowers with an orange colour as a result of activity of the maize A1 gene (Meyer *et al.*, 1987). Seven transgenic plants with single gene inserts were examined with regard to methylation of the CCGG sites within the CaMV35S promoter using the isoschisomer endonucleases *Msp*1/*Hpa*ll to determine the extent of methylation in the transgenic plants. Of the seven plants studied, three showed methylation of these sequences. Interestingly, the four non-methylated plants had red-coloured flowers while the three methylated plants had white flowers suggesting a close correlation with expression of the gene and its methylation status. In plants carrying multiple copies of the DFR gene, the methylation patterns were very complex and the flowers were either white or variegated (Linn *et al.*, 1990). Fluctuations in environmental conditions in the field such as extremes of temperature tend to lead to variable degrees of methylation and gene expression (Saedler, personal communication). From the point of view of manipulating secondary metabolism in culture by altering expression of key enzymic steps in a metabolic pathway, it appears likely that low or single copy numbers of foreign genes are desirable, perhaps in conjunction with a continuous or periodic selection pressure and/or treatment with azacytidine which has been shown to reduce or eliminate the effects of methylation on expression of foreign genes in plants (Van Slogteren *et al.*, 1984; Hepburn *et al.*, 1987).

5.6 Conclusions and future prospects

It is clear that recent advances in transformation methodology and molecular genetics have had an impact upon the study of secondary metabolism in plants and in organs and tissues grown *in vitro*. Although significant advances in understanding the biosynthesis of certain secondary metabolites have been possible using disorganised cell culture, the use of fast growing, productive and stable transformed organ cultures has greatly extended this approach and such cultures are proving to be a convenient system for biosynthetic and genetic manipulation experiments aimed at understanding and increasing flux through metabolic pathways. The number of genes isolated and characterised which code for enzymes specific to secondary metabolism continues to grow and this will enable control to be studied at the molecular level. In addition, such genes will be of value for experiments aimed at perturbing secondary metabolism by direct gene transfer. Recently reported phenomena such as co-suppression and inactivation of foreign gene activity by methylation, and the use of antisense technology to control specific gene expression, all have implications for experiments aimed at manipulating secondary metabolism in plants and in tissues cultured *in vitro*. Also, it is clear that we are still ignorant of the factors controlling the regulation of secondary metabolic pathways in the majority of cases. Recent advances in elucidating the control of phenylpropanoid secondary metabolism, particularly with flavonoid biosynthesis, have indicated the existence of sophisticated molecular mechanisms governing the overall expression of these pathways. It is apparent that this area of secondary metabolism is subject to rigorous developmental and/or environmental regulation at the level of the co-ordinated expression of structural genes encoding enzymes in the biosynthetic pathway. In many cases specific DNA sequences in the 5′ region of the genes, or even in the coding region of the structural gene itself, such as the 4-hydroxycinnamate CoA ligase gene from parsley (Douglas *et al.*, 1991), are required for optimal expression of the gene. These structural genes are regulated by genes encoding DNA binding proteins, the production of which is itself highly regulated (reviewed by Hahlbrock and Scheel, 1989; Dixon and Lamb, 1990; Van Tunen *et al.*, 1990; Martin *et al.*, 1991; Martin and Gerats, 1992).

It will be of considerable interest to ascertain whether other secondary metabolic pathways have such highly regulated control mechanisms as the phenylpropanoid biosynthetic pathway. In the light of experiments noted earlier, it is clear that many secondary metabolic pathways are subject to developmental regulation so it will not be too surprising if similar mechanisms operate in other secondary metabolic pathways. The design of experiments aimed at manipulating secondary metabolism in culture will benefit greatly from detailed knowledge regarding the regulation of pathways at the molecular level during the life cycle of the plant. It is possible that manipulation of regulatory genes controlling the expression of entire metabolic pathways may be more appropriate than manipulation of the expression of structural genes encoding individual enzymic steps in biochemical pathways. The extent to which primary and secondary metabolism interact is also an area where many questions remain to be answered.

Clearly the next few years promise to be interesting as these questions are addressed experimentally.

Acknowledgements

Research in the laboratories of the authors is supported by the Australian Research Council (J.D.H.) and the Agricultural and Food Research Council of the UK (M.J.C.R.). We are very grateful to Mrs J. Elliston for her skill and patience in the typing of the manuscript.

References

Aird, E.L.H., Hamill, J.D. and Rhodes, M.J.C. (1988a) Cytogenetic analysis of hairy root cultures from a number of plant species transformed by *Agrobacterium rhizogenes. Plant Cell Tiss. Org. Cult.* **15 :** 47–57.

Aird, E.L.H., Hamill, J.D., Robins, R.J. and Rhodes, M.J.C. (1988b) Chromosome stability in transformed hairy root cultures and the properties of variant lines of *Nicotiana rustica* hairy roots. In *Manipulating Secondary Metabolism in Culture* (Robins, R.J. and Rhodes, M.J.C., eds.), Cambridge University Press, Cambridge, pp. 137–144.

Ambros, P.F., Matzke, M.A. and Matzke, A.J.M. (1986) Detection of a 17 kb unique sequence (T-DNA) in plant chromosomes by *in situ* hybridization. *Chromosoma (Berlin)* **94**: 11–18.

Anderson, L.A., Keene, A.T. and Philipson, J.D. (1982) Alkaloid production by leaf organ, root organ and cell suspension cultures of *Cinchona ledgeriana. Planta Medica* **46** : 25–27.

Anderson, L.A., Hay, C.A., Roberts, M.F. and Phillipson, J.D. (1986) Studies on *Ailanthus altissimia* cell suspension cultures. Precursor feeding of L-[methylene^{14}C] tryptophan and L-tryptophan. *Plant Cell Rep.* **5**: 387–390.

Asamizu, T., Akiyama, K. and Yasuda, I. (1988) Anthoquinone production by hairy root culture in *Cassia obtusifolia. Yakagaku Zasshi* **108**: 1215–1218.

Bailey, J.E. (1991) Toward a science of metabolic engineering. *Science* **252**: 1668–1675.

Banerjee-Chattopadhyay, S., Schwemmin, A.M. and Schwemmin, D.J. (1985) A study of karyotypes and their alterations in cultured and *Agrobacterium* transformed roots of *Lycopersicon peruvianum. Theor. Appl. Genet.* **71**: 258–262.

Barz, W., Daniel, S., Winderer, W., Koster, J., Olto, C. and Tiemann, K. (1988) Elicitation and metabolism of phytoalexins in plant cell cultures. In *CIBA Foundation Symp. 137: Applications of Plant Cell and Tissue Culture* (Yamada, Y., ed.), Wiley and Sons, New York, pp. 178–191.

Bassez, T., Paris, J., Omilli, F., Dorel, C. and Osborne, H.B. (1990) Post-transcriptional regulation of ornithine decarboxylase in *Xenopus laevis* oocytes. *Development* **110**: 955–962.

Bell, E. and Malmberg, L. (1990) Analysis of cDNA encoding arginine decarboxylase from oat reveals similarity to the *Escherichia coli* arginine decarboxylase and evidence of protein processing. *Mol. Gen. Genet.* **224**: 431–436.

Berlin, J. and Witte, L. (1982) Metabolism of phenylalanine and cinnamic acid in tobacco cell lines with high and low yields of cinnamoyl putrescines. *J. Nat. Prod. (Lloydia)* **45**: 88–93.

Berlin, J., Bedorf, N., Mollenschott, C., Wray, V., Sasse, F. and Höfle, G. (1988) On the podophyllotoxins of root cultures of *Linium flavum. Plant Medica* **54**: 204–206.

Berlin, J., Dietze, P., Fecker, L., Goddijn, O. and Hoge, J.H.C (1991) Production of high levels of serotonin in *Peganum* cultures by expression of a foreign plant tryptophan decarboxylase. *Abstracts 3rd Int. Cong. Plant Mol. Biol.* (Hallick, R.B., ed.), Tucson, October 1991, p. 1091.

Bevan, M. (1984) Binary *Agrobacterium* vectors for plant transformation. *Nucl. Acids Res.* **12**: 8711–8721.

Bey, P., Darzin, C. and Jung, M. (1987) Inhibition of basic amino acid decarboxylase involved in polyamine biosynthesis. In *Inhibition of Polyamine Metabolism* (McCann, P.P., Pegg, A.E. and Sjoerdsma, A., eds.), Academic Press, Orlando.

Brodelius, P., Collinge, M.A., Funk, C., Gugler, K. and Marques, I. (1989) Studies on alkaloid formation in plant cell cultures after treatment with a yeast elicitor. In *Primary and Secondary Metabolism of Plant Cell Cultures* (Kurz, W.G.W., ed.), Springer-Verlag, Berlin, pp. 191–199.

Cannon, M., Platz, J., O'Leary, M., Sookdeo, C. and Cannon, F. (1990) Organ-specific modulation of gene expression in transgenic plants using antisense RNA. *Plant. Mol. Biol.* **15**: 39–47.

Charlwood, B.V. and Moustou, C, (1988) Essential oil accumulation in shoot-proliferation cultures of *Pelargonium* spp. In *Manipulating Secondary Metabolism in Cultures* (Robins, R.J. and Rhodes, M.J.C., eds.), Cambridge University Press, Cambridge, pp. 187–194.

Charlwood, B.V., Brown, J.T., Moustou, C., Morris, G.S. and Charlwood, K.A. (1988a) The accumulation of isoprenoid flavour compounds in plant cell cultures. In *Bioflavour '87* (Schreier, P., ed.), Walter de la Gruyter, Berlin, pp. 303–314.

Charlwood, K.A., Brown, S. and Charlwood, B.V. (1988b) The accumulation of flavour compounds by cultures of *Zingiber officinale*. In *Manipulating Secondary Metabolism in Cultures* (Robins, R.J. and Rhodes, M.J.C., eds.), Cambridge University Press, Cambridge, pp. 195–200.

Chavadez, S., Brisson, N. and De Luca, U. (1991) Production of transgenic plants which accumulate tryptamine through expression of tryptophan decarboxylase. *Abstracts 3rd Int. Plant Mol. Biol.* (Hallick, R.B., ed.), Tucson, October 1991, p. 1098.

Christen, P., Roberts, M.F., Phillipson, J.D. and Evans, W.C. (1989) High yield production of tropane alkaloids by hairy-root cultures of a *Datura candida* hybrid. *Plant Cell Rep.* **8**: 75–77.

Cline, S.D. and Coscia, C.J. (1988) Stimulation of sanguinarine production by combined fungal elicitation and hormonal deprivation in cell suspension cultures of *Papaver bracteatum*. *Plant Physiol.* **86**: 161–165.

Coen, E.S. and Meyerowitz, E.M. (1991) The war of the whorls: genetic interactions controlling flower development. *Nature* **353**: 31–37.

Constabel, C.P. and Towers, G.H.N. (1988) Thiarubrine accumulation in hairy root cultures of *Chaenactis douglasii*. *J. Plant Physiol.* **133**: 67–72.

Croes, A.F., Van den Berg, A.J.R., Bosveld, M., Breteler, H. and Wullems, G.J. (1989) Thiophene accumulation in relation to morphology in roots of *Tagetes patula*: Effects of auxin and transformation by *Agrobacterium*. *Planta* **179**: 43–50.

D'Amato, F. (1985) Cytogenetics of plant cell and tissue culture and their regenerates. *CRC Crit. Rev. Plant Sci.* **3**: 73–112.

Dawson, R.F. (1942) Nicotine synthesis in excised tobacco roots. *Am. J. Bot.* **29**: 813–815.

Deno, H., Yamagata, T., Emoto, T., Yoshioka, T., Yamada, Y. and Fijita, Y. (1987) Scopalamine production by root cultures of *Duboisia myoporoides*. II. Establishment of a hairy root culture by infection with *Agrobacterium rhizogenes*. *J. Plant Physiol.* **131**: 315–323.

Deus-Neumann, B. and Zenk, M.H. (1984) Instability of indole alkaloide production in *Catharanthus roseus* cell suspension cultures. *Planta Medica*. **50**: 427–431.

De Vries-Uijtewaal, E., Gilssen, L.J.W., Flipse, E., Sree Ramulu, K. and de Groot, B. (1988) Characterisation of root clones obtained after transformation of monohaploid and diploid potato genotypes with hairy root inducing strains of *Agrobacterium*. *Plant Science* **58**: 192–202.

Dixon, R.A. (1986) The phytoalexin response: elicitation, signalling and control of gene expression. *Biol. Rev.* **61**: 239–291.

Dixon, R.A. and Lamb, C.J. (1990) Molecular communications in interactions between plants and microbial pathogens. *Ann. Rev. Plant Physiol.* **41**: 339–367.

Dixon, R.A., Blyden, E.R., Dron, M., Harrison, M.J., Lamb, C.J., Lawton, M.A. and Mavandad, M. (1989) Regulation of gene expression in biologically stressed bean cell cultures. In *Primary and Secondary Metabolism of Plant Cell Cultures* (Kurz, W.G.W., ed.), Springer-Verlag, Berlin, pp. 266–273.

Douglas, C.J., Hauffe, K.D., Ites-Morales, M.E., Ellard, M., Passkowski, U., Hahlbrock, K. and Dangl, J.L. (1991) Exonic sequences are required for elicitor and light activation of a plant defense gene, but promoter sequences are sufficient for tissue specific expression. *EMBO J.* **10**:1767–1775.

Eckes, P., Schmitt, P., Daub, W. and Wengenmayer, F. (1989) Overproduction of alfalfa glutamine synthetase in transgenic tobacco plants. *Mol. Gen. Genet.* **217**. 263–268.

Eilert, U. (1989) Elicitor induction of secondary metabolism in dedifferentiated and differentiated, *in vitro* systems of *Ruta graveolens*. In *Primary and Secondary Metabolism of Plant Cell Cultures* (Kurz, W.G.W., ed.), Springer-Verlag, Berlin, pp. 219–228.

El-Dabbas, S.W. and Evans, W.C. (1982) Alkaloids of the genus *Datura*, Section Burgmansia. X. Alkaloid content of *Datura* hybrids. *Plant Medica* **44**: 184–185.

Elkind, Y., Edwards, R., Mavandad, M., Hedrick, S.A., Ribak, O., Dixon, R.A. and Lamb, C.J. (1990) Abnormal plant development and down-regulation of phenylpropanoid biosynthesis in transgenic tobacco containing a heterologous phenylalanine ammonia-lyase gene. *Proc. Natl Acad. Sci. USA* **87**: 9057–9061.

Endo, T., Goodbody, A. and Misawa, M. (1987) Alkaloid production in root and shoot cultures of *Catharanthus roseus*. *Planta Medica* **53**: 479–482.

Flores, H.E. (1987) Use of plant cell and organ culture in the production of biological chemicals. In *Application of Biotechnology to Agricultural Chemistry* (Mumma, R.O., Lebaron, H., Mumma, R.O., Honeycutt, R.C. and Duesing, J.H., eds.), Am. Chem. Soc. Symposium series, Washington DC, **334**: 67–86.

Flores, H.E. and Filner, P. (1985) Metabolic relationships of putrescine, GABA and alkaloids in cell and root cultures of Solanaceae. In *Primary and Secondary Metabolism of Plant Cell Cultures* (Neumann, K.H., Barz, W. and Reinhard, E., eds.), Springer-Verlag, Berlin, pp. 174–185.

Flores, H.E., Pickard, J.J. and Hoy, M.W. (1988) Production of polyacetylenes and thiophenes in heterotrophic and photosynthetic root cultures of Asteraceae. In *Chemistry and Biology of Naturally Occurring Acetylenes and Related Compounds (NOARC). Bioactive Molecules* (Lam, J., Breheler, H., Arnason, T. and Hansen. L., eds.),**7**: 233–254.

Fonzi, W.A. and Sypherd, P.S. (1987) The gene and primary structure of ornithine decarboxylase from *Saccharomyces cerevisiae*. *J. Biol. Chem.* **262**: 10127–10135.

Fujita, Y. (1990) The production of industrial compounds. In *Plant Tissue Culture: Applications and Limitations* (Bhojwani, S.S., ed.), Elsevier, Amsterdam, pp. 259–275.

Fujita, Y. and Tabata, M. (1987). Secondary metabolites from plant cells—pharmaceutical applications and progress in commercial production. In *Plant Tissue and Cell Culture* (Green, C.E. *et al.*, eds.), Alan R. Liss, New York, pp. 169–186.

Fujita, Y., Takahashi, S. and Yamada, Y. (1984) Selection of cell lines of high productivity of shikonin derivatives through protoplasts of *Lithospermum erythrorhizom*. *Proc. Third European Cong. Biotechnol.* **1**:161–166.

Furner, I.J., Huffman, G.A., Amasino, R.M., Garfinkel, D.J., Gordon, M.P. and Nester, E.W. (1986) An *Agrobacterium* transformation in the evolution of the genus *Nicotiana*. *Nature* **319**: 422–427.

Furze, J.M., Rhodes, M.J.C., Parr, A.J., Robins, R.J., Whitehead, J.M. and Threlfall, D.R. (1991) Abiotic factors elicit sesquiterperiod phytoalexin production but not alkaloid production in transformed roots of *Datura stramonium*. *Plant Cell Rep.* **10**: 111–114.

Goring, D.R., Thomson, L. and Rothstein, S.J. (1991) Transformation of a partial nopaline synthase gene into tobacco suppresses a resident wild-type gene. *Proc. Natl Acad. Sci. USA* **88**:1770–1774.

Grierson, D., Fray, R.G., Hamilton, A.J., Smith, C.J.S. and Watson, C.F. (1991) Does co-suppression of sense genes in transgenic plants involve antisense RNA? *Trends Biotechnol.* **9**: 122–123.

Guerche, P., Jouanin, L., Tepfer, D. and Pelletier, G. (1987) Genetic transformation of oilseed rape (*Brassica napus*) by the Ri T-DNA of *Agrobacterium rhizogenes* and analysis of inheritance of the transformation phenotype. *Mol. Gen. Genet.* **206**: 382–386.

Gugler, K., Funk, C. and Brodelius, P. (1988) Elicitor-induced tyrosine decarboxylase in berberine-synthesising cell suspension cultures of *Thalictrum rugosum*. *Eur. J. Biochem.* **170**: 661–669.

Hagemann, M-L. and Grisebach, H. (1984) Enzymatic rearrangement of flavonone to isoflavone. *FEBS Lett.* **175**: 199–202.

Hagemann, M-L., Heller, W. and Grisebach, H. (1984) Induction and characterisation of a microsomal flavonoid 3′-hydroxylase from parsley cell cultures. *Eur. J. Biochem.* **142**: 127–131.

Hahlbrock, K. and Scheel, D. (1989) Physiology and molecular biology of phenylpropanoid metabolism. *Ann. Rev. Plant Physiol. Plant Mol. Biol.* **40**: 347–369.

Hahlbrock, K., Chappell, J. and Kuhn, D.N. (1984) In *The Genetic Manipulation of Plants and its Application to Agriculture* (Lia, P.J. and Stewart, G.R., eds.), Oxford University Press, London and New York, pp. 171–82.

Hahn, M.G. and Albersheim, P. (1978) Host-Pathogens Interactions XIV. Isolation and partial characterisation of an elicitor from yeast extract. *Plant Physiology* **62**: 107–111.

Hain, R., Birseler, B., Kindl, W., Schroder, G. and Stocker, R. (1990) Expression of a stilbene synthase gene in *Nicotiana tabacum* results in the synthesis of the phytoalexin resveratrol. *Plant Mol. Biol.* **15**: 325–335.

Hamill, J.D., Parr, A.J. Robins, R.J. and Rhodes, M.J.C. (1986) Secondary product formation by cultures of *Beta vulgaris* and *Nicotiana rustica* transformed with *Agrobacterium rhizogenes*. *Plant Cell Rep.* **5**: 111–114.

Hamill, J.D., Robins, R.J. and Rhodes, M.J.C. (1989) Alkaloid production by transformed root cultures of *Cinchona ledgeriana*. *Planta Medica* **55**: 354–357.

Hamill, J.D., Robins, R.J., Parr, A.J., Evans, D.M., Furze, J.M. and Rhodes M.J.C. (1990a) Over-expressing a yeast ornithine decarboxylase gene in transgenic roots of *Nicotiana rustica* can lead to enhanced nicotine accumulation. *Plant Mol. Biol.* **15**: 27–38.

Hamill, J.D., Robins, R.J., Parr, A.J., Evans, D.M., Furze, J.M., Bent, E.G. and Rhodes, M.J.C. (1990b) The effects of over-expressing the yeast ornithine decarboxylase gene upon the nicotine and polyamine levels in transformed roots of *Nicotiana rustica*. In *Progress in Plant Cellular and Molecular Biology: Current Plant Science and Biotechnology in Agriculture* (Nijkamp, H.J.J., Van der Plas, L.H.W. and Van Aartrijk, J., eds.), Kluwer Academic Publishers, pp. 732–737.

Hanisch ten Cate, C.H., Sree Ramulu, K., Dijkhuis, P. and de Groot, B. (1987) Genetic stability of cultured hairy roots by *Agrobacterium rhizogenes* on tubers of potato cv Bintje. *Plant Sci.* **49**: 217–222.

Hanisch ten Cate, C.H., Ennik, E., Roest, S., Sree Ramulu, K., Dukhuis, P. and de Groot, B. (1988) Regeneration and characterisation of plants from potato root lines transformed by *Agrobacterium rhizogenes*. *Theor. Appl. Genet.* **75**: 452–459.

Harborne, J.B. (1988) *Introduction to Ecological Biochemistry*, Academic Press, New York.

Hartmann, T. and Toppel, G. (1987) Senecionine *N*-oxide, the primary product of pyrrolizidine alkaloid biosynthesis in root cultures of *Senecio vulgaris*. *Phytochemistry* **26**: 1639–1643.

Hartmann, T., Sander, H., Adolph, R. and Toppel, G. (1988) Metabolic links between the biosynthesis of pyrrolizidine alkaloids and polyamines in root cultures of *Senecio vulgaris*. *Planta* **175**: 82–90.

Hartmann, T., Ehmke, A., Eilert, U., Von Borstel, K. and Theuring, C. (1989) Sites of synthesis, translocation and accumulation of pyrrolizidine alkaloid *N*-oxides in *Senecio vulgaris*. *Planta* **177**: 98–107.

Hashimoto, T., Hayashi, A., Amano, Y., Kohno, J., Iwanari, H., Usuda, S. and Yamada, Y. (1991) Hyoscyamine 6β-hydroxylase, an enzyme involved in tropane alkaloid biosynthesis, is localised at the pericycle of the root. *J. Biol. Chem.* **266**: 4648–4653.

Hay, C.A., Anderson, L.A., Roberts, M.F. and Phillipson, J.D. (1986) *In vitro* cultures of Cinchona species: Precursor feeding of *C. ledgeriana* root organ suspension cultures with tryptophan. *Plant Cell Rep.* **5**: 1–4.

Hepburn, A.G., Belanger, F.C. and Mattheis, J.R (1987) DNA methylation in plants. *Dev. Genet.* **8**: 475–493.

Hermann, F.B. (1991) A unique opportunity for plant cell culture. *Agricell Report* **16**: 34.

Herminghaus, S., Schreier, P.H., McCarthy, J.E.G., Landsmann, J., Botterman, J. and Berlin, J. (1991) Expression of a bacterial lysine decarboxylase gene and transport of the protein into chloroplasts of transgenic tobacco. *Plant Mol. Biol.* **17**: 475–486.

Himeno, H. and Sano, K. (1987) Synthesis of crocin, picrocrocin and safranol by saffron stigma-like structures poliferated *in vitro*. *Agric. Biol. Chem.* **57**: 2393–2400.

Hunter, C.S., McCalley, D.V. and Barraclough, A.J. (1982) Alkaloids produced by cultures of *Cinchona ledgeriana*. In *Plant Tissue Culture* (Fujiwara, A., ed.), Japanese Association for Plant Tissue Culture, Tokyo, pp. 317–318.

Jorgensen, R. (1991) Beyond antisense—how do transgenes interact with homologous plant genes? *Trends Biotechnol.* **9**: 266–267.

Jung, G. and Tepfer, D. (1987) Use of genetic transformation by the Ri T-DNA of *Agrobacterium rhizogenes* to stimulate biomass and tropane alkaloid production in *Atropa belladonna* and *Calystegia sepium* roots grown *in vitro*. *Plant Science* **50**: 145–151.

Kacser, H. (1987) Control of metabolism. In *The Biochemistry of Plants* **11**: 39–67.

Kamada, H., Okamura, N., Satake, M., Harada, M. and Shimomura, K. (1986) Alkaloid production by hairy root cultures in *Atropa belladonna*. *Plant Cell Rep.* **5**: 239–242.

Kameji, T. and Pegg, A.E. (1987) Inhibition of translation of mRNAs for ornithine decarboxylase and *S*-adenosylmethionine decarboxylase by polyamines. *J. Biol. Chem.* **262**: 2427–2430.

Kay, R., Chan, A., Daly, M. and McPherson, J. (1987) Duplication of CaMV35S promoter sequences creates a stronger enhancer for plant genes. *Science* **236**: 1299–1302.

Koblitz, H., Koblitz, D., Schmauder, H.P. and Groger, D. (1983) Studies on tissue cultures of the genus *Cinchona* L. Alkaloid production in cell suspension cultures. *Plant Cell Rep.* **2**: 122–125.

Koromilas, A.E. and Kriakidis, D.A. (1988) The existence of ornithine decarboxylase-antizyme complex in germinating barley seeds. *Physiol. Plantanum* **72**: 718–724.

Leube, J. and Grisbach, H. (1983) Further studies on induction of enzymes of phytoalexin synthesis in soybean and cultured soybean cells. *Z. Naturforsch.* **38C**: 730–735.

Linn, F., Heidmann, I., Saedler, H., Meyer, P. (1990) Epigenetic changes in the expression of the maize A1 gene in *Petunia hybrida*: Role of numbers of integrated gene copies and state of methylation. *Mol. Gen. Genet.* **222**. 329–336.

Lockwood, G.B. and Essa, A.K. (1984) The effect of varying hormonal and precursor supplementations on levels of nicotine and related alkaloids in cell cultures of *Nicotiana tabacum*. *Plant Cell Rep.* **3**: 109–111.

Luckner, M. (1990) *Secondary Metabolism in Micro-organisms, Plants and Animals*, 3rd edn, Springer-Verlag, Berlin.

Mano, Y., Ohkawa, H. and Yamada, Y. (1989) Production of tropane alkaloids by hairy root cultures of *Duboisia leichhardtii* transformed by *Agrobacterium rhizogenes*. *Plant Science* **59**: 191–201.

Mano, Y., Nabeshima, S., Matsui, C. and Ohkawa, H. (1986) Production of tropane alkaloids by hairy root cultures of *Scopolia japonica*. *Agric. Biol. Chem.* **50**: 2715–2722.

Marchant, Y.Y. (1988) *Agrobacterium rhizogenes*–transformed root cultures for the study of polyacetylene metabolism and biosynthesis. In: *Chemistry and biology of naturally occurring acetylenes and related compounds (NOARC)*: *Bioactive molecules* **7**: 217–231.

Martin, C. and Gerats, A.G.M. (1992) The control of flower coloration. In *Molecular Biology of Flowering* (Jordan, B., ed.), CAB International (in press).

Martin, C., Prescott, A., Mackay, S., Bartlett, J. and Vrijlandt, E. (1991) Control of anthocyanin biosynthesis in flowers of *Antirrhinum majus*. *Plant Journal* **1**: 37–49.

Matsuda, J., Okabe, S., Hashimoto, J. and Yamada, Y. (1991) Molecular cloning of hyoscyamine 6 β-hydroxylase, a 2-oxoglutarate-dependent dioxygenase from cultured roots of *Hyoscyamus niger*. *J. Biol. Chem*. **266**: 9460–9464.

McConlogue, L., Gupta, M., Wu, L. and Coffino, P. (1984) Molecular cloning and expression of the mouse ornithine decarboxylase gene. *Proc. Natl Acad. Sci. USA* **81**: 540–544.

Medford, J.I., Horgan, R., El-Sawi, Z. and Klee, H.J. (1989) Alterations of endogenous cytokinins in transgenic plants using a chimeric isopentenyl transferase gene. *Plant Cell* **1**: 403–413.

Meyer, P., Heidmann, I., Forkmann, G. and Saedler, H. (1987) A new *Petunia* flower colour generated by transformation of a mutant with a maize gene. *Nature* **330**: 677–678.

Minocha, S.C. and De Scenzo, R.A. (1991) Expression of a murine ornithine decarboxylase (ODC) gene in transgenic *Nicotiana tabacum*. *Abstracts 3rd Int. Cong. Plant Mol. Biol.* (Hallick, R.B., ed.), Tucson, October 1991, p. 812.

Mol, J., van Blockland, R. and Kooter, J. (1991) More on co-suppression. *Trends Biotechnol*. **9** 182–183.

Moore, R.C. and Boyle, S.M. (1990) Nucleotide sequence and analysis of the *Spe*A gene encoding biosynthetic arginine decarboxylase in *Escherichia coli*. *J. Bacteriol*. **172**: 4631–4640.

Nabeshima, S., Maro, Y. and Ohkawa, H. (1986) Production of tropane alkaloids by hairy root cultures of *Scopolia japonica*. *Symbiosis* **2**:11–18.

Nahrstedt, A. (1989) The significance of secondary metabolites for interactions between plants and insects. *Planta Medica* **55**: 333–338.

Napoli, C., Lemieux, C. and Jorgensen, R. (1990) Introduction of a chimeric chalcone synthase gene into *Petunia* results in reversible co-suppression of homologous genes *in trans*. *Plant Cell* **2**: 279–289.

Ooms, G., Bains, A., Burrell, M., Karp, A., Twell, D. and Wilcox, E. (1985a) Genetic manipulation of cultivars of oilseed rape (*Brassica negus*) using *Agrobacterium*. *Theor. Appl. Genet.* **7**: 325–329.

Ooms, G., Karp, A., Burrell, M.M., Twell, D. and Roberts, J. (1985b) Genetic modification of potato development using Ri T-DNA. *Theor. Appl. Genet.* **70**: 440–446.

Parr, A.J. (1989) The production of secondary metabolites by plant cell culture. *J. Biotechnol*. **10**: 1–26.

Parr, A.J. and Hamill, J.D. (1987) Relationship between *Agrobacterium rhizogenes* transformed hairy roots and intact uninfected *Nicotiana* plants. *Phytochemistry* **26**: 3241–3245.

Parr, A.J., Peerless, A.C.J., Hamill, J.D., Walton, N.J., Robins, R.J. and Rhodes, M.J.C. (1988) Alkaloid production by transformed root cultures of *Catharanthus roseus*. *Plant Cell Rep*. **7**: 309–312.

Parr, A.J., Payne, J., Eagles, J., Chapman, B.T., Robins, R.J. and Rhodes, M.J.C. (1990) Variation in tropane alkaloid accumulation within the solanaceae and strategies for its exploitation. *Phytochemistry* **29**: 2545–2550.

Payne, J., Hamill, J.D., Robins, R.J. and Rhodes, M.J.C. (1987) Production of hyoscyamine by hairy root cultures of *Datura stramonium*. *Planta Medica* **53**: 474–478.

Phillips, M.A., Coffino, P. and Wang, C.C. (1987) Cloning and sequencing of the ornithine decarboxylase gene from *Trypanosoma brucei*. Implications for enzyme turnover and selective difluoromethylornithine inhibition. *J. Biol. Chem*. **262**: 8721–8727.

Ramsay, G., and Kumar, A. (1990) Transformation of *Vicia faba* cotyledon and stem tissues by *Agrobacterium rhizogenes*: Infectivity and cytological studies. *J. Exptl. Bot*. **41**: 841–847.

Rhodes, M.J.C., Hilton, M., Parr, A.J., Hamill, J.D. and Robins, R.J. (1986) Nicotine production by hairy root cultures of *Nicotiana rustica*: fermentation and product recovery. *Biotech. Lett.* **8**:415–420.

Rhodes, M.J.C., Robins, R.J., Aird, E.L.H., Payne, J., Parr, A.J. and Walton, N.J. (1989) Regulation of secondary metabolism in transformed root cultures. In *Primary and Secondary Metabolism of Plant Cell Cultures* (Kurz, W.G.W., ed.), Springer-Verlag, Berlin, pp. 58–72.

Rhodes M.J.C., Robins, R.J., Hamill, J.D., Parr, H.J., Hilton, M.G. and Walton, N.J. (1991) Properties of transformed root cultures. In *Secondary Products from Plant Cell Culture* (Charlwood, B.V. and Rhodes, M.J.C., eds.), Oxford University Press, Oxford, pp. 201–226.

Rhodes, M. J. C., Spencer, A., Hamill, J. D. and Robins, R. J. (1992) Flavour improvement through plant cell culture. In *Bioformation of Flavours* (Chorlwood, B. V., Patteson, R. W., McLeod, G. and Williams, A. A., eds.), Royal Society of Chemistry, London, pp. 42–64.

Robins, R.J., Parr, A.J., Walton, N.J. and Rhodes, M.J.C. (1990) Factors regulating tropane-alkaloid production in a transformed root culture of a *Datura candida* × *D. aurea* hybrid. *Planta* **181**: 414–422.

Robins, R.J., Parr, A.J. and Walton, N.J. (1991a) Studies on the biosynthesis of tropane alkaloids in *Datura stramonium* L. transformed root cultures. On the relative contribution of L-arginine and L-ornithine to the formation of the tropane ring. *Planta* **183**: 196–201.

Robins, R.J., Parr, A.J., Bent, E.G. and Rhodes, M.J.C. (1991b) Studies on the biosynthesis of tropane-alkaloids in *Datura stramonium* L. transformed root cultures. I. The kinetics of alkaloid production and the influence of feeding intermediate metabolites. *Planta* **183**: 185–195.

Robins, R.J., Bent, E.G. and Rhodes, M.J.C. (1991c) Studies on the biosynthesis of tropane alkaloids by *Datura stramonium* transformed root cultures. 3. The relationship between morphological integrity and alkaloid biosynthesis. *Planta* **185**: 385–390.

Robins, R. J., Bachmann, P., Robinson, T., Rhodes, M. J. C. and Yamada, Y. (1991d) The formation of 3α- and 3β-acetoxytropanes by *Datura stramonium* transformed root cultures involves two acetyl-CoA-dependent acyltransferases. *FEBS Letts.* **292** : 293–297.

Robinson, C. (1991) Targeting of proteins to chloroplasts and mitochondria In *Plant Biotechnology, Vol. I* (Grierson, D., ed.), Blackie, Glasgow, pp. 179–198.

Rogers, S., Wells, R. and Rechsteiner, M. (1986) Amino acid sequences common to rapidly degraded proteins: The PEST hypothesis. *Science* **234**: 364–368.

Ryan, C.A. and Farmer, E.E. (1991) Oligosaccharide signals in plants: A current assessment. *Ann. Rev. Plant Physiol.* **42**: 651–674.

Saito, K., Murakoshi, I., Inze, D. and Van Montagu, M. (1989) Biotransformation of nicotine alkaloids by tobacco shooty teratomas induced by a Ti plasmid mutant. *Plant Cell Rep.* **7**: 607–610.

Saito, K., Yamazaki, M., Shimorua, K., Yoshimatsu, K. and Murakoshi, T. (1990) Genetic transformation of foxglove (*Digitalis purpurea*) by chimeric foreign genes and production of cardioactive glycosides. *Plant Cell Rep.* **9**: 121–124.

Sano, K. and Himeno, H. (1987) *In vitro* proliferation of saffron (*Crocus sativus* L.) stigma. *Plant Cell, Tissue and Organ Culture* **11**: 159–166.

Sasse, F., Buchholz, M. and Berlin, J. (1983) Selection of cell lines of *Catharanthus roseus* with increased tryptophan decarboxylase activity. *Z. Naturforsch.* **38c**: 916–922.

Sauerwein, M., Yamazaki, T. and Shimomura, K. (1991) Hernandulcin in hairy root cultures of *Lippia dulcis*. *Plant Cell Rep.* **9**: 579–581.

Schallenberg, J. and Berlin, J. (1979) 5-methyltryptophan resistant cells of *Catharanthus roseus*. *Z. Naturforsch* **34c**: 541–545.

Schumacher, H.M., Gundlach, H., Fiedler, F. and Zenk, M.H. (1987) Elicitation of benzophenanthridine alkaloid synthesis in *Eschscholtzia* cell cultures. *Plant Cell Rep.* **6**: 410–413.

Shah, D.M., Horsch, R.B., Klee, H.J., Kishore, G.M., Winter, J.A., Turner, N.E., Hironaka, C.M., Sanders, P.R., Gasser, C.S., Aykert, S., Siegel, N.R., Rogers, S.G. and Fraley, R.T. (1986) Engineering herbicide tolerance in transgenic plants. *Science* **233**: 478–481.

Sharp, J.M. and Doran, P.M. (1990) Characteristics of growth and tropane alkaloid synthesis in *Atropa belladonna* roots transformed by *Agrobacterium rhizogenes*. *J. Biotechnol.* **16**: 171–186.

Shimomura, K., Sudo, H., Saga, H. and Kamada, H. (1991) Shikonin production and secretion by hairy root cultures of *Lithospermum erythrorhizon*. *Plant Cell Rep.* **10**: 282–285

Signs, M.W. and Flores, H.E. (1989) Elicitation of sesquiterpene phytoalexin biosynthesis in transformed root cultures of *Hyoscyamus muticus*. *Plant Physiol.* **89** sup.: 135.

Smith, C.J.S., Watson, C.F., Ray, J., Bird, C.R., Morris, P.C., Schuch, W. and Grierson, D. (1988) Antisense RNA inhibition of polygalacturonase gene expression in transgenic tomatoes. *Nature* **334**: 724–726.

Smith, C.J.S., Watson, C.F., Bird, C.R., Ray, J., Schuch, W. and Grierson, D. (1990a) Expression of a truncated tomato polygalacturonase gene inhibits expression of the endogenous gene in transgenic plants. *Mol. Gen. Genet.* **224**: 477–481.

Smith, C.J.S., Watson, C.F., Morris, P.C., Bird, C.R., Seymour, G.B., Gray, J.E., Arnold, C., Tucker, G.A., Schuch, W., Harding, S. and Grierson, D. (1990b) Inheritance and effect on ripening of antisense polygalacturonase genes on transgenic tomatoes. *Plant Mol. Biol.* **14**: 369–379.

Songstad, D.D., De Luca, V., Brisson, N., Kurz, W.G.W. and Nessler, C.L. (1990) High levels of tryptamine accumulation in transgenic tobacco expressing tryptophan decarboxylase. *Plant Physiol.* **94**: 1410–1413.

Spanier, K., Schell, J. and Schreier, P.H. (1989) A functional analysis of T-DNA gene 6b: The fine tuning of cytokinin effects on shoot development. *Mol. Gen. Genet.* **219**: 209–216.

Spencer, A., Hamill, J.D. and Rhodes, M.J.C. (1990a) Production of terpenes by differentiated shoot cultures of *Mentha citrata* transformed with *Agrobacterium tumefaciens* T37. *Plant Cell Rep.* **8**: 601–604.

Spencer, A., Hamill, J.D., Reynolds, J. and Rhodes, M.J.C. (1990b) Production of terpenes by transformed differentiated shoot cultures of *Mentha piperita citrata* and *M. piperita vulgaris*. In *Progress in Plant Cellular and Molecular Biology: Current Plant Science and Biotechnology in Agriculture* (Nijkamp, H.J.J., Van der Plas, L.H.W. and Van Aartrijk, J., eds.), Kluwer Academic Publishers, pp. 619–624.

Spencer, J.A. (1991) The development of shooty teratomas in *Mentha* species by genetic manipulation and studies on their growth and terpene production *in vitro*. PhD thesis, University of East Anglia.

Spencer, A., Hamill, J.D. and Rhodes, M.J.C. (1992) Transformation in *Mentha*. In *Biotechnology in Agriculture and Forestry. Protoplast and Genetic Engineering II* (Bajaj, Y.P.S., ed.), Springer-Verlag, Berlin (in press).

Tepperman, J.M. and Dunsmuir, P. (1990) Transformed plants with elevated levels of chloroplastic SOD are not more resistant to superoxide toxicity. *Plant Mol. Biol.* **14**: 501–511.

Threlfall, D.R. and Whitehead, I.M. (1988) Co-ordinated inhibition of squalene synthetase and induction of enzymes of sesquiterpenoid phytoalexin biosynthesis in cultures of *Nicotiana tabacum*. *Phytochemistry* **27**: 2567–2580.

Tiburcio, A.F. and Galston, A.W. (1986) Arginine decarboxylase as the source of putrescine for tobacco alkaloids. *Phytochemistry* **25** 107–110.

Tinland, B., Huss, B., Paulus, F., Bonnard, G. and Otten, L. (1989) *Agrobacterium tumefaciens* 6b genes are strain specific and effect the activity of auxin as well as cytokinin genes. *Mol. Gen. Genet.* **219**: 217–224.

Toivonen, L., Balsevich, J. and Kurz, W.G.W. (1989) Indole alkaloid production by hairy root cultures of *Catharanthus roseus*. *Plant Cell, Tissue and Organ Culture* **18**: 79–93.

Toppel, G., Witte, L., Riebesehl, B., Borstel, K.V. and Hartmann, T. (1987) Alkaloid patterns and biosynthetic capacity of root cultures from some pyrrolozidine alkaloid producing Senecio species. *Plant Cell Rep.* **6**: 466–469.

Tran Thanh Van, M., Thi Dien, N. and Chylah, A. (1974) Regulation of organogenesis in small explants of superficial tissue of *Nicotiana tabacum* L. *Planta* **119**: 149–159.

Trypsteen, M., Van Lijsebettens, M., Van Severen, R. and Van Montagu, M. (1991) *Agrobacterium rhizogenes*-mediated transformation of *Echinacea purpurea*. *Plant Cell Rep.* **10**: 85–89.

Tyler, R.T., Eilert, U., Rijndere, C.O.M., Roewer, I.A., McNabb, C.K. and Kurz, W.G.W. (1989) Studies on benzophenanthridine alkaloid production in elicited cell cultures of *Papaver somniferum*. In *Primary and Secondary Metabolism of Plant Cell Cultures* (Kurz, W.G.W., ed.), Springer-Verlag, Berlin, pp. 201–207.

Ulbrich, B., Wiesner, W., Arens, H. (1985) Large-scale production of rosmarinic acid from plant cell cultures of *Coleus blumei Benth*. In *Primary and Secondary Metabolism of Plant Cell Cultures* (Neumann, K-H., Barz, W. and Reinhard, E., eds.), Springer-Verlag, Berlin, pp. 293–303.

Van der Heijden, R., Verpoorte, R. and Harkes, P.A.A. (1988a) Effects of elicitors on the production of secondary metabolism in plant cell cultures of *Cinchona, Rubia, Morinda* and *Tabernaemontana*. In *Manipulating Secondary Metabolism in Culture* (Robins, R.J. and Rhodes, M.J.C., eds.), Cambridge University Press, Cambridge, pp.217–224.

Van der Heijden, R., Verheij, E.R., Schripsema, J., Baerheim Svendsen, A., Verpoorte, R. and Harkes, P.A.A. (1988b) Induction of triterpene biosynthesis by elicitors in suspension cultures of *Tabernaemontana* species. *Plant Cell Rep.* **7**: 51–54.

Van der Krol, A.R., Lenting, P.E., Veerstra, J., Meer, I.M., Van der Koes, R.E., Gerats, A.G.M., Mol, J.N.M. and Stuitje, A.R. (1988) An antisense chalcone synthase gene in transgenic plants inhibits flower pigmentation. *Nature* **333**: 866–869.

Van der Krol, A.R., Mur, L.A., de Lange, P., Mol, J.N.M. and Stuitje, A.R. (1990a) Inhibition of flower pigmentation by antisense CHS genes: promoter and minimal sequence requirements for the antisense effect. *Plant Mol. Biol.* **14**: 457–466.

Van der Krol, A.R., Mur, L.A., Beld, M., Mol, J.N.M. and Stuitje, A.R. (1990b) Flavonoid genes in *Petunia*: Addition of a limited number of gene copies may lead to a suppression of gene expression. *Plant Cell* **2**: 291–299.

Van Slogteren, G.M.S., Hooykaas, P.J.J. and Schilperoort, R.A. (1984). Silent T-DNA genes in plant lines transformed by *Agrobacterium tumefaciens* are activated by grafting and by 5-azacytidine treatment. *Plant Mol. Biol.* **3**: 333–336.

Van Tunen, A.J. and Mol, J.N.M. (1991) Control of flavonoid synthesis and manipulation of flower colour. In *Plant Biotechnology, Vol. II* (Grierson, D., ed.), Blackie, Glasgow, pp. 95–130.

Walton, N.J., Robins, R.J. and Rhodes, M.J.C. (1988) Perturbation of alkaloid production by cadaverine in hairy root cultures of *Nicotiana rustica*. *Plant Science* **54**: 125–131.

Walton, N.J., Robins, R.J. and Peerless, A.C.J. (1990) Enzymes of *N*-methylputrescine biosynthesis in relation to hyoscyamine formation in transformed root cultures of *Datura stramonium* and *Atropa belladonna*. *Planta* **182**: 136–141.

Westcott, R. (1988) Thiophene production from 'Tagetes' hairy roots. In *Manipulating Secondary Metabolism in Culture* (Robins, R.J. and Rhodes, M.J.C., eds.), Cambridge University Press, Cambridge, pp. 233–237.

Widholm, J.M. (1972a) Cultured *Nicotiana tabacum* cells with an altered anthranilate synthetase which is less sensitive to feedback inhibition. *Biochem. Biophys. Acta* **261**: 52–58.

Widholm, J.M. (1972b) Anthranilate synthetase from 5-methyltryptophan susceptible and resistant cultured *Daucus carota* cells. *Biochem. Biophys. Acta* **279**: 48–57.

Wink, M. (1987) Why do lupin cell cultures fail to produce alkaloids in large quantities? *Plant Cell, Tissue and Organ Cultures* **8**: 103–111.

Wink, M. (1988) Plant breeding: importance of plant secondary metabolites for protection against pathogens and herbivores. *Theor. Appl. Genet.* **75**: 225–233.

Yamada, Y., Hashimoto, T. and Sato, F. (1991) Enzymology and molecular biology in alkaloid biosynthesis. *Abstracts 3rd Int. Cong. Plant Mol. Biol.* (Hallick, R.B., ed.), Tucson, October 1991, p.5.

Yonemitsu, H., Shimomura, K., Satake, M., Mochida, S., Tanaka, M., Endo, T. and Kaji, A. (1990) Lobeline production by hairy root culture of *Lobelia inflata* L. *Plant Cell Rep.* **9**: 307–310.

Yoshikawa, T. and Furuya, T. (1987) Saponin production by cultures of *Panax ginseng* transformed with *Agrobacterium rhizogenes*. *Plant Cell Rep.* **6**: 449–453.

Zenk, M.H. (1985) Enzymology of isoquinoline alkaloid biosynthesis. In *The Chemistry and Biology of Isoquinoline Alkaloids* (Phillipson, J.D., Roberts M.F. and Zenk, M.H., eds.), Springer-Verlag, Berlin, pp. 240–256.

Zenk, M.H. (1988) Biosynthesis of alkaloids using plant cell cultures. In *Recent Advances in Phytochemistry Plant Nitrogen Metabolism* (Poulton, J.E., Romeo, J.T. and Conn, E.E., eds.), **23**: 429–457.

Zenk, M.H., El-Shagi, H., Arens, H., Stockigt, J., Weiler, E.W. and Deus, B. (1977) Formation of the indole alkaloids serpentine and ajmalicine in cell suspension cultures of *Catharanthus roseus*. In *Plant Tissue Culture and its Biotechnological Application* (Barz, W., Reinhard, E. and Zenk, M.H., eds.), Springer-Verlag, Berlin, Heidelberg, New York, pp. 27–43.

Zenk, M.H., Bueffer, M., Amann, M., Deus-Neumann, B. and Nagakura, N. (1985) Benzylisoquinoline biosynthesis by cultivated plant cells and isolated enzymes. *J. Nat. Prod. (Lloydia)* **48**: 725–738.

6 Structure, function and applications of ricin and related cytotoxic proteins

M. R. HARTLEY and J. M. LORD

6.1 Introduction

Ricin is the best known example of a family of proteins known as ribosome-inactivating proteins or RIPs. These are widely distributed in Angiosperms and also occur in several species of fungi and bacteria. A commonly used criterion for the inclusion of proteins within this group is that they catalytically and irreversibly inactivate the 60S subunit of eukaryotic ribosomes, rendering it incapable of binding elongation factor 2 (Stirpe and Barbieri, 1986). Although this definition was proposed before the nature of the RIP-mediated modifications to ribosomes were known, it remains a useful working definition but should be broadened to take into account the recent finding that some RIPs are also active on prokaryotic ribosomes. Protein toxins which act by the catalytic inactivation of protein synthesis accessory factors such as diphtheria toxin and *Pseudomonas* exotoxin are not included within this group.

It is axiomatic that because of the extreme cytotoxicity of RIPs, the ribosomes of the organisms producing them are either resistant to their action or are prevented from coming into contact with the RIP by sequestration of the latter. Both of these protective strategies are known to occur.

A major application of RIPs, particularly ricin, is in the construction of immunotoxins in which specific monoclonal antibodies are conjugated with potent toxins in order to target the toxin to specific cells *in vivo* and destroy them. Targets include neoplastic cells, lymphocytes involved in autoimmune reactions and virally-infected cells. RIPs are also being used increasingly to probe ribosome structure and function, particularly the role of ribosomal RNA (rRNA) in protein synthesis.

This chapter reviews recent work on the cell and molecular biology of RIPs and discusses their proposed physiological roles. There are several older reviews on these topics (Olsnes and Pihl, 1982; Jiminez and Vasquez, 1985) and a recent issue of *Seminars in Cell Biology* is devoted to RIP structure and the applications of RIPs in immunotoxin constructs (Lord, 1991). Here a more 'ribocentric' view of RIPs is taken.

6.2 Nomenclature and classification of RIPs

TThe most widely adopted classification of RIPs is that proposed by Stirpe and Barbieri (1986) who designated those RIPs existing in nature as single chain proteins or glycoproteins as type I and those consisting of an A (catalytically active) chain covalently linked to a B (cell binding) chain with lectin properties as type II. With the discovery of the mode of RIP action on ribosomes, a distinction should be made between the plant and bacterial RIPs, which exhibit sequence homologies in their enzymatic chains and possess a rRNA: *N*-glycosidase activity, and the fungal RIPs, which are unrelated in sequence to the former and act as specific rRNA endonucleases. The nomenclature of RIPs can be confusing, even to the initiated, and is based on several different criteria. The most commonly used system is to name the RIP after the generic name of its plant of origin. Thus ricin, abrin and dianthin are the RIPs from *Ricinus communis, Abrus precatorius* and *Dianthus caryophyllus* respectively. Ambiguities can arise here for the following reasons:

(i) Many plants produce several RIP isoforms of varying relatedness, either present within the same organ, or in different organs. For example, leaves of the soapwort (*Saporania officianalis*) contain at least six chromatographically-distinguishable RIPs termed saporin 1–6 (Stirpe *et al*., 1983) and *R. communis* seeds contain two RIPs: a dimeric RIP (ricin) which exhibits high cytotoxicity but relatively low haemaggtination activity and a related tetrameric RIP (*R. communis* agglutinin or RCA) which exhibits strong agglutination activity but poor cytotoxicity (Olsnes *et al*., 1974)

(ii) Several genera are known in which individual species produce distinctive RIPs. For example, the tubers of *Trichosanthes* spp. contain high concentrations of type I RIPs (trichosanthins) which are thought to be responsible for the ability of tuber extracts to induce abortion, treat choriocarcinomas and ectopic pregnancies and act as inhibitors of human immunodeficiency virus (HIV) (Chow *et al*., 1990). *T. kirilowii* tubers contain three RIPs of which two are known to differ in amino acid sequence and show different anti-HIV activities (see section 6.8). *T. cucumeroides* tubers produce another RIP, termed β-trichosanthin (Ng *et al*., 1991). RIPs have also been named to signify one of their biological activities, i.e. their ability to inhibit virus replication, which in many cases was discovered before the active proteins responsible were purified and characterised. Examples include pokeweed antiviral protein (PAP) from *Phytolacca americana* and *Mirabilis* antiviral protein (MAP). It is unfortunate that the acronym MAP has also recently been given to another type I RIP from *Momordica charantia* (Lee-Huang *et al*., 1990).

6.3 Distribution of RIPs

Type I RIPs are widely distributed in Angiosperms and screening surveys have shown them to occur in some sixty species belonging to fifteen different families (Barbieri and Stirpe, 1982; Stirpe and Barbieri, 1986; Merino *et al*., 1990). They occur in different plant organs (seeds, leaves, latices, roots and tubers) in concentrations ranging from a few micrograms to several hundred milligrams per 100 g

Table 6.1 Some examples of type 1 RIPs.

Plant source	Inhibitor	M_r	Glycosylated
Phytolacca americana (pokeweed) seeds or leaves	Pokeweed antiviral proteins (PAP)	30 000	No
Triticum aestivrum (wheat) seeds	Tritin	30 000	No
Gelonium multiflorum seeds	Gelonin	30 000	Yes
Momordica charantia (bittergourd) seeds	Momordin	31 000	Yes
Saponaria officinalis (soapwort) seeds	Saporin	29 500	No
Dianthus caryophyllus (carnation) leaves	Dianthin	30 000	Yes
Zea mays (maize) seeds	Maize RIP	16 500 18 500	?

of tissue. With only one known exception, they occur as monomeric proteins of around 30 kDa, are frequently, but not always, *N*-glycosylated and are often strongly basic (pIs ≥ 9.5). The exception is the active RIP in maize kernels which consists of two non-covalently associated polypeptides of 16.5 and 8.5 kDa (Walsh *et al.*, 1991). Some examples of type I RIPs are shown in Table 6.1. Type II RIPs of plant origin are less common than their type I counterparts and have been identified in five species from four different families (Table 6.2). In all cases the

Table 6.2 Plant type 2 RIPs.

Plant source	Inhibitor	M_r	Glycosylated
Ricinus communis (castor bean) seeds	Ricin	65 000	Yes
	A-chain	32 000	Yes
	B-chain	34 000	Yes
Abrus precatorius (jequirity bean) seeds	Abrin	65 000	Yes
	A-chain	30 000	No
	B-chain	36 000	Yes
Adenia digitata roots	Modeccin	63 000	Yes
	A-chain	28 000	
	B-chain	31 000	
Adenia volkensii roots	Volkensin	62 000	Yes
	A-chain	29 000	
	B-chain	36 000	
Viscum album (mistletoe leaves)	Viscumin	60 000	Yes
	A-chain	29 000	Yes
	B-chain	32 000	Yes

A-chain is joined through a single disulphide bond to a galactose-binding B-chain, also of around 30 kDa. Some enteric bacteria which cause diarrhoeal and/or dysenteric disease produce high levels of type II RIPs referred to as Shiga toxin or Shiga-like toxins (SLTs). Shiga toxin is produced by *Shigella dysenteriae* serotype I, responsible for epidemic outbreaks of bacillary dysentry. The chromosomally encoded toxin is composed of a single catalytic A-subunit of 32 kDa in non-covalent association with a pentamer of 7.7 kDa B-subunits which interact with a membrane glycolipid, globotriaosylceramide or GB_3(reviewed by O'Brien and Holmes, 1987). Shiga-like toxins are produced by certain strains of *E. coli* associated with haemorrhagic colitis and haemolytic uraemic syndrome in humans. These phage-encoded toxins are similar in structure and receptor specificity to Shiga toxin. Plant type I RIPs and plant and bacterial type II RIPs all act by the inhibition of protein synthesis through their rRNA : *N*-glycosidase activity. Although type I and type II RIPs show a similar effectiveness in the inhibition of protein synthesis in cell-free systems, type I RIPs are considerably less toxic to cells and animals (by a factor of 10^2–10^4) than type 11 RIPs since the former do not bind to and enter cells (Olsnes and Pihl, 1982).

Three RIPs of fungal origin are known, all from the genus *Aspergillus*. The best-characterised of these is α-sarcin (so named because of its anti-sarcoma activity) secreted by *A. giganteus* and first identified in 1956 in a screening programme by the Michigan Department of Health (reviewed by Wool, 1984). The toxin is composed of a single, basic, non-glycosylated polypeptide of 17 kDa, the amino acid sequence of which is highly homologous (86%) to the *A. restrictus* toxins mitogillin and restrictocin (Lopez-Otin *et al.*, 1984). All of these fungal toxins inhibit protein synthesis by making a single, endonucleolytic cleavage in 23S/26S/28S rRNAs. Although they are relatively non-toxic to animals and intact cells, the recent finding that an 18 kDa antigen found in the urine of patients suffering from aspergilliosis suggests they could have a role in pathogenesis (Lamy *et al.*, 1991).

Little is known about the evolutionary relationships between the various plant and bacterial RIPs. It seems probable that the ancestral RIP resembled an extant type I toxin. Whether this arose in prokaryotes or higher plants is uncertain since it has been proposed that horizontal gene transfer has occurred between pro- and eukaryotes and *vice versa* (Doolittle *et al.*, 1990). The heterodimeric structure of type II RIPs has been suggested to be the result of a fusion between a gene encoding a single-chain RIP (analogous to a type I RIP) and a lectin gene. It is possible that such an event has occurred more than once since the species in which type II RIPs are found are not closely related taxonomically. This notion is also supported by the finding that *Gelonium multiflorum*, which is in the same family as *R. communis* (Euphorbiaceae), produces a type I RIP (Stirpe *et al.*, 1980). Ready *et al.* (1988) have shown that some plant RIPs are homologous at the polypeptide domain level to the *pol* gene product of retroviral reverse transcriptases and to RNAse H from *E. coli*. These proteins may therefore have arisen from an ancient folding unit capable of catalysing RNA hydrolysis.

6.4 Structure and function of RIPs

6.4.1 The structure of ricin and related RIPs

To date, ricin is the only RIP for which a detailed X-ray crystallographic structure is known. The recent report of extensive sequence homologies between the A- and B-chains respectively of ricin and abrin (Wood *et al.*, 1991) and the known homologies between the A-chains of these toxins and type I RIPs (see below) together with their common enzymatic activity suggest basic similarities in tertiary and quaternary structure for this class of proteins. For detailed information on the structure of ricin, the reader should consult the several excellent articles by Robertus and co-workers (Rutenber *et al.*, 1991; Katzin *et al.*, 1991; Rutenber and Robertus, 1991; Robertus, 1991) and only an overview will be given here. A ribbon representation of the ricin backbone is shown in Figure 6.1. The A-chain, comprising 267 amino acids, is arbitrarily divided into three folding domains. The amino-terminal domain (residues 1–111) is dominated by a five-stranded β-sheet, whereas the second domain (residues 118–210) is dominated by five α-helices and contains the only free sulphydryl group (Cys 171) in the holotoxin. Domain 3 (residues 211–267) is compact and disk-like and interacts both with domains 1 and 2 and the

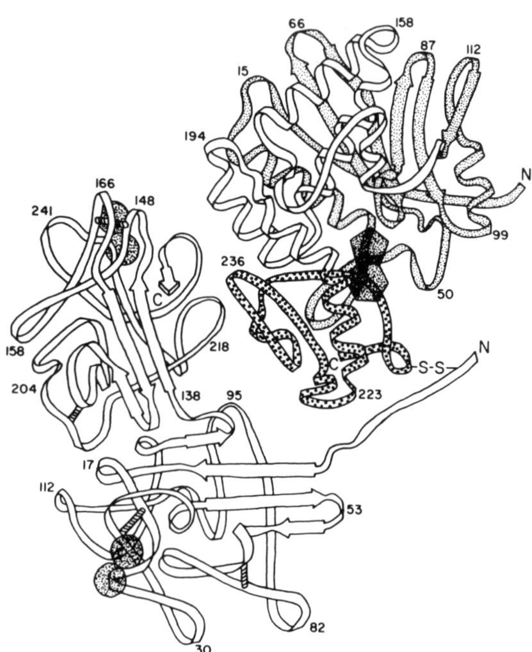

Figure 6.1 Ribbon representation of the ricin backbone. The A-chain is on the upper right, and the B-chain is on the lower left. The two lactose molecules bound to the B-chain are each represented as pairs of discs. From Montfort *et al.* (1987).

two lobes of the B-chain. The A-chain contains two *N*-glycosylation sites (Asn-x-Ser/Thr) of which one, or both, may be occupied (Foxwell *et al.*, 1985). Glycosylation is not required for activity of the A-chain since recombinant ricin A-chain purified from *E. coli* is as toxic to ribosomes as native A-chain (O'Hare *et al.*, 1987). The active site of the A-chain is believed to lie in a cleft created at the interface of all three domains (Montfort *et al.*, 1987).

Ricin B-chain (262 amino acids) is a galactose-specific lectin which forms two distinct globular domains with identical folding topologies (Montfort *et al.*, 1987). Each domain contains two internal disulphide bonds, one glycosylation site, which is usually occupied, and one sugar-binding site which lies in a pocket formed, in part, by a kink in the polypeptide chain by the tripeptide Asp-Val-Arg (Rutenber *et al.*, 1987). These authors have suggested that ricin B-chain is the product of a gene duplication and that each of the galactose-binding domains is in turn composed of three copies of the products of an ancestral gene coding for a galactose-binding peptide of about forty residues. Only two of the six potential galactose-binding peptides retain sugar-binding activities in the present-day ricin B-chain molecule. Recent work in which the galactose-binding tripeptides were mutated in one or other or both of the sites has shown that only when both sites were changed simultaneously was the resulting B-chain devoid of lectin activity (Wales *et al.*, 1991). It was also shown that when the Asn residues of the two *N*-glycosylation sites were changed to Gln the non-glycosylated mutant B-chain rapidly aggregated when expressed in *Xenopus* oocytes, suggesting that the presence of the *N*-linked oligosaccharides is required to stabilise the molecule.

6.4.2 Structure–function relationships

Katzin *et al.* (1991) have proposed a putative model for the active site region of ricin A-chain. It contains the residues Tyr80, Tyr123, Glu177, Arg180 and Trp211 which are invarient in all of the *N*-glycosidase RIPs sequenced to date (Figure 6.2). In addition to these invarient residues, Asn78, Arg134, Gln173, Glu208 and Asn209 are highly conserved active site residues. Site-directed mutagenesis studies from several laboratories have shown that the substitution of Arg180 and Glu177 for one of several other residues reduces the ribosome-inactivating activities of the mutant polypeptides by a factor of ≥ 100 (Frankel *et al.*, 1990; Schlossman *et al.*, 1989; Ready *et al.*, 1991).

Ready *et al.* (1991) have proposed a mechanism for catalysis in which Glu177 stabilises an oxocarbonium ion on the ribose attached to the target adenine in the rRNA substrate (see section 6.7.1) and Arg180 is involved in binding to the sugar-phosphate backbone through electrostatic interaction. Hydrolysis of the *N*-glycosidic bond is envisaged as being facilitated by the partial protonation of the target adenine through its hydrogen bonding with an unspecified residue and the positive charge development on the ribose (Figure 6.3).

```
                 10             20          30          40          50          60
RTA       IFPKQYPIINFTTA    GATVQSYTNFIRAVRGRLTTGADVRHEIPVLPNRVGLPINQRFILV
ATA          EDRPI KFSTE    GATSQSYKQFIEALRERLRGGL I HDIPVLPDPTTLQERNRYITV
TCS               DVSFRLS   GATSSSYGVFISNLRKALPNERKL YDIPLL  RSSLPGSQRYALI
MAP         APTLETIASLD LNNPT  TYLSFITNIRTKVADKTEQ CTIQKI  SKTF  TQRYSYI
SO6               VTSITLD LVNPTAGQYSSFVDKIRNNVKDPNLK YGGTDI  AVIGPPSKEKFLR
BPSI   AAKMAKNVDKPLFTATFNVQASSADYATFIAGIRNKLRNPAHFSHNEPVLPPVEPNVPPSRWFHV
SLTA1             KEFTLDFS    TAKTYVDSLNVIRSAIGTPLQTISSGGTSLLMIDSGSGDNLFAV
SLTA2             KEFTIDFS    TQQSYVSSLNSIRTEISTPLEHISQGTTSVSVINHTHGS YFAV

                  70            80             90         100        110
RTA       ELSNHAELS VTLALD    VTNAYVVGYRAGNS        AYFFHPDNQEDAEAITHLF TDVQ
ATA       ELSNSDTES IEVGID    VTNAYVVAYRAGTQ        SYFLRDAPSS   ASDYLF TGTD
TCS       HLTNYADET ISVAID    VTNVYIMGYRAGDT        SYFFNEASATE  AAKYVF KDAM
MAP       DLIVSSTQK ITLAID    MADLYVLGYSDIANNKGR AFFFKDVTEAV  ANNF FPGATG
SO6       INFQSSRGT VSLGLK    RDNLYVVAYLAMDNTNVNRAYYF  RSEITSAESTALFPEATT
BPSI      VLKASPTSAGLTLAIR    ADNIYLEGFKSSDG      TWWELT                 PGLI
SLTA1     DVRGIDPEEGRFNNLRLIVERNNLYVTGFVNRTNNVFYRFADF               SHVTFPGTT
SLTA2     DIRGLDVYQARFDHLRLIIEQNNLYVAGFVNTATNTFYRFSDF               THISVPGVT

                120         130          140          150          160          170
RTA        NRYTFAFGGNYDRLEQLAGNLRENIELGNGPLEEAISALYYY   STGGTQLPT LARS FII
ATA        QH SLPFYGTYGDLERWAHQSRQQIPLGLQALTHGIS     FF   SGGNDNEE KART LIV
TCS        RKVTLPYSGNYERLQTAAGKIRENIPLGLPALDSAITTLFYY         NANSA   ASA LMV
MAP        TNRIKLTFTGSYGDLEKNGG LRKDNPLGIFRLENSIVNIYGK    AGDVKKQ   ALF FLL
SO6        ANQKALEYTEDYQSIEKNAQITQGDQSRKELGLGIDLLSTSMEAVNKKARVVKD EARF LLI
BPSI       PGATYVGFGGTYRDLLGDTDKL TNVALGRQQLEDAVTALHGRTKADKASGPKQQQAREAVTT
SLTA1      A  VTLSGDSSYTTLQRVAGISRTGMQINRHSL     TTSYLDLMSHSGTSLTQSVAR AMLR
SLTA2      T  VSMTTDSSYTTLQRVAALERSGMQISRHSL     VSSYLALMEFSGNIMTRDASR AVLR

               180           190           200          210         220
RTA       CIQMISEAARFQ     YIEGEMRTRIRY NRRSAPDPS VITLENSWGRLSTAIQE   SNQGAF
ATA       IIQMVAEAARFR     YISNRVRVSIQTG TAFQPDAA MISLENNWDNLRG VQE   SVQDTF
TCS       LIQSTSEAARYK     FIEQQIGKRVDKT    FLPSLA IISLENSWSALSKQIQIASTNNGQF
MAP       AIQMVSEAARFK     YISDKIPSEKYEE    VTVDEY MTALENNWAKLSTAVYNSKPS TTT
SO6       AIQMTAEAARFR     YIQNLVIKNFPNK    FNSENK VIQFEVNWKKISTAIYG DAKNGVF
BPSI      LLLMVNEATRFQTVSGFVAGLL HPKAVEKKSGKIGNE MKAQVNGWQDLSAALLKTDVKPPPG
SLTA1     FVTVTAEALRFR     QIQRGFRTTLDDLSGRSYVMTAEDVDLTLNWGRLSSVLPD   YHGQDS
SLTA2     FVTVTAEALRFR     QIQREFRQALSE TAPVYTMTPGDVDLTLNWGRLDAALGE   YRGEDG

            230          240          250          260
RTA       ASPIQLQRRNGSKFSVYDVS  ILIPIIALMVYRCAPPPSSQF
ATA       PNQVTLTNIRNEPVIVDSLSH PTVAVLALMLFVCNPPN
TCS       ETPVVLINAQNQRVMITNVDAGVVTSNIALLLNRNNMA
MAP       ATKCQLATSPVTISPWIFKTVEEIKLVMGLLKSS
SO6       NKDYDFGFGKVRQVKDLQMGLLMY    LGKPKSSNEAN
BPSI      KSPAKFTEKM      GVRTAEQAAATLGILLFVEVPGGLTVAKALELFHASGGKPI
SLTA1     VRVGRISF  GSINAILGSVALILNCHHHASRVARMA  SDEFRSMCPADGRVRGITHNKILW
SLTA2     VRVGRISF  NNISAILGTVAVILNCHHQGARSVRAV  NEESQPECQITGDRPVIKINNTLW

SLTA1     DSSTLGAILMRRTISS
SLTA2     ESNTAAAFLNRKSQFLYTTGK
```

Figure 6.2 Sequence comparison of ribosome inactivating proteins. The amino acid sequences of proteins from five dicotyledonous plants, one monocot and two strains of *Escherichia coli*, are aligned using the one-letter code. The numbers along the top refer to the ricin A-chain sequence. The shaded residues are invarient. RTA is ricin A-chain (Lamb *et al.*, 1985); ATA is abrin A-chain (Funatsu *et al.*, 1988), TCS is trichosanthin (Collins *et al.*, 1990); MAP is mirabilis antiviral protein (Habuka *et al.*, 1989); SO6 is saporin-6 (Benatti *et al.*, 1989); BPS1 is the barley seed inhibitor (Asano *et al.*, 1986); SLTA I is one form of Shiga-like toxin (Calderwood *et al.*, 1987) while SLTA II is an antigenically-distinct form (Jackson *et al.*, 1987). From Robertus (1991).

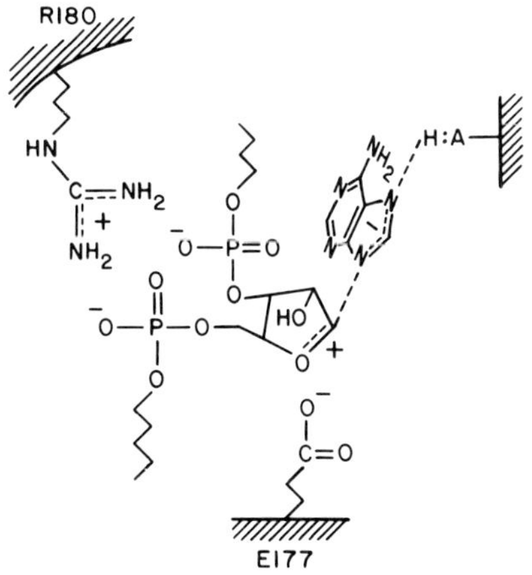

Figure 6.3 Schematic diagram of the proposed ricin A-chain mechanism. The substrate adenine is shown in the *syn* conformation. H:A represents an unspecified residue that hydrogen-bonds to the leaving adenine. This partial protonation facilitates N–C bond cleavage. For further details, see text. From Ready *et al.* (1991).

6.5 Synthesis of RIPs

To date, the RIPs for which the most is known about genomic sequence organisation, regulation of expression, post-translational modifications and targeting, are ricin and the maize seed RIP (a type I RIP also known as b-32).

Southern blot hybridisation studies using a ricin cDNA probe have shown that the *R. communis* genome contains about eight members of a ricin/*R. communis* agglutinin (RCA) multigene family (Tregear and Roberts, 1992). At least three members of this family are pseudogenes. One gene member analysed in detail represents a functional ricin gene similar in coding sequence to the published cDNA sequence (Lamb *et al.*, 1985). It contains typical eukaryotic DNA consensus sequences and two separate motifs resembling the CATGCATG RY repeat described for legume seed storage protein genes (Dickinson *et al.*, 1988) in its 5′-flanking region. The ricin/RCA genes lack introns, a feature of all RIP genes analysed to date. Transcripts of the ricin/RCA gene family accumulate maximally in the endosperm around the onset of testa formation but are undetectable in the dry seed, leaves, roots and the endosperm of germinating seeds (Tregear and Roberts, 1992).

Ricin and RCA are synthesised simultaneously in equivalent amounts in the endosperm during and after testa formation when protein storage bodies are being rapidly formed (Roberts and Lord, 1981). The primary translation product of both mRNAs is an inactive preproprotein containing both A- and B-chain sequences joined by a 12 amino acid linking peptide and a 35 amino acid presequence preceding the A-chain (Butterworth and Lord, 1983; Lamb *et al.*, 1985). An N-terminal signal sequence, believed to lie within the first 22 residues of the presequence, is cleaved during contranslational translocation of the preproprotein across the endoplasmic reticulum (ER) membrane. At the same time the four intrachain and one interchain disulphide bonds are formed, and appropriate Asn residues are core glycosylated. Proricin then travels to the Golgi and finally to the protein bodies by vesicles which fuse with these membrane-bound organelles. The location of the sequence information required to target proricin to the protein bodies is unknown, but does not reside in the N-terminal presequence remaining after signal peptide cleavage (Westby, 1991). This scheme of synthesis and processing ensures that the toxic A-chain does not come into contact with the endosperm cell ribosomes.

Ricin is proteolytically degraded during seed germination and in this respect, together with its location in protein bodies and accumulation during seed maturation, resembles a 'conventional' seed storage protein.

Knowledge about the biosynthesis of the maize RIP (also known as b-32) has been arrived at by a circuitous route and provides a good example of the convergence of formerly diverse interests. A monomeric 32 kDa protein (b-32), now known to be synonymous with the RIP, was first identified in the albumin fraction of developing kernels by Soave *et al.* (1981). The protein was absent in kernels homozygous for the recessive allele *opaque*-2 (O_2) in which accumulation of zeins is greatly reduced. It was subsequently shown that the O_2 gene encodes a 48 kDa polypeptide with properties resembling those of the leucine-zipper class of trans-acting factors (Lohmer *et al.*, 1991), which serves to regulate the transcription of the b-32 gene. This was shown by the co-transformation of tobacco leaf protoplasts with plasmids containing a constitutively-expressed O_2 cDNA and a β-glucuronidase reporter gene fused to a genomic promoter sequence extending from −1283 to +4 positions of the b-32 gene. The sequence responsible for the expression of GUS activity was shown to reside in a 440 bp region flanking the b-32 gene containing five O_2 protein binding sites (GATGAPyPuTGPu), two of which were embedded in the 'endosperm box' motif thought to be involved in endosperm-specific expression (Forde *et al.*, 1985). The identity of the b-32 protein only became apparent when its amino acid sequence, derived from a cDNA clone (Hartings *et al.*, 1990), was found to be 96% homologous with that of a maize kernel RIP cloned independently by Walsh *et al.* (1991). The maize kernel RIP is unique among type I RIPs in that it is synthesised and stored in the kernel as an inactive propeptide of 34 kDa and is converted to an active form during germination, as judged by its ability to inhibit protein synthesis in the reticulocyte lysate system, by the removal of 25 amino acids from the centre of the propeptide (Walsh *et al.*, 1991). The resulting fragments remain tightly associated by non-covalent interactions.

Comparisons of the N- and C-terminal amino acid sequences of several other mature type I RIPs with the corresponding gene sequences have revealed the likelihood of signal sequences involved in targeting precursors to the ER and possibly the vacuole. For example, the translation of a cDNA sequence for α-trichosanthin revealed that the mature protein is preceded by 23 amino acids, which resemble a consensus secretory protein signal peptide, and also contains a C-terminal extension of 19 residues (Chow *et al.*, 1990). Similar observations have also been made for *Mirabilis* antiviral protein (Kataoka *et al.*, 1991), saporins-2 and -6 (Fordham-Skelton *et al*, 1991; Benatti *et al.*, 1991) and dianthin-30 (Legname *et al.*, 1991). In the case of saporins-2 and -6 and dianthin-30 the putative C-terminal extensions have potential N-linked glycosylation sites (Asn-Ser-Thr). Similar C-terminal extensions are known to be required for the vacuolar targeting of several plant proteins including barley lectin, the vacuolar forms of β-glucanases and wheat germ agglutinin (Bednarek *et al.*, 1990; Raikhel and Williams, 1987). Thus there is the possibility that some type I RIPs are targeted to the vacuole. However, in cytological studies using immunogold labelling, PAP and type I RIPs from *Dianthus barbatus*, *Spinacea oleracea* and *Chenopodium amaranticolor* were all shown to be localised in the cell wall matrix of leaf mesophyll cells (Ready *et al.*, 1986; Frotschl *et al.*, 1990).

6.6 Entry of cytotoxic lectins into cells

6.6.1 Cell surface binding

Cytotoxic lectins bind opportunistically to the surface of target cells by reversible interaction between the B-chain and carbohydrate moieties containing terminal galactose residues occurring on both glycoproteins and glycolipids. Mammalian cells contain an abundance of such binding sites, ensuring a high concentration of bound toxin. HeLa cells, for example, possess 3×10^7 binding sites per cell for ricin (Sandvig *et al.*, 1976). Binding is reduced by the presence of free galactosides and increased by prior treatment of cells with neuraminidase to remove terminal sialic acid residues (Rosen and Hughes, 1977).

6.6.2 Internalisation

A proportion of the surface-bound toxin is internalised by the target cell. The entry of ricin into mammalian cells has been extensively studied, primarily by electron microscopy using ricin linked to an appropriate marker for visualisation. From these studies it is clear that ricin enters cells by endocytosis primarily, but not exclusively (Moya *et al.*, 1985), via clathrin coated pits and vesicles (van Deurs *et al.*, 1985). A large proportion of the ricin taken into cells is either recycled back to the cell surface or is delivered into lysosomes where it is presumably degraded. In addition a small proportion of ricin avoids recycling or degradation and it is from this pool that ricin A-chain crosses an intracellular membrane to reach its ribosomal substrates in the cytosol. The intracellular compartment from which ricin A-chain translocates is not known at present, but it is clear that endocytosis continues beyond the endosomes before translocation takes place (van Deurs *et al.*, 1986). Several electron microscopic studies have shown that endocytosed ricin is first delivered to the endosomes and a fraction subsequently appears within the Golgi complex, in particular the trans Golgi network (van Deurs *et al.*, 1986). Indeed it has been proposed that ricin A-chain is translocated into the cytosol from the Golgi complex (Yoshida *et al.*, 1991), although the possibility that retrograde transport carries ricin further into the endomembrane, perhaps back to the endoplasmic reticulum, cannot be dismissed (Lord *et al.*, 1991).

6.7 The action and specificities of RIPs on ribosomes

6.7.1 Action of RIPs on ribosomes

Early studies on the effects of the A-chains of ricin, abrin and modeccin on the inhibition of protein synthesis in cell-free systems showed that the 60S ribosomal subunit alone was affected and that one A-chain molecule could inactivate *c.* 1500 ribosomes per minute (reviewed by Olsnes and Pihl, 1982). The finding that ribosome inactivation occurred in simple buffer solutions suggested that there was no cofactor requirement. Despite considerable efforts over about fifteen years, the nature of the modification to the 60S subunit remained elusive (reviewed by Jiminez and Vasquez, 1985). Armed with the knowledge that α-sarcin acts as a specific rRNA endonuclease (see section 6.7.5) and the similarities between the partial reactions of protein synthesis inhibited by α-sarcin and ricin and related RIPs, Endo *et al.* (1987) investigated the possibility that ricin may also act on rRNA. The first indication that this may be the case was the observation that 28S rRNA isolated from rat liver ribosomes treated with ricin A-chain co-migrated marginally more slowly on non-denaturing gels than 28S rRNA from control ribosomes. Furthermore a 550 nucleotide fragment generated from the 3′ end of 28S rRNA by the action of endogenous nuclease also had a decreased mobility, suggesting that the site of action in 28S rRNA was located close to the 3′ end. Further experiments revealed the precise nature of the ricin-catalysed modification in 28S rRNA (Endo *et al.*, 1987; Endo and Tsurugi, 1987). Adenine at position 4324 (numbered from the 5′ end of 28S rRNA) had been removed, leaving the sugar-phosphate backbone intact. Thus ricin acts as a highly specific hydrolase, cleaving a single *N*-glycosidic bond among *c.* 7000 nucleotide residues in rRNA. It is highly significant that adenine (A) 4324 lies near the centre of a 14 nucleotide sequence that is the most strongly conserved structural feature of the large rRNA from the large ribosomal subunit (LSU rRNA) and which has remained largely unchanged throughout evolution (Raué *et al.*, 1988). This is shown in Table 6.3. Furthermore α-sarcin cleaves the LSU rRNA at the phosphodiester bond linking guanine (G) 4325 and adenine (A) 4326 (Wool, 1984) suggesting that the structural integrity of this sequence is of crucial importance to the functioning of the ribosome. Thus the organisms producing the *N*-glycosidase and endonuclease RIPs, which are unrelated in sequence, appear to have independently evolved activities which damage the ribosome at its Achilles heel.

Two methods have been reported for the assay of the *N*-glycosidase activities of RIPs on their ribosome substrates. One of these makes use of the fact that the

Table 6.3 Nucleotide sequences in rRNA surrounding the modification sites for ricin A-chain and α-sarcin. The target sites for ricin A-chain (R) and α-sarcin (S) in rat 28S rRNA are indicated by arrows. The number after each sequence gives the distance, in nucleotides, from the ricin A-chain target adenine to the 3′ end of the rRNA.

Ribosomal RNA	Sequence	Reference
	R ↓ S ↓	
Rattus norvegicus 28S	A G U A C G A G A G G A A C 393	Chan *et al.* (1983)
Citrus limon 26S	A G U A C G A G A G G A A C 360	Kolosha and Fodor (1990)
Saccharomyces cerevisiae 26S	A G U A C G A G A G G A A C 368	Veldman *et al* (1981)
Escherichia coli 23S	A G U A C G A G A G G A C C 244	Brosius *et al.* (1980)
Nicotiana tabacum chloroplast 23S	A G U A C G A G A G G A C C 133	Takaiwa and Sugiura (1982)

phosphodiester bonds on either side of the depurinated ribose are sensitive to amine-catalysed hydrolysis at acidic pH by a β-elimination reaction (Peattie, 1979). The resulting fragments (4324 and 394 nucleotides in the case of rat 28S rRNA according to the sequence data of Chan *et al.* (1983)) are then separated on a polyacrylamide or agarose gel. The extent of depurination can then be calculated following densitometric scanning of the ethidium bromide-stained gel using 5.8S rRNA as an internal standard (Osborn, 1990). In the second method, the released adenine is converted into a highly fluorescent 1-N^6-etheno derivative through a reaction with chloroacetaldehyde (Zamboni *et al.*, 1989). The fluorescent derivitive is measured either directly, or following separation by HPLC. The latter method has the advantage that the release of adenine can be measured in ribosome preparations in which the rRNA is non-specifically degraded (a problem commonly encountered with plant ribosomes) but suffers from the drawback that a variable amount of chloroacetaldehyde-reactive material is also released from control incubations of ribosomes.

Endo and co-workers subsequently went on to show that momordin, gelonin, saporin, PAP, PAP II, PAP S, *Phoradendron californicum* lectin, barley protein synthesis inhibitor, Shiga toxin and Shiga-like toxin II all act on rat liver ribosomes in an analogous manner to ricin A chain (Endo *et al.*, 1988, a,b, c, 1989). Stirpe *et al.* (1988) showed that the 28S rRNA of rabbit reticulocyte ribosomes and the 26S rRNA of yeast ribosomes also produced a diagnostic aniline-labile site with ricin A-chain and a variety of type I RIPs. Site-specific depurination has also been observed following the injection of ricin, shiga toxin and shiga-like toxin II into *Xenopus* oocytes (Saxena *et al.*, 1989).

In the majority of the studies on RIP action on ribosomes *in vitro* the reactions were performed in a simple Tris/Mg^{2+}/KCl solution hence giving credence to the observations based on protein synthesis inhibition which suggested that the majority of RIPs have no cofactor requirements. This conclusion is almost certainly an over-simplification. The first indication that some RIPs require cofactors for activity was reported by Coleman and Roberts (1981), who found that the inhibition of protein synthesis in cell-free extracts from Acites cells by tritin (a type I RIP from wheat seeds) had a marked requirement for ATP and a loosely bound ribosomal factor(s) which could be removed by washing the ribosomes in 0.5 M NH_4Cl. Work in which the RNA:*N*-glycosidase activity of tritin was assayed on reticulocyte and yeast ribosomes has confirmed this observation (A. Massiah and M. Hartley, unpublished). Sperti *et al.* (1991) found that the release of adenine from *Artemia salina* ribosomes by gelonin required ATP and a high M_r factor present in rabbit reticulocyte post-ribosomal supernatant. In the absence of these factors the kinetic constants for the reaction were $K_m = 4.35$ μM and $K_{cat} = 0.1$ min^{-1}, and in their presence $K_m = 1.05$ μM and $K_{cat} = 108$ min^{-1}. The authors suggest that the phosphorylation of a ribosomal protein may be required for gelonin activity. It is possible that the supernatant factor requirement is for a ribosomal protein kinase.

The cofactor requirements demonstrated for tritin and gelonin activities could well represent extreme examples of a more general, but less pronounced, dependence shown by most RIPs. For example, Ready *et al.* (1991) have shown that ricin A-chain, previously thought not to show cofactor dependence, was stimulated in its action on *Artemia salina* ribosomes by ATP and a post-ribosomal fraction. The K_m dropped three-fold and the K_{cat} increased three-fold in the presence of these factors.

From the foregoing discussion, it becomes apparent that it is not yet possible to rank RIPs in any meaningful hierarchical order of their relative *N*-glycosidase activities on ribosomes. Ribosomes from different sources are known to vary considerably in their susceptibility to a given RIP under identical assay conditions (see below) and the precise method of ribosome isolation, including the composition of buffers, may influence their susceptibility to RIP action. In this context, it is interesting to note that EDTA-treatment of ribosomes, or their dialysis against Mg^{2+} -free buffer, irreversibly damages them such that they are inactive in protein synthesis and as substrates for modification by gelonin, even when Mg^{2+} is restored in the assay (Zamboni *et al.*, 1989). It is likely that ribosomes undergo irreversible conformational changes which either prevent the binding of the RIP and/or inhibit catalysis. This is surprising in view of the findings that both naked (deproteinised) rRNA and synthetic oligoribonucleotides serve as substrates for RIP modification (see section 6.7.4).

6.7.2 The susceptibility of plant and E.coli *ribosomes to RIPs*

Since all known eukaryotic 80S and eubacterial 70S ribosomes possess the conserved RIP target site in their LSU rRNAs it might be expected that under the appropriate conditions they would all be susceptible to modification by RIPS. This is not the case. Cawley *et al.* (1977) found that wheat germ ribosomes required a 5000-fold higher concentration of ricin A-chain to inhibit their activity in poly(U)-directed Phe polymerisation than for a comparable inhibition of rat liver ribosomes, and *E. coli* ribosomes were completely refractory to the toxin. The binding of

[^{3}H]-labelled ricin A-chain to the ribosomes, as measured after gel filtration and sedimentation, was appreciable and comparable in the rat and wheat germ ribosomes, but was undetectable with *E. coli* ribosomes. These results indicate that the binding of the RIP alone is not sufficient for ribosome inactivation. Work in our laboratory has confirmed and extended these findings. Osborn (1990) showed that wheat germ and tobacco leaf 80S ribosomes are *c*. 1000-fold more resistant to the action of recombinant ricin A-chain than rabbit reticulocyte ribosomes as measured by the aniline assay, and Wood (1991) has made similar observations for recombinant abrin A-chain. The notion emerged from the earlier studies that plant cytoplasmic ribosomes are much less sensitive to RIPs in general than are mammalian ribosomes and that *E. coli* ribosomes are completely refractory. More specifically, experiments were interpreted to suggest that the ribosomes from individual plant species producing RIPs were uniquely resistant to their homologous RIP (Battelli *et al*., 1984; Stirpe and Hughes, 1989). This is almost certainly an over-simplification since it is now known for several species that the ribosomes isolated from RIP-producing tissue are depurinated during their extraction. For example, Taylor and Irvin (1990) found that the 25S rRNA from pokeweed leaf was specially depurinated, both in RNA preparations from isolated ribosomes and following the direct extraction of leaf RNA using guanidine hydrochloride. Work in our laboratory has confirmed and extended this observation. We have found that the 25S rRNA extracted directly from the leaves of *Phytolacca americana, P. dodecandra* and *Dianthus caryophyllus* by both the phenol-detergent method of Kirby (1968) and the guanidinium isothiocyanate method was specifically depurinated (B. Stott and M. Hartley, unpublished). In contrast unmodified rRNA could be extracted directly from the endosperm of *Ricinus communis* seeds, although the RNA from isolated ribosomes was modified (K. Wood and M. Hartley, unpublished). The results obtained with the rRNA from isolated ribosomes from RIP-producing tissue are easily interpreted by the RIP coming into contact with sensitive ribosomes when the tissue is homogenised. The observed modification of the rRNA extracted directly from leaves is more difficult to interpret. Either the protein denaturants used do not entirely inhibit RIP activity, or the RNA is modified *in situ*. Although the latter suggestion seems unlikely, it cannot be ruled out entirely since we have observed that viable yeast transformants expressing a mutant form of ricin A-chain cytoplasmically have a significant proportion of their ribosomes depurinated, yet the cells divide at a rate similar to that of non-transformed cells (Gould *et al*., 1991; see also section 6.7.3). In the case of the rRNA extracted directly from the leaves of *Phytolacca dodecandra* samples from fully expanded leaves on the point of senescence were more heavily depurinated than those from young, expanding leaves. It is therefore conceivable that ribosome inactivation *in situ* is part of the physiological mechanism to inhibit protein synthesis during senescence.

The earlier ideas about ribosome resistance (Battelli *et al*., 1984) probably arose because the activities of the ribosomes, as measured by cell-free protein synthesis, were very low even prior to adding the RIP, and presumably represented the residual activities of depurinated ribosomes. The only documented example of resistance to the endogenous RIP is that of wheat germ ribosomes to seed tritin (Taylor and Irvin, 1990) although wheat germ ribosomes are sensitive to several other type I RIPs, including an isoform of tritin present in wheat leaves (A. Massiah and M. Hartley, unpublished). This is consistent with the reports that the seed forms of RIPs in maize and barley are unusual in that they are not synthesised as precursors containing N-terminal signal sequences (Leah *et al.*, 1991; Walsh *et al.*, 1991) and presumably accumulate in the cytosol in contact with ribosomes.

As previously stated, it is a widely accepted view that prokaryotic ribosomes are insensitive to plant RIPs. In keeping with this *E. coli* has been used successfully as a host for the production of biologically active recombinant ricin A-chain (O'Hare *et al.*, 1987; Piatak *et al.*, 1988) and abrin A-chain (Wood *et al.*, 1991). These proteins were produced cytoplasmically where they accounted for up to 10% of the total bacterial protein without affecting bacterial growth. It was therefore surprising when attempts to express the type I RIP *Mirabilis* antiviral protein (MAP) cytoplasmically in *E. coli* resulted in severely inhibited growth of the host caused by the recombinant protein, the yield of which was very low (Habulka *et al.*, 1989). MAP was subsequently shown to inhibit protein synthesis by *E. coli* ribosomes (Habuka *et al.*, 1990). This inhibition was probably the result of the depurination of 23S rRNA, since it has recently been shown that A2660 in *E. coli* 23S rRNA (the analagous base to A4324 in rat ribosomes) is removed by four other type I RIPs (Hartley *et al.*, 1991).

From the foregoing discussion, the reader may be left with the impression that because of the widely differing sensitivites of ribosomes from different sources to RIPs it is difficult to make generalisations on this subject. However, the following generalisations are held to be true at the time of writing.

i) Ribosomes from mammals (and probably most other vertebrates) and *Saccharomyces cerevisiae* are highly sensitive to most RIPs.
ii) Higher plant 80S ribosomes are moderately resistant to the A-chains of the toxic lectins ricin and abrin, but show a sensitivity comparable with that of mammalian ribosomes to several type I RIPs. Cereal seed ribosomes are resistant to their homologous RIPs.
iii) *E. coli* ribosomes are completely refractory to the A-chains of ricin and abrin, but are moderately sensitive to several type 1 RIPs.

In making these generalisations, it should be borne in mind that the possibility of cofactor requirements has not been systematically studied, and could affect the outcome.

6.7.3 The isolation of mutant cells with RIP-resistant ribosomes

One approach which may improve our understanding of the molecular basis of the binding of RIPs to ribosomes and subsequent catalysis is to isolate mutant cells whose ribosomes show resistance to a given RIP and then attempt to analyse biochemically the difference(s) between the sensitive, wildtype ribosomes and the resistant, mutant ribosomes. A similar approach has been used with great success for the characterisation of antibiotic-resistant ribosomes in prokaryotes. The fact that naturally occurring resistance to RIPs is known would suggest that this strategy is feasible. With this aim Sallustio and Stanley (1990) isolated four Chinese hamster ovary (CHO) cell mutants whose ribosomes showed increased cross-resistance to ricin and abrin, as measured by protein synthesis inhibition and the generation of a fragment from 28S rRNA upon aniline treatment. Three of the mutant phenotypes were recessive in somatic cell hybrids whereas a fourth, termed LEC17, was dominant. The concentrations of toxins causing a 50% inhibition in protein synthesis by LEC17 ribosomes were shown to be 22-, 9- and 56-fold higher for abrin holotoxin, ricin holotoxin and ricin A-chain respectively when compared with the parental cell ribosomes. No difference was observed in the sensitivity to modeccin, also a dimeric RIP. The authors reasoned that the mutation in LEC17 was likely to reside in a ribosomal protein rather than rRNA, since the genes for the latter are extensively reiterated. However, the selective removal by ethanol of acidic ribosomal proteins which are thought to be involved in the ricin-sensitive steps of protein synthesis including EF-2-dependent GTPase activity (see section 6.6.6) did not abrogate the LEC17 phenotype. This was despite the fact that both the parental and LEC17 ribosomes were considerably more resistant to ricin following ethanol treatment than control, untreated ribosomes. Thus the nature of the mutation in LEC17 was not resolved.

Other work has taken advantage of yeast genetics and screening procedures in an attempt to isolate and characterise ricin A-chain-resistant mutants. Intact yeast cells are resistant to ricin holotoxin because they lack galactose receptors on their cell surfaces. Therefore yeast transformants containing integrated copies of a galactose-regulated ricin A-chain expression plasmid were constructed (Gould *et al.*, 1991). Mutants which survived the induction of ricin A-chain expression by galactose were selected, and their ribosomes tested *in vitro* for resistance to ricin A-chain. None of the mutants analysed had resistant ribosomes, suggesting that they contained mutations in ricin A-chain. Mutations in yeast genes which confer resistance to ricin A-chain must therefore occur at a very low frequency, if at all, in relation to the frequency of mutations in ricin A-chain which lead to a reduction in, or a complete loss of, toxicity. We do not have a satisfactory explanation to account for the relative ease of isolation of toxin-resistant CHO cells and the difficulties with yeast, apart from the possibility that the levels of ribosome resistance observed in CHO cells, had they occurred in yeast, may not have been sufficiently high to allow the recovery of viable mutants.

6.7.4 RNA identity elements required for recognition and catalysis by ricin A-chain

Endo *et al.* (1987) and Endo and Tsurugi (1988) found that ricin A-chain could depurinate naked 28S rRNA with an identical specificity to that of 28S rRNA in native ribosomes. They also reported that the 553-nucleotide fragment generated

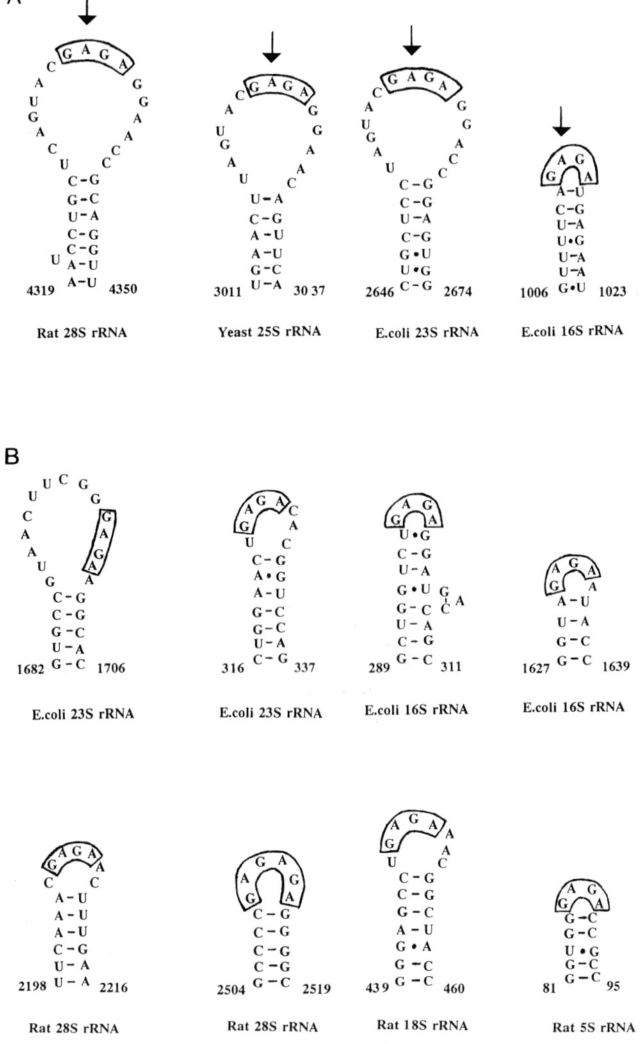

Figure 6.4 The structure near the ricin A-chain site in rRNA. A, the ricin A-chain sensitive structures; B, the ricin A-chain-insensitive structures. From Endo and Tsurugi (1988).

from the 3′ end of 28S rRNA could serve as a substrate following its deproteinisation. The only additional requirement for these substrates was that they retained their secondary structure. An analysis of the kinetic parameters for the depurination of ribosomes and naked 28S rRNA revealed that the K_m values were similar (2.6 μM and 5.8 μM respectively) whilst the turnover number (K_{cat}) differed by a factor of *c*. 10^5 (1777 min^{-1} and 0.02 min^{-1} respectively). These results are important for several reasons: (i) they show that ribosomal proteins are not required for the

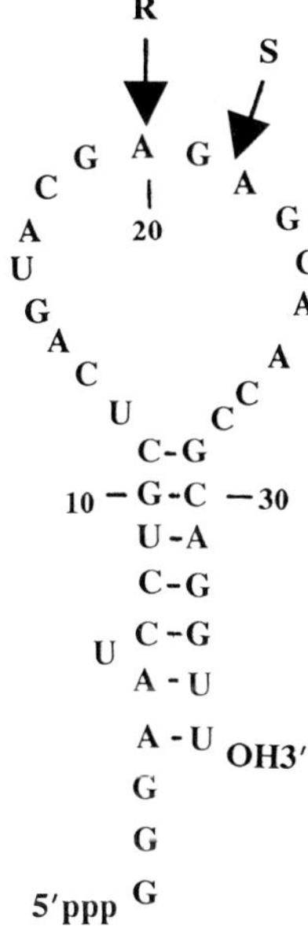

Figure 6.5 The structure of the synthetic oligonucleotide that mimics the ricin/A-sarcin domain in rat 28S rRNA. The ricin A-chain (R) and α-sarcin (S) modification sites are shown by arrows. A20 corresponds to A4324 in rat 28S rRNA. The three guanosine residues at the 5′ end of the 35-mer are remnants of the T7 promoter. They make no contribution to the identification of the substrate by the toxins. From Endo *et al.* (1991).

sequence specificity of the modification site; (ii) they show that intact 28S rRNA is not required; (iii) they suggest that ricin A-chain has a similar affinity for both ribosomes and rRNA (as evidenced by the similarities in K_m values) and (iv) they suggest that the native conformation of rRNA in the ribosome, which is probably dependent on ribosomal proteins, is required for efficient catalysis (as evidenced by the large difference between the K_{cat} values). Endo and Tsurugi (1987) also showed that naked *E. coli* rRNA served as a substrate for ricin A-chain modification (cf. the insensitivity of the native ribosome) but here depurination occurred at two sites; A2660 in 23S rRNA (corresponding to A4324 in rat 28S rRNA) and A1014 in 16S rRNA. Inspection of the assumed structures around these depurination sites (Figure 6.4) reveals that they all contain the sequence GAGA present in a single-stranded region which is subtended by a double-helical stem structure. However, this structural motif occurs at a further four sites in both *E. coli* and rat rRNAs but these are not modified by ricin A-chain in either ribosomes or naked rRNA (Endo and Tsurugi, 1988).

The detailed requirements for recognition and catalysis were studied by using a synthetic oligoribonucleotide (35 mer) that mimics the primary and secondary structure (a helical stem, bulged nucleotide and 17-membered single-stranded loop) of the rat 28S rRNA target site (Endo *et al.*, 1991). This is shown in Figure 6.6. A comparison of the effects of ricin A-chain on these 'wildtype' and 'mutant' oligonucleotides revealed the following: (i) there is an absolute requirement for A^* in the sequence GA^*GA; (ii) the helical stem is essential, although the length of the stem can be reduced from seven to three base pairs, suggesting that the stem is necessary only to tether the ends of the loop. However, the identities of the base pairs in the stem can affect catalysis both negatively, as seen when the G11-A29 base pair was replaced by U11-A29, or positively, as seen when U9-A31 was changed to G9-C31; (iii) the omission of the bulged nucleotide did not affect catalysis; (iv) activity was abolished by altering the nucleotides in the 'universal' sequence surrounding A20 and by changing the position in the loop of GA (ricin) GA (although these findings are at variance with the ability to depurinate *E. coli* 16S rRNA at A1014, shown in Figure 6.6); (v) following cleavage of the phosphodiester bond between G21 and A22 by α-sarcin, ricin A-chain was unable to remove A20. Since ricin A-chain is able to remove A4324 in rat liver ribosomes following α-sarcin cleavage, the authors suggest that the free ends of the RNA may somehow be bridged by ribosomal proteins.

6.7.5 The action of α-sarcin on ribosomes and rRNA

α-Sarcin inhibits protein synthesis in cell-free systems from a variety of eukaryotes, *E. coli* and the Archebacterium *Sulpholobus solfataricus* (reviewed by Jiminez and Vasquez, 1985). It does so by hydrolysing a single phophodiester bond on the 3′ side of G4325 in rat ribosomes or G2661 in *E.coli* ribosomes present in the universal sequence GAG (α-sarcin) A. The 26S rRNA in the ribosomes of the α-sarcin producing mould are also sensitive to the toxin *in vitro* (Miller and Bodley, 1988). Unlike ricin holotoxin it has little effect on intact cells, presumably because it fails to enter them (Ackerman *et al.*, 1988). Initial experiments by Schindler and Davies (1977) showed that α-sarcin was highly effective on yeast ribosomes, as assayed by the cleavage of 26S rRNA. However, rat ribosomes suspended in a simple buffer were refractory. Paradoxically, α-sarcin was highly effective in inhibiting protein synthesis in rat liver extracts (reviewed by Wool, 1984). The apparent discrepancy between these data is resolved by the finding that only those ribosomes engaged in protein synthesis can serve as substrates for α-sarcin cleavage. Ackerman *et al.* (1988) showed that an amount of α-sarcin sufficient to cause a substantial inhibition of protein synthesis when injected into *Xenopus* oocytes resulted in only a small proportion of 28S rRNA cleavage and suggested that this resides in the *c.* 2% of oocyte ribosomes which are active. Experiments by Wool *et al.* (1990) using antibiotic inhibitors of protein synthesis have shown that reticulocyte ribosomes are sensitive to α-sarcin only when peptidyl tRNA is in the ribosomal A site prior to translocation, suggesting that the α-sarcin site alternates between open and closed states during translation.

An early report (Endo *et al.*, 1983) on the action of α-sarcin on naked 28S rRNA showed that its unique specificity for hydrolysing the phosphodiester bond on the 3′ side of G4325 was lost; instead cleavage occurred in the 3′ side of all purine residues. However, Miller and Bodley (1988) found that specificity was retained on naked yeast 26S and *E. coli* 23S rRNA, even at very high concentration of α-sarcin, and that the α-sarcin concentration-dependence of the cleavage was comparable to that with ribosomes (cf. ricin A-chain). The authors suggest the specificity of α-sarcin action on naked rRNA is lost upon storage and/or further purification.

The substrate specificity of α-sarcin has been assessed by using the synthetic oligonucleotide which reproduces the sequence and secondary structure of the ricin modification site (Endo *et al.*, 1991; Figure 6.6). In general, the sequence requirements for binding and catalysis are similar to those for ricin A-chain (section 6.7.4) but with the following differences: (i) there is a strong, but not absolute, preference for G at position 21, corresponding to G4325 in rat 28S rRNA; (ii) A20 (the base removed by ricin) can be substituted by G, C or U without loss of specificity or affecting the extent of cleavage.

6.7.6 The effects of rRNA modifications by RIPs on the functional activities of ribosomes

The target site for the action of RIPs lies in the centre of the most strongly conserved structural feature of LSU rRNA, implying that the structural integrity of this stem-loop structure is necessary in a functional ribosome. However, intact LSU rRNA *per se* is not necessary for ribosome function, since mild treatment of ribosomes with nuclease which causes many nicks in rRNA does not affect their protein synthesising activities (Cahn *et al.*, 1970). Also, some organisms physiologically divide their 28S rRNA into domains (White *et al.*, 1986). It is now generally recognised that rRNA is responsible for the basic biochemistry of protein synthesis, including peptide bond formation, translocation, and the binding of aminoacyl-tRNA and supernatant factors (Wool *et al.*, 1990). It has even been suggested that rRNA plays a direct catalytic role, the ribosomal proteins merely facilitating the optimal structure of the rRNA (Noller, 1991).

Moazed *et al.* (1988) have provided direct evidence for the involvement of the ricin/α-sarcin loop in protein synthesis. They showed that when the two bacterial elongation factors EF-Tu (involved in binding the aminoacyl-tRNA/GTP complex to ribosomes) and EF-G (translocase) were bound to *E. coli* ribosomes a number of specific rRNA nucleotides were protected from modification by dimethyl sulphate and kethoxal, as monitored by primer extension. Both factors protected G2655, A2660 and G2661 while EF-Tu also protected A2665 and EF-G also protected A1067 and A1069. With the exception of the last two bases, the others all lie within the conserved single-stranded loop; A2660 is the base removed by single chain *N*-glycosidase RIPs and G2661 is the base on the 5′ side of the α-sarcin cleavage site. The authors conclude that the ricin/α-sarcin loop is the binding site for both factors. This is substantiated by the finding that the pre-binding of EF-G to *E. coli* ribosomes in the presence of fusidic acid to stabilise the association prevents α-sarcin cleavage (Miller and Bodley, 1991). Furthermore, the treatment of *E. coli* ribosomes with α-sarcin prevents the subsequent binding of both EF-Tu and EF-G (Hausner *et al.*, 1987). It might be expected from the almost universal conservation of the sequence of the α-sarcin/ricin target site and its importance in binding elongation factors, that mutations in this sequence would lead to inactive ribosomes and hence be lethal. The only study on this was reported by Tapprich and Dahlberg (1990) working on *E. coli*. G2661 was changed to cytosine, and expressed from a high copy-number plasmid containing the entire *rrn*B operon, from which 65% of the total cell rRNA was derived. The rRNA containing this mutation was correctly assembled into ribosomes which were fully functional in an otherwise wildtype background. However, ribosomes with a second mutation in ribosomal protein S12 showed a greatly reduced affinity for the EF-Tu/tRNA/GTP ternary complex. More work is obviously required in this area.

There is an extensive literature on the partial reactions of protein synthesis inhibited by α-sarcin and the *N*-glycosidase RIPs in eukaryotic systems (Olsnes and Pihl, 1982). While there is general agreement that α-sarcin inhibits both the EF-1 (equivalent to *E. coli* EF-Tu) and EF-2 (equivalent to *E. coli* EF-G) dependent reactions, the situation with the *N*-glycosidase RIPs is more controversial. The controversy surrounds whether the EF-1-dependent binding of aminoacyl tRNA and associated GTPase is inhibited in addition to the well-established inhibition of translocation. Some authors (Fernandez-Puentes *et al.*, 1977; Olsnes *et al.*, 1975) have shown a strong inhibition of factor-dependent binding of aminoacyl tRNA to salt-washed ribosomes translating poly (U), whereas others (Sperti and Montanaro, 1979; Brigotti *et al.*, 1989) have shown no effect. It has been suggested that this could be due to different concentrations of EF-1 in the assays (Olsnes and Pihl, 1982) or to contaminating amounts of EF-2 in the ribosome preparations (Sperti and Montanaro, 1979). Using an unfractionated reticulocyte lysate system translating globin mRNA, Osborn and Hartley (1990) showed that the N-terminal dipeptide (Met-Val) was formed on ricin-modified ribosomes and that the subsequent translocation of this dipeptide was blocked, showing that translocation was the step showing most inhibition.

A possible explanation to account for the differences in the partial reaction of protein synthesis inhibited by α-sarcin and the *N*-glycosidase RIPs has come from the observations by Teraro *et al.* (1988) that the two classes of toxins alter the conformation of rat liver ribosomes at neighbouring, but different, sites. Using a double-labelling technique in which ribosomes were treated with the toxins and then reacted with either [^{3}H] or [^{14}C]- labelled *N*-ethylmaleimide it was found that the labelling of ribosomal proteins L14 was specifically reduced by treatment with ricin A-chain, and that of L3 and L4 by α-sarcin. L3 and L4 have been located in the A and P sites respectively and can be crosslinked to EF-2 (Uchiumi *et al.*, 1986).

Indirect evidence on changes in the association between 5.8S rRNA and 28S rRNA in rat liver ribosomes caused by α-sarcin treatment suggest that the effects of rRNA cleavage on ribosome structure may be very pronounced. 5.8S rRNA is believed to be associated with 28S rRNA through hydrogen bonding interactions involving *c.* 40 base pairs (Wool, 1984). In rRNA samples from control ribosomes complete dissociation of the 28S:5.8S RNA complex requires a urea concentration of 5 to 6 M, whereas after α-sarcin treatment the complex dissociates in the absence of urea (Wool, 1984). The author concludes that cleavage at the α-sarcin site causes a collapse in the structure of 28S rRNA.

The more extensive alteration in conformation seen after α-sarcin treatment presumably reflects the greater disruption that occurs when the rRNA backbone is cleaved when compared with the smaller effect of the removal of one base by ricin. The finding that high concentrations of Mg^{2+} can partially overcome the inhibition of protein synthesis in both α-sarcin and ricin-modified ribosomes (Cawley *et al.*, 1979; Teraro *et al.*, 1988) could mean that the alterations in conformation can be partially reversed by this ion, presumably through the formation of salt bridges between neighbouring phosphate groups in the rRNA backbone.

In addition to the well-studied effects of the *N*-glycosidase RIPs on the elongation cycle of protein synthesis, there are two reports that initiation is also inhibited. Skorve *et al.* (1977) found that abrin A-chain reduced the formation of the 80S initiation complex in wheat germ extracts and Osborn and Hartley (1990) made a similar observation for reticulocyte lysate. Since the only involvement of the 60S subunit in initiation is in the final step, i.e. the formation of the 80S initiation complex from the 40S preinitiation complex, this is most probably the step inhibited. These observations may account for the finding by Gould *et al.* (1991) that yeast cells expressing mutant ricin A-chain in which serine 203 was changed to asparagine grew at a rate comparable to that of parental cells even though a significant proportion of their 26S rRNA was depurinated. Although the mutant ricin A-chain was *c.* 100-fold less active than the wildtype in the inhibition of protein synthesis, it showed an altered specificity in that it caused the preferential modification in the rRNA of free 60S subunits; 26S rRNA in monosomes and polysomes was barely affected. In this situation the toxin-modified 60S subunits may be preferentially excluded from entering polysomes, thereby allowing protein synthesis to continue.

6.8 Antiviral properites of RIPs

It has long been known that proteinaceous extracts from certain plants inhibit plant virus infection when the extract and virus are mixed and rubbed onto the surface of an appropriate test plant (Kassanis and Kleczkowski, 1948). Subsequently, it has been shown that several purified *N*-glycosidase RIPs, especially single chain RIPs including PAP, MAP, dianthins and saporins, are potent inhibitors of plant viruses, at concentrations as low as 25 ng/ml (Irvin, 1975; Kubo *et al.*, 1990; Stevens *et al.*, 1981; Frotschl *et al.*, 1990). A striking feature of the antiviral activity of RIPs is that it is entirely non-specific with regard to the identities of the viruses inhibited. For example Chen *et al.* (1992) showed that PAP inhibited plant viruses from seven groups (five RNA virus groups and two DNA virus groups) when assayed on the leaves of appropriate indicator plants. Physical association between the virus and the RIP is not required for antiviral activity. This was demonstrated by Kumon *et al.* (1990) who showed that although there was a weak association between TMV and PAP at low ionic strength, at high ionic strength (300 mM KCI) the association was completely lost yet the antiviral effect was unaltered. Furthermore, there is no direct effect of the RIP on the virus when the two are mixed since Tomlinson *et al.* (1974) showed that cucumber mosaic virus mixed with PAP regained full infectivity on *Chenopodium quinoa* when the virus was separated by centrifugation. These findings suggest that the antiviral effect of the RIP acts on the host cell, presumably through ribosome inactivation, thus preventing virus replication. This interpretation is strengthened by studies on the inhibition by RIPs of animal virus infection of cultured cells. For example Ussery *et al.* (1977) showed that the treatment of poliovirus-infected HeLa cells with PAP prevented the subsequent release of infectious virus, with a concomitant inhibition of protein synthesis. Protein synthesis was not affected by PAP in control cells. These data support the generally accepted view that viral infection permeabilises mammalian cells to protein toxins (Fernandez-Puentes and Carrasco, 1980).

The mechanism of antiviral action of RIPs described above cannot, however, account for the recently described inhibitory effects of RIPs on human immunodeficiency virus (HIV-1) gene expression in acutely and chronically-infected macrophages and T-cells. McGrath *et al.* (1989) showed that the single chain RIP trichosanthin selectively inhibited both HIV RNA and protein accumulation without affecting host cell gene expression. When freshly-drawn blood samples from HIV-infected patients were treated with a single 3 h exposure to trichosanthin, HIV replication was blocked for at least five days in subsequently cultured monocyte/macrophages. However, doubts were cast over the efficacy of treating HIV-positive patients with trichosanthin because of its well-documented cytotoxic side effects (Palacca, 1990). In a search for RIPs with more efficaceous anti-HIV properties Lee-Huang *et al.* (1991) found that TAP29 (another single chain RIP from the root tubes of *Trichosanthes kirilowii*) had a 'therapeutic' index (defined as the ratio of the minimum concentration of the RIP required to inhibit HIV gene expression to that exhibiting cytotoxicity in uninfected cells) which was 100–1000 fold higher than for trichosanthin. Surprisingly the IC^{50} (the concentration required for 50% inhibition) values for both RIPs in inhibiting protein synthesis in the reticulocyte lysate system were identical. The authors conclude that at least two unique domains must be involved in the anti-HIV and cytotoxic activities of these RIPs. If this proves to be the case, the anti-HIV activity must operate even more efficiently than the ribosome inactivating activity. Clearly detailed structure–function comparisons of the two RIPs should be made in order to shed light on their exciting therapeutic potential.

6.9 Physiological roles of plant RIPs

The extreme cytotoxicity of the heterodimeric RIPs, such as ricin, together with their widespread distribution and the highly efficient ribosome inactivating activities of the single chain RIPs, have prompted the widely held view that RIPs play a defensive role in plants, protecting them against predators and pathogens. This defensive role could operate through two general strategies; the ribosomes of the RIP-producing plant could be resistant to the homologous RIP whereas those of its predators and cellular pathogens are sensitive or the ribosomes of the producing plant are sensitive to the homologous RIP but the two are separated by sequestration of the RIP in an organelle or secreted extracellulary. In the latter strategy damage caused to plant cells by the pathogen, or the pathogen's vector in the case of viruses, could release the RIP into the cytosol, thereby inhibiting protein synthesis and causing local 'suicide' which prevents the pathogens' growth or replication.

Ricin, abrin and modeccin are particularly toxic to mammals when injected and are highly toxic even after oral ingestion (reviewed by Olsnes and Pihl, 1982). There is considerable variation in toxicity between animal species. For example, on a weight basis the guinea pig is more sensitive to ricin than the mouse, and the horse is particularly sensitive. It is therefore likely that potential grazing herbivores and rodents which ingested *Ricinus* seeds would suffer or die in consequence, and their populations may have 'learned' to avoid such seeds. In a recent study Gatehouse *et al.* (1990) showed that ricin and saporin, incorporated into artificial diets, were extremely toxic to the larvae of two Coleopteran species with average LD^{50} values of less than 10^{-2}% (expressed in terms of dry weight of the diet). The larvae of two Lepidopteran species were considerably more resistant to the RIPs and this correlated with the ability of their digestive enzymes to hydrolyse and so inactivate the RIPs in the gut, thus abrogating their toxic effects. The reason why saporin, a single chain RIP lacking the cell-binding B-chain, should be almost as toxic as ricin to the Lepidopteran larvae is unclear.

Cereal seed RIPs have been shown to exhibit antifungal properties which may protect the seeds against fungal attack. Roberts and Selitrennikoff (1986) showed that the IC^{50} for the inhibition of protein synthesis by *Neurospora* ribosomes by barley seed RIP was some 10-fold lower than for Ascites cell ribosomes, and that barley seed RIP was a potent inhibitor of mycelial growth of *Trichoderma reesei* on agar plates. They suggested also that the RIP may act synergistically with other antifungal agents, some of which appear to act by permeabilising the fungal plasma membrane (Albersheim and Valent, 1978). This idea was followed up by Leah *et al.* (1991) who found that a combination of barley seed (1,3)-β-glucanase, chitinase and barley seed RIP acted synergistically to inhibit myclcial growth of *T. reesei* and *Fusarium sporotrichioidies*.

The most widely accepted physiological role of single chain RIPs is that they are antiviral agents. This view was strengthened with the discovery that in several plants RIPs are located in the cell wall matrix (see section 6.5) and are known to be active on their own ribosomes. Ready *et al.* (1986) proposed that the local wounding caused by vectors which transmit viruses to plants (e.g. aphids and nematodes) would release the RIP into the cytosol, resulting in local suicide at the site of infection. Although this hypothesis has its attractions two facts conspire against it. Firstly, organs containing high concentrations of RIPs in several species are susceptible to virus infection. For example, the leaves of *Phytolacca americana* (containing PAP) and *Dianthus caryophyllus* (containing dianthins) are highly susceptible to infection by pokeweed mosaic virus and carnation mottle virus respectively (Shepherd *et al.*, 1969; Hollings and Stone, 1970). Secondly, studies on the inhibitory effects of RIPs on the mechanical transmission of plant viruses have consistently shown that RIPs are effective only against heterologous plants and do not protect the homologous plant (Gendron and Kassanis, 1954). For example, purified PAP does not protect the pokeweek plant from local lesion formation caused by TMV (Grasso and Shepherd, 1978). The explanation traditionally offered for this observation is that the host cell ribosomes are uniquely resistant to the action of the homologous RIP (Stirpe and Hughes, 1989). Such a strategy would appear to have limited use in nature and recent findings that RIPs do act on homologous ribosomes would appear to negate it. Thus the antiviral role of RIPs is highly questionable, although it is quite possible that the expression of RIPs in transgenic plants could protect such plants. Work with this aim is currently in progress at Rothamsted (R. White and J. Antoniw, personal communication) and in our own laboratory.

Finally, it will not have escaped the reader's notice that several plant organs containing type 1 RIPs are eaten by humans, sometimes in an uncooked state. For example, we have shown that wheat germ purchased from a local health food shop is an excellent source of tritin, which accounts for about 1% of the total protein. We have shown that under optimal assay conditions *in vitro* the turnover number for tritin on reticulocyte ribosomes is *c.* 1000 min^{-1}. A calculation involving Avagadro's number reveals that a liberal sprinkling (say 20 g) of wheat germ on a bowl of breakfast cereal contains about 10^{18} mols of tritin which could depurinate 10^{21} ribosomes per minute. If we assume that the adult human possesses 10^{13} cells each of which contains 3×10^{6} ribosomes (Alberts *et al.*, 1989), then all 3×10^{19} ribosomes could be inactivated in about two seconds. Bon appetit!

References

Ackerman, E.J., Saxena, S.K. and Ulbrich, N. (1988) α-sarcin causes a specific cut in 28S rRNA when microinfected into *Xenopus* occytes *J. Biol. Chem.* **263**: 17076–17083.

Albersheim, P. and Valent, B.S. (1978) Host–pathogen interactions in plants. *J. Cell Biol.* **78**: 627–643.

Alberts, B., Bray, D., Lewis, J., Raff, M., Roberts, K. and Watson, J.D. (1989) *Molecular Biology of the Cell*, 2nd edn, Garland, New York.

Asano, K., Svenson, B., Svensden, I., Poulsen, P., Roepstorff, P. (1986) The complete primary structure of protein synthesis inhibitor II from barley seeds. *Carlsberg Res. Commun.* **51**: 129–141.

Barbieri, L. and Stirpe, F. (1982) Ribosome-inactivating proteins from plants: properties and possible uses. *Cancer Surveys* **1**: 489–520.

Battelli, M.G., Enzo, L., Stirpe, F., Cella, R. and Parisi, B. (1984) Differential effect of ribosome inactivating proteins on plant ribosome activity and plant cell growth. *J. Exp. Bot.* **35**: 882–889.

Bednarek, S.V., Williams, T.A., Dombrowski, J.E. and Raikhel, N.V. (1990) A carboxyl-terminal propeptide is necessary for proper sorting of barley lectin to tobacco. *The Plant Cell* **2**: 1145–1155.

Benatti, L., Saccardo, M.B., Dani, M., Nitti, G., Sassano, M., Lorenzetti, R., Lappi, D.P and Soria, M. (1989) Nucleotide sequence of cDNA coding for saporin-6, a type-1 ribosome-inactivating protein from *Saponaria officinalis*. *Eur. J. Biochem.* **183**: 465–470.

Benatti, L., Nitti, G., Solinas, M., Valsasina, B., Vitale, A., Ceriotti, A. and Soria, M.R. (1991) A saporin-6 cDNA coding for a carboxyl-terminal extension. *FEBS Lett.* **291**: 285–288.

Brigotti, M., Rambelli, F., Zamboni, M., Montanaro, L. and Sperti, S. (1989) Effect of α-sarcin and ribosome-inactivating proteins on the interaction of elongation factors with ribosomes. *Biochem. J.* **257**, 723–727.

Brosius, J., Dull, T.J. and Noller, H.F. (1980) Complete nucleotide sequence of a 23S ribosomal RNA gene from *Escherichia coli*. *Proc. Natl Acad. Sci. USA* **77**: 202–204.

Butterworth, A.G. and Lord, J.M. (1983) Ricin and *Ricinus communis* agglutinin subunits are all derived from a single-size polypeptide precursor. *Eur. J. Biochem.* **137**: 57–65.

Cahn, F., Schachter, E.M. and Rich, A. (1970) Polypeptide synthesis with ribonuclease-digested ribosomes. *Biochim. Biophys. Acta* **209**: 512–520.

Calderwood, S.B., Auclair, F., Donohue-Rolfe, A., Keush, G.T., Mekalano, J.T. (1987) The nucleotide sequence of the shiga-like toxin genes of *Escherichia coli*. *Proc. Natl Acad. Sci. USA* **84**: 4364–4368.

Cawley, D.B., Hedblom, M.L., Hoffman, E.J. and Houston, L.L. (1977) Differential sensitivity of rat liver and wheat germ ribosomes to polyuridylic acid translation. *Arch. Biochem. Biophys.* **182**: 690–695.

Cawley, D.B., Hedblom, M.L. and Houston, L.L. (1979) Protection and rescue of ribosomes from the action of ricin A chain. *Biochemistry* **18**: 2648–2654.

Chan, Y.-L., Olvera, J. and Wool, I.G. (1983) The structure of rat 28S ribosomal ribonucleic acid inferred from the sequence of nucleotides in a gene. *Nucleic Acids Res.* **11**: 7819–7831.

Chen, Z.C., White, R.F., Antoniw, J.F. and Lin, Q. (1991) Effect of pokeweed antiviral protein on the infection of plant viruses. *Plant Pathology* **40** : 612–620.

Chow, T.P., Feldman, R.A., Lovett, M. and Piatak, M. (1990) Isolation and DNA sequence of a gene encoding α-trichosanthin, a type 1 ribosome-inactivating protein. *J. Biol. Chem.* **265**: 8670–8674.

Coleman, W.H. and Roberts, W.K. (1981) Factor requirements for the tritin inactivation of animal cell ribosomes. *Biochem. Biophys. Acta* **654**: 57–66.

Collins, E.J., Robertus, J.D., LoPreti, M., Stone, K.L., Williams, K.R., Wu, P., Hwang, K. and Piatak, M. (1990) Primary amino acid sequence of α-trichosanthin and molecular models for abrin A-chain and α-trichosanthin. *J. Biol. Chem.* **265**: 8665–8669.

van Deurs, B., Pederson, O.W., Sundan, A., Olsnes, S. and Sandvig, K. (1985) Receptor-mediated endocytosis of a ricin-colloidal gold conjugate in Vero cells. *Exp. Cell Res.* **159**: 287–304.

van Deurs, B., Tonnesson, T.O., Peterson, O.W., Sandvig, K. and Olsnes, S. (1986) Routing of internalized ricin and ricin conjugates to the Golgi complex. *J. Cell. Biol.* **102**: 37–47.

Dickinson, C.D., Evans, R.P. and Nielson, N.C. (1988) RY repeats are conserved on the 5-flanking regions of legume seed protein genes. *Nucleic Acids Res.* **16**: 371.

Doolittle, R.F.,Feng, D.F.,Anderson, K.L. and Alberro, M.R. (1990) A naturally occurring horizontal gene transfer from a eukaryote to a prokaryote. *J. Mol. Evol.* **31**:383–388.

Endo, Y. and Tsurugi, K. (1987) RNA N-glycosidase of ricin A chain. Mechanism of action of the toxic lectin ricin on eukaryotic ribosomes. *J. Biol. Chem.* **262**: 8128–8130.

Endo, Y. and Tsurugi, K. (1988). The RNA *N*-glycosidase activity of ricin A-chain: the characteristics of the enzymatic activity of ricin A-chain with ribosomes and rRNA. *J. Biol. Chem.* **263**: 8735–8739.

Endo, Y.,Huber, P.W. and Wool, I.G. (1983) The ribonuclease activity of the cytotoxin α-sarcin. *J. Biol. Chem.* **258**: 2662–2667.

Endo, Y.,Mitsui, K.,Motizuki, M. and Tsurugi, K. (1987) The mechanism of action of ricin and related toxin on eukaryotic ribosomes. *J. Biol. Chem.* **262**: 5908–5912.

Endo, Y.,Tsurugi, K. and Lambert, J.M. (1988a) The site of action of six different ribosome-inactivating proteins from plants on eukaryotic ribosomes: The RNA *N*-glycosidase activity of the proteins. *Biochem. Biophys Res. Comm.* **150**: 1032–1036.

Endo, Y.,Tsurugi, K. and Ebert, R.F. (1988b) The mechanism of action of barley toxin: a type I ribosome inactivating protein with RNA *N*-glycosidase activity. *Biochim. Biophys. Acta.* **954** : 224–226.

Endo, Y.,Tsurugi,K.,Yutsudo, T.,Tukeda, Y.,Ogasawara, K. and Igarashi, K. (1988c) Site of action of Vero toxin (VT2) from *E. coli* 0157:H7 and of Shiga toxin on eukaryotic ribosomes. *Eur. J. biochem.* **131**: 45–50.

Endo, Y., Oka, T., Tsurugi, K. and Franz, H. (1989) The mechanism of action of the cytoxic lectin from *Phoradendron californicum*: the RNA *N*-glycosidase activity of the protein. *FEBS Lett.* **248**: 115–118.

Endo, E., Gluck, A., Chan, Y.-L., Tsurugi, K. and Wool, I.G. (1990) RNA-protein interaction. *J. Biol. Chem.* **265**: 2216–2222.

Endo, Y.,Gluck, A. and Wool, I.G. (1991) Ribosomal RNA identity elements for ricin A-chain recognition and catalysis. *J. Mol. Biol.* **221**: 193–207.

Fernandez-Puentes, C. and Carrasco, L. (1980) Viral infection permeabilizes mammalian cells to protein toxins. *Cell* **20**: 769–775.

Fernandez-Puentes, C., Carrasco, L. and Vazquez, D. (1977) Site of action of ricin on ribosomes. *Biochemistry* **15**: 4364–4369.

Forde, B.G., Heyworth, A., Pywell, J. and Kreis, M. (1985) Nucleotide sequence of a B1 hordein gene and the identification of possible upstream regulatory elements in endosperm storage protein genes from barley, wheat and maize. *Nucleic Acids Res.* **13**: 7327–7339.

Fordham-Skelton, A.P., Taylor, P.N., Hartley, M.R. and Croy, R.D.D. (1991) Characterisation of saporin genes; *in vitro* expression and ribosome inactivation. *Mol. Gen. Genet.* **229**: 460–466.

Foxwell, B.M.J., Donovan, T.A., Thorpe, P.E. and Wilson, G. (1985) Removal of carbohydrates from ricin with endoglycosidase H, endoglycosidase F, endoglycosidase D and α-mannosidase. *Biochim. Biophys. Acta* **840**: 193–203.

Frankel, A., Welsh, P., Richardson, J. and Robertus, J.D. (1990) The role of arginine 180 and glutamic acid 177 of ricin toxic A chain in the enzymatic inactivation of ribosomes. *Mol. Cell Biol.* **10**: 6257–6263.

Frotschl, R., Schonfelder, M., Mundry, K.W. and Adam, G. (1990) Functional studies and subcellular distribution of RIPs from plants with antiviral activity. *VIIIth International Congress of Virology*, Berlin, August 1990, Abstracts p. 485.

Funatsu, G., Taguchi, Y., Kamenosona, M. and Yanaka, M. (1988) The complete amino acid sequence of the A-chain of abrin-a, a toxin protein from the seeds of *Abrus precatorius*. *Agric. Biol. Chem.* **52**: 1095–1097.

Gatehouse, A.M.R., Barbieri, L., Stirpe, F. and Croy, R.R.D. (1990) Effects of ribosome inactivating proteins on insect development—differences between Lepidoptera and Coleoptera. *Entemol. Exp. Appl.* **54**: 43–51.

Gendron, Y. and Kassanis, B. (1954) The importance of the host species in determining the action of virus inhibitors. *Ann. Appl. Bot.* **41**: 183–188.

Gould, J.H., Hartley, M.R., Welsh, P.C., Hoshizaki, D.K., Frankel, A., Roberts, L.M. and Lord, J.M. (1991) Alteration of an amino acid outside the active site of ricin A chain reduces its toxicity towards yeast ribosomes. *Mol. Gen. Genet.* **230**: 81–90.

Grasso, S. and Shepherd, R.J. (1978) Isolation and partial characterizations of virus inhibitors from plant species taxanomically related to *Phytolacca*. *Phytopathology* **68**: 199–205.

Habuka, N., Murakami, Y., Noma, M., Kudo, T. and Horikoshi, K. (1989) Amino acid sequence of *Mirabilis* antiviral protein, total synthesis of its gene and expression in *Escherichia coli*. *J. Biol. Chem.* **264**: 6629–6637.

Habuka, K., Akiyama, K, Tsuge, H., Miyano, M., Matsumoto, T. and Noma, M. (1990) Expression and secretion of *Mirabilis* antiviral protein in *Escherichia coli* and its inhibition *in vitro* of eukaryotic and prokaryotic protein synthesis. *J. Biol. Chem.* **265**: 10988–10992.

Hartings, H., Lazzaroni, N., Marsan, P.A., Aragay, A., Thompson, R., Salamini, F., DiFonzo, N., Palau, J. and Motto, M. (1990) The b-32 protein from maize endosperm: characterization of genomic sequences encoding two alternative central domains. *Plant Molec. Biol.* **14**: 1031–1040.

Hartley, M.R., Legname, G., Osborn, R.W., Chen, Z. and Lord, J.M. (1991) Single-chain ribosome inactivating proteins from plants depurinate *Escherichia coli* 23S ribosomal RNA. *FEBS Lett* **290**: 65–68.

Hausner, T.P., Atmadja, J. and Nierhaus, K.K. (1987) Evidence that the G_{2661} region of 23S rRNA is located at the ribosomal binding sites for both elongation factors. *Biochimie* **69**: 911–923.

Hollings, M. and Stone, O.M. (1970) Carnation mottle virus. C.M.I./A.A.B. Descriptions of plant viruses No. 7.

Irvin, J.D. (1975) Purification and partial characterization of a protein from *Phytolacca americana* which inhibits eukaryotic protein synthesis. *Arch. Biochem. Biophys.* **169**: 522–528.

Jackson, M.P., Neill, R.J., O'Brien, A.D., Holmes, R.K. and Newland, J.W. (1987) Nucleotide sequence analysis and comparison of the structural genes for Shiga-like toxin I and Shiga-like toxin II encoded by bacteriophages from *Escherichia coli* 933. *Microbiol. Lett.* **44**: 109–114.

Jiminez, A. and Vasquez, D. (1985) Plant and fungal proteins and glycoproteins inhibiting eukaryotic protein synthesis. *Ann. Rev. Microbiol.* **39**: 649–672.

Kassanis, B. and Kleczkowski, A. (1948) The isolation and some properties of a virus-inhibiting protein from *Phytolacca esculenta*. *J. Gen. Microbiol.* **2**: 143–153.

Kataoka,J.,Habuku, N., Furono, M., Miyano, M., Takanami, Y. and Koiwai, A. (1991) DNA sequence of *Mirabilis* antiviral protein (MAP), a ribosome-inactivating protein with an antiviral property from *Mirabilis jalapa* L. and its expression in *Escherichia coli*. *J. Biol. Chem.* **266**: 8426–8430.

Katzin, B.J., Collins, E.J. and Robertus, J.D. (1991) Structure of ricin A-chain at 2.5 Å. *Proteins* **10**: 251–259.

Kirby, K.S. (1968) Isolation of nucleic acids with phenolic solvents. In *Methods in Enzymology Vol. XIIB* (Grossman, L. and Moldave, K.,eds), Academic Press, New York, pp. 87–100.

Kolosha, V.O. and Fodor, I. (1990) Nucleotide sequence of *Citrus limon* 26S rRNA gene and secondary structure model of its RNA. *Plant Molec. Biol.* **14**: 147–161.

Kubo, S., Ikeda, T., Imaizumi, S., Takanami, Y. and Mikami, Y. (1990) A potent plant virus inhibitor found in *Mirabilis jalapa* L. *Ann. Phytopath. Soc. Japan* **56**: 481–487.

Kumon, K., Sasaki, J., Sejima, M., Takeuchi, Y. and Hayashi, Y. (1990) Interaction between tobacco mosaic virus, pokeweed antiviral proteins and tobacco cell wall. *Phytopathology* **80**: 636–641.

Lamb, F.I., Roberts, L.M. and Lord, J.M. (1985) Nucleotide sequence of cDNA coding for preproricin. *Eur. J. Biochem.* **148**: 265–270.

Lamy, B., Moutaouakil, M., Latge, J.-P. and Davies, J. (1991) Secretion of a potential virulence factor, a fungal ribonucleotoxin, during human aspergillosis infection. *Molecular Microbiology* **5**: 1811–1815.

Leah, R., Tommerup, H., Svendsen, I. and Mundy, J. (1991) Biochemical and molecular characterization of three barley seed proteins with antifungal properties. *J. Biol. Chem.* **266**: 1564–1573.

Lee-Huang, S., Huang, P.L., Nara, P.L., Chen., H.-C., Kung, H., Huang, P., Huang, H.I. and Huang, P.L. (1990) MAP 30: a new inhibitor of HIV-1 infection and replication. *FEBS Lett.* **272**: 12–18.

Lee-Huang, S., Huang, P.L., Kung, H.-F., Li, B.-Q., Huang, P.L., Huang, P., Huang, H.I. and Chen, H.-C. (1991) TAP 29: An anti-human immunodeficiency virus protein from *Trichosanthes kirilowii* that is non-toxic to intact cells. *Proc. Natl. Acad. Sci. USA* **88**: 6570–6574.

Legname, G., Bellosta, P., Gromo, G., Modena, D., Keen, J.N., Roberts, L.M. and Lord, J.M. (1991) Nucleotide sequence of cDNA coding for dianthin 30, a ribosome inactivating protein from *Dianthus caryophyllus. Biochim. Biophys. Acta* **1090**: 119–122.

Lohmer, S., Maddaloni, M., Motto, M., DiFonzo, N., Hartings, H., Salamini, F. and Thompson,R.D. (1991) The maize regulatory locus *opaque*-2 encodes a DNA-binding protein which activates the transcription of the b-32 gene. *EMBO Journal* **10**: 617–624.

Lopez-Otin, C., Barber, D., Fernandez-Luna, J.L., Soriano, F. and Mendez, E. (1984) The primary structure of the cytotoxin restrictonin. *Eur. J. Biochem.* **143**: 621–634.

Lord, J.M. (1991) (ed) *Redirecting Nature Toxins, Seminars in Cell Biology Vol. 2*, Saunders, Philadelphia.

Lord, J.M., Hartley, M.R. and Roberts, L.M. (1991) Ribosome inactivating proteins of plants. *Sem. Cell. Biol* **2**: 15–22.

McGrath, M.S., Huang, K.M., Caldwell, S.E., Gaston, I., Luk, K.-G., Wu, P., Ng, V.L., Crowe,S., Daniels, J., Marsh, J., Deinhart, T., Lekas, P.V., Vennari, J.C., Yeung, H.-W. and Lifson, J.D. (1989) GLQ223: An inhibitor of human immunodeficiency virus replication in acutely and chronically infected cells of lymphocyte and mononuclear phagocyte lineage. *Proc. Natl. Acad. Sci. USA* **86**: 2844–2848.

Merino, M.J., Ferneras, J.M., Munoz, R., Ingelias, R. and Girbes, T. (1990) Plant species containing inhibitors of eukaryotic protein synthesis. *J. Exp. Bot.* **41**: 67–70.

Miller, S.P. and Bodley, J.W. (1988) α-sarcin cleaves ribosomal RNA at the α-sarcin site in the absence of ribosomal proteins. *Biochem. Biophys. Res. Commun.* **154**: 404–410.

Miller, S.P. and Bodley, J.W. (1991) α-sarcin cleavage of ribosomal RNA is inhibited by the binding of elongation factor G or thiostrepton to the ribosome. *Nucleic Acids Res.* **19**: 1657–1660.

Moazed, D., Robertson, J.M. and Noller, H.F. (1988) Interaction of elongation factors EF-G and EF-Tu with a conserved loop in 23S rRNA. *Nature* **331**: 362–364.

Montfort, W., Villafranca, J.E., Monzaningo, A.F., Ernst, S., Katzin, B., Rutenber, E., Xuong,N.H., Hamlin, R. and Robertus, J.D. (1987) The three-dimensional structure of ricin at 2.8 Å. *J. Biol. Chem.* **262**: 5398–5403.

Moya, M., Dantry-Varsat, A., Goud, B., Louvard, D. and Boquet, P. (1985) Inhibition of coated pit formation in Hep2 cells blocks the cytotoxicity of diphtheria toxin but not that of ricin. *J. Cell Biol.* **101**: 548–559.

Ng, T.B., Feng, Z., Li, W.W. and Yeung, H.W. (1991) Improved isolation and further characterization of beta-trichosanthin, a ribosome-inactivating protein and abortifacient from *Trichosanthes cucumeroides. Int. J. Biochem.* **23**: 561–567.

Noller, H.F. (1991) Drugs and the RNA world. *Nature* **353**: 302–303.

O'Brien, A.D. and Holmes, R.K. (1987) Shiga and shiga-like toxins. *Microbiol. Rev.* **51**: 206–220.

O'Hare, M., Roberts, L.M., Thorpe, P.E., Watson, G.J., Prior, B. and Lord, J.M. (1987) Expression of ricin A chain in *Escherichia coli. FEBS Lett.* **216**: 73–78.

Olsnes, S. and Pihl, A. (1982) Toxin lectins and related proteins. In *Molecular Action of Toxins and Viruses* (Cohen, P. and vanHeyringen, J. eds) Elsevier, Amsterdam, pp. 51–105.
Olsnes, S., Saltvedt, E. and Pihl, A. (1974) Isolation and comparison of galactose-binding lectins from *Abrus precatorius* and *Ricinus communis*. *J. Biol. Chem.* **249**: 803–810.
Olsnes, S., Fernandez-Puentes, C., Carrasco, L. and Vazquez, D. (1975) Ribosome inactivation by the toxic lectins abrin and ricin. *Eur. J. Biochem.* **60**: 281–288.
Osborn, R.W. (1990) The action of ricin A chain on eukaryotic ribosomes. PhD thesis, University of Warwick.
Osborn, R.W. and Hartley, M.R. (1990) Dual effects of the ricin A chain on protein synthesis in rabbit reticulocyte lysate. *Eur. J. Biochem.* **193**: 401–407.
Palacca, J. (1990) Trials and tribulations of AIDS drug testing. *Science* **247**: 1406.
Peattie, D.A. (1979) Direct chemical method for sequencing RNA. *Proc. Natl. Acad. Sci. USA* **76**: 1760–1764.
Piatak, M., Lane, J.A., Laird, W., Bjorn, M.J., Wang, A. and Williams, M. (1988) Expression of fully functional ricin A-chain in *Escherichia coli* is temperature-sensitive. *J. Biol. Chem.* **263**: 4837–4843.
Raikhel, N.V. and Williams, T.A. (1987) Isolation and characterization of a cDNA clone encoding wheat germ agglutinin. *Proc. Natl. Acad. Sci. USA* **84**: 6745–6749.
Raué, H.A., Klootwijk, J. and Musters, W. (1988) Evolutionary conservation of structure and function in high molecular weight ribosomal RNA. *Prog. Biophys. Molec. Biol.* **51**: 77–129.
Ready, M.P., Brown, D.T. and Robertus, J.D. (1986) Extracellular localization of pokeweed antiviral protein. *Proc. Natl Acad. Sci USA* **83**: 5053–5056.
Ready, M.P., Katzin, B.J. and Robertus, J.D. (1988) Ribosome inhibiting proteins, retroviral reverse transcriptases and RNAse H share common structural elements. *Proteins* **3**: 53–59.
Ready, M.P., Kim, Y. and Robertus, J.D. (1991) Site directed mutagenesis of ricin A chain and implications for the mechanism of action. *Proteins* **10**: 270–278.
Roberts, L.M. and Lord, J.M. (1981) Protein biosynthetic capacity in the endosperm of ripening castor bean seeds. *Planta* **152**: 420–487.
Roberts, W.K. and Selitrennikoff, C.P. (1986) Isolation and characterization of two antifungal proteins from barley. *Biochim. Biophys. Acta* **880**: 161–170.
Robertus, J.D. (1991) The structure and action of ricin, a cytotoxic *N*-glucosidase. In *Redirecting Nature's Toxins* (Lord, J.M. ed), *Seminars in Cell Biology*, vol.2. 23–30.
Rosen, S.W. and Hughes, R.C. (1977) Effects of neuraminidase on lectin binding by wild-type and ricin-resistant strains of hamster fibroblasts. *Biochemistry* **16**: 4908–4915.
Rutenber, E. and Robertus, J.D. (1991) Structure of ricin B-chain at 2.5 Å resolution. *Proteins* **10**: 260–269.
Rutenber, E., Ready, M. and Robertus, J.D. (1987) Structure and evolution of ricin B chain. *Nature* **326**: 624–626.
Rutenber, E., Katzin, B.J., Ernst, S., Collins, E.J., Mesna, D., Ready, M.P. and Robertus, J.D. (1991) Crystallographic refinement of ricin at 2.5 Å. *Proteins* **10**: 240–250.
Sallustio, S. and Stanley, P. (1990) Isolation of Chinese hamster ovary ribosomal mutants differentially resistant to ricin, abrin and modeccin. *J. Biol. Chem.* **265**: 582–588.
Sandvig, K., Olsnes, S. and Pihl, A. (1976) Kinetics of binding of the toxic lectins abrin and ricin to surface receptors on human cells. *J. Biol. Chem.* **251**: 3977–3984.
Saxena, S.K., O'Brien, A.O. and Ackerman, E.J. (1989) Shiga toxin, Shiga-like toxin II variant and ricin are all single site RNA glycosidases of 28S RNA when microinjected into *Xenopus*. oocytes. *J. Biol. Chem.* **264**: 596–601.
Schindler, D.G. and Davies, J.E. (1977) The specific cleavage of ribosomal RNA caused by α-sarcin. *Nucleic Acids Res.* **4**: 1097–1110.
Schlossman, D., Withers, D., Welsh, P., Alexander, A., Robertus, J.D. and Frankel, A. (1989). Expression and characterization of mutants of ricin toxin A chain in *Escherichia coli*. *Mol. Cell Biol.* **9**: 5012–5021.
Shepherd, R.J., Fulton, J.P. and Wakeman, R.J. (1969) Properties of a virus causing pokeweed mosaic. *Phytopathology* **59**: 19–222.
Skorve, J.,Abraham, A.K., Olsnes, S. and Pihl, A. (1977) Effect of abrin on peptide chain initiation. *Eur. J. Biochem.* **79**: 559–564.
Soave, C., Tardani, I., DiFonzo, N. and Salamini, F. (1981) Regulation of zein level in maize endosperm by a protein under control of *opaque*-2 and *opaque*-6 loci. *Cell* **27**: 403–410.
Sperti, S. and Montanaro, L. (1979) Ricin and modeccin do not inhibit the elongation factor 1-dependent binding of aminoacyl tRNA to ribosomes. *Biochem. J.* **179**: 233–236.

Sperti, S., Brigotti, M., Zamboni, M., Carnecelli, D. and Montanaro, L. (1991) Requirements for the inactivation of ribosomes by gelonin. *Biochem.J.* **277**: 281–284.

Stevens, W.A., Spurdon, C., Ongon, L.J. and Stirpe, F. (1981) Effects of inhibitors of protein-synthesis from plants on tobacco mosaic virus infection. *Experientia* **37**: 257–259.

Stirpe, F. and Barbieri, L. (1986) Ribosome inactivating proteins up to date. *FEBS Lett* **195**: 1–8.

Stirpe, F. and Hughes, C.R. (1989) Specificity of ribosome-inactivating proteins with RNA-N glycosidase activity. *Biochem. J.* **262**: 1001–1002.

Stirpe, F., Olsnes, S. and Pihl, A. (1980) Gelonin, a new inhibitor of protein synthesis non-toxic to intact cells. *J. Biol. Chem.* **255**: 6947–6955.

Stirpe, F., Gasperi-Campani, A., Barbieri, L., Falaska, A., Abbondanza, A. and Stevens, W.A. (1983) Ribosome-inactivating proteins from the seeds of *Saponaria officinalis* L. (soapwort) of *Agrostemma githago* L. (corncockle) and of *Asparagus officinalis* L.(asparagus) and from the latex of *Hura* crepitans L. (sandbox) tree. *Biochem. J.* **216**: 617–625.

Stirpe, F., Bailey, S., Miller, S.P. and Bodley, J.W. (1988) Modification of ribosomal RNA by ribosomes inactivating proteins from plants. *Nucleic Acids Res.* **16**: 1349–1357.

Takaiwa, F. and Sugiura, M. (1982) The complete nucleotide sequence of a 23S rRNA gene from tobacco chloroplasts. *Eur. J. Biochem.* **124**: 13–19.

Tapprich, W.E. and Dahlberg, A.E. (1990) A single base mutation at position 2661 in *E. coli* 23S ribosomal RNA effects the binding of ternary complex to the ribosome. *EMBO J.* **9**: 2649–2655

Taylor, B.E. and Irvin, J.D. (1990) Depurination of plant ribosomes by pokeweed antiviral protein. *FEBS Lett.* **273**: 144–146.

Teraro, K., Uchiumi, T., Endo, Y. and Ogata, K. (1988) Ricin and α-sarcin alter the conformation of 60S ribosomal subunits at neighbouring but different sites. *Eur. J. Biochem.* **174**: 459–463.

Tomlinson, J.A., Walker, V.M., Flewett, T.H. and Barclay, G.R. (1974) The inhibition of infection by cucumber mosaic virus and influenza virus by extracts from *Phytolacca americana*. *J. Gen. Virol.* **22**: 225–232.

Tregear, J.W. and Roberts, L.M. (1992) The lectin gene family of *Ricinus communis*: Cloning of a functional ricin gene and three lectin pseudogenes. *Plant Mol. Biol.* (in press).

Uchiumi, T., Kikuchi, M. and Ogata, K. (1986) Cross-linking study on protein neighbourhoods at the subunit interface of rat liver ribosomes with 2-iminothiolane. *J. Biol. Chem.* **261**: 9663–9667.

Ussery, M.A., Irvin, J.D. and Hardesty, B. (1977) Inhibition of poliovirus replication by a plant antiviral peptide. *Ann. N.Y. Acad. Sci.* **284**: 431–440.

Veldman, G.M., Klootwijk, J., DeRegt, V.C.H.F. and Planta, R.J. (1981) The primary and secondary structure of yeast 26S rRNA. *Nucleic Acids Res.* **9**: 6935–6952.

Wales, R., Richardson, P.T., Roberts, L.M., Woodland, H.R. and Lord, J.M. (1991) Mutational analysis of recombinant ricin B chain. *J. Biol. Chem.* **266**: 19172–19179.

Walsh, T.A., Morgan, A.E. and Hey, T.E. (1991) Characterization and molecular cloning of a proenzyme form of a ribosome-inactivating protein from maize. *J. Biol. Chem.* **266**: 23422–23427.

Westby, M. (1991) Processing and trafficking of the ricin precursor. PhD thesis, University of Warwick.

White, T.C., Rudenko, G., and Borst, P. (1986) Three small RNAs within the 10 kb trypanosome transcription unit are analogous to domain VII of other eukaryotic 28S rRNAs. *Nucleic Acids Res.* **14**: 9471–9849.

Wool, I.G. (1984) The mechanism of action of the cytotoxic nuclease α-sarcin, and its use to analyse ribosome structure. *Trends in Biochemical Sciences* Sept. 1984: 14–17.

Wool, I.G., Endo, Y., Chan, Y.-L. and Gluck, A. (1990) Structure, function and evolution of mammalian ribosomes. In *The Ribosome: Structure, Function and Evolution (Hill, W.E. ed.), American Society for Microbiology, Washington D.C. pp. 203–217.*

Wood, K.A. (1991) Preproabrin: Genomic cloning and expression of the constituent polypeptides in heterologous systems. PhD thesis, University of Warwick.

Wood, K.A., Lord, J.M., Wawrzynczak, E.J. and Piatak, M. (1991) Preproabrin: genomic cloning, characterization and the expression in *Escherchia coli*. *Eur. J. Biochem.* **198**: 723–732.

Yoshida, T., Chen, C., Zhang, M. and Wu, H.C. (1991) Disruption of the Golgi complex by brefeldin A inhibits the cytotoxicity of ricin, modeccin and *Pseudomonas* toxin. *Exp. Cell Res.* **192**: 389–395.

Zamboni, M., Brigotti, M., Rambelli, F., Montanaro, L. and Sperti, S. (1989) High-pressure-liquid-chromatographic and fluorimetric methods for determination of adenine released from ribosomes by ricin and gelonin. *Biochem. J.* **259**:6 39–643.

Index